t i m e

Catalogue

Edited by A.J. Turner
Texts by H.F. Bienfait, E. Dekker, W. Dijkhuis, V. Icke, and A.J. Turner

The Hague 1990

Time, you old gypsy man,
Will you not stay,
Put up your caravan
Just for one day?

Ralph Hodgson (1917)

t i m e

DESIGN
SDU ontwerpgroep (Moniek Voorhout)

TRANSLATION
The authors, TT&T, Marguerite Rothengatter-Lake

TYPESETTING
Interpoint, Hapert

PRINTING
Ten Brink Meppel bv

ISBN 90 12 06863 0

Contents

The 'Nieuwe Kerk', Amsterdam

Exhibition Organisation

Images of Time is an initiative of the Foundation 'Tijd voor Tijd'.

Board
A. Stikker (chairman)
S.E. Huyzer (secretary, treasurer)
W. Van Dieren
E. Veen

Exhibition Committee
A.J. Turner (chairman), History of Time and Time-Measurement
H.F. Bienfait, Natural Time
E. Dekker, Modern Precision
W. Dijkhuis, Social Time
V. Icke, Time and Physics

Collection Curator, A. Aerts
Horologists, M. Spaander, H. Mulder

Concept and organisation
Mindz Associates B.V.
Van Diemenstraat 130
1013 CN Amsterdam

Sponsors

Rabobank Nederland
Digital Equipment bv
Schaap en Citroen, jewellers
Landis & Gyr B.V.
Koninklijke PTT Nederland NV
Nederlands Meetinstituut
Public Information Centre for Science, Technology and the Humanities
City of Amsterdam
Christie's South Kensington
Electrogig Nederland b.v.

Lenders to the Exhibition

Amsterdam
Economisch - Historische Bibliotheek
Joods Historisch Museum
Junghans Nederland B.V.
Koninklijk Instituut voor de Tropen, Tropenmuseum
Rijksmuseum 'Nederlands Scheepvaart Museum'
Rijksmuseum
Spaarpottenmuseum

Asten
Nationaal Beiaardmuseum

Barcelona
Eduard Farré-Olivé

Besançon
Musée du Temps et de l'Histoire

Brussels
Koninklijke Musea voor Kunst en Geschiedenis

Cambridge
Whipple Museum of the History of Science

La Chaux-de-Fonds
Musée International d'Horlogerie

Clermont-Ferrand
Société Michelin

De Bilt
K.N.M.I.

Delft
Technische Universiteit Delft, Faculteit Mijnbouwkunde en Petroleumwinning

Den Haag
Haags Gemeentemuseum
Koninklijke Bibliotheek
Museon
Rijksmuseum Meermanno-Westreenianum
Stichting Het Nederlandse PTT Museum

Edinburgh
Messrs Hamilton & Inches
Royal Museum of Scotland

Elburg
Gemeentemuseum

Genève
Cartier S.A.

Groningen
André Wegener Sleeswyk

Haarlem
Teylers Museum

Leicester
Dr. A.A. Mills

Leiden
Bibliotheek der Rijksuniversiteit te Leiden
Museum Boerhaave
Nationaal Natuurhistorisch Museum
Rijksmuseum voor Volkenkunde

Leidschendam
PTT Research

London
National Maritime Museum Greenwich
David Penney
Rogers Turner Books Ltd.
The Science Museum

Louviers
J.-C. Sabrier

Munich
Deutsches Museum

Oegstgeest
Marine Electronisch en Optisch Bedrijf

Oxford
Museum of the History of Science

Paris
Cité des Sciences et de l'Industrie
Dudley Barnes
l'Institut du Monde Arabe

Rotterdam
Historisch Museum Rotterdam

St. Nicolas d'Aliermont
Musée d'Horlogerie Aliermontaise

Schoonhoven
Goud-, Zilver- en Klokkenmuseum

Strasbourg
Musée de l'Oeuvre Notre-Dame

Utrecht
Universiteitsmuseum

Woodstock
Oxfordshire County Museum

Zaandam
Zaans Uurwerkenmuseum

Our thanks also to the various private lenders who wish to remain anonymous.

Acknowledgements

All exhibitions are co-operative ventures, and the organisers thank the following for their invaluable help and advice :

d'Arpiani, M.
Arps, C.E.S.
Bennett, J.A.
Bernold, Th.
Bertels, C.O.
Bessis, Ph.
Bloemen, E.S.A.
Boekschoten, G.J.
Van Brakel, J.H.
Breugelmans, R.
Van den Broek, A.J.
Brongers, A.
Bull, S.
Bryden, B.J.
Cardinal, C.
De Clerq, P.
De Clerq, S.
Collins, J.
Daan, S.
Dahan, A.
Van Dalen, B.
Darius, J.
Dawson, D.
Deiman, J.
Dobbs, A.
Douma, J.
Draaisma, D.
Drye, E.
Dubelaar, C.W.
Ebbinge, E.
Elverding, J.P.
Enthoven, D.
Eppink, A.
Gerkema, M.P.
Gillan, M.
Grapperhaus, F.
Groult, F.
Groult, P.
Hackmann, W.D.
Haspels, J.J.L.
Haverkamp, F.
Van 't Hof, A.
Hogeveen, F.
Van Hoorn, M.
Van Houten, H.J.
Huisman, P.
Hutchinson, B.
Jager, G.
De Jong, G.
Karig, J.S.
Kerkhof, G.A.
Keus, A.A.
Kist, J.B.
Klaassens, S.J.
De Koning, M.
Korteweg, A.
Korving, R.
Kroes, P.
Kruit, P.
Van Laarhoven, K.
Latcham, M.
Lehr, A.
Lemaire, A.
Leopold, J.H.
Maddison, F.R.
Martin, E.
Mauerhan, J.
Mertz, P.
Meyerman, A.M.
Moene, M.
Morrison-Low, A.
Morzer-Bruyns, W.
Mouliérac, J.
North, J.D.
Van Noten, F.
Nussbaum, E.
Offerhaus, J.
Querido, J.
Rathier, J.C.
Rijling, J.
Rogers, P.J.
Roux, A.
Schuub, K.P.J.
Sevenster, A.
Simons, J.S.
Simpson, A.D.C.
Smid, S.
Stender, N.
Van Tooren, M.
Varichon, A.
Vaughan, D.
Veerman, A.
Vermeulen, T.
Vickers, M.
Vos, K.
De Vries, D.J.
Waale, L.R.
Weiss, E.E.J.
Wijnen, G.
Wisse, P.
Woldringh, C.L.
Zwart, B.

We also wish to thank the following organisations for their support

Apollo B.V.
Coopers & Lybrand Dijker Van Dien Subsidy Advisors
Fides
Intertel Productions B.V.
Jenny Spits Type- en Uitzendbureau
Fa. A. de Jong T.H. BV
Koninklijke Nederlandse Akademie voor Wetenschappen
Luhlf Studios
Omniversum
Rijksvoorlichtingsdienst

Introduction

The invention of the clock is said to have alienated mankind from nature. This rather strong observation clearly contains an element of truth.
The exhibition 'Images of Time' of which the following pages provide a record, offers the visitor a wealth of information on mankind's perception of time and its philosophical, cultural, technological and social implications. It may therefore serve to bring time back into a broader perspective. This is important as the clock has a tendency to lock us up in a time-prison, from which we would like to be liberated.
This liberation can come about by understanding the way time has entered our world-view and the many different faces time has.

The presentation of the subject in this book has been kept as simple as possible with the Essays in Part I offering broad syntheses as a complement to the more fragmented picture that emerges from the catalogue. Here, in the descriptions of individual items, the authors have not hesitated to write quite technical descriptions if an object justified this, either because of its importance or because it was not previously published.

The exhibition itself consists of four main themes and their interconnection :

- time-measurement
- time in physics
- time in the natural world
- time in society

The organizers are greatly indebted to Anthony Turner who has prepared the exhibition and catalogue with great skill and dedication. In turn Mr. Turner was able to draw from his many colleagues and friends who supported this project with enthusiastic response and ready and unfailing help. The names of all those listed in the aknowledgements and the list of lenders should be considered to be written in letters of gold.

I wish the reader of the catalogue and the visitor to the exhibition an enrichment of their world-view through 'Images of Time'.

Allerd Stikker
Chairman Foundation Time for Time

Vincent Icke

Time Scales

Introduction

Human experience of time is limited, even though the pace of events around us may vary enormously. A human heartbeat takes about one second, but by the time we celebrate our first birthday our heart has already beat thirty million times. When we consider the world outside our immediate experience, we find an even larger variety of characteristic durations (we shall call these *time scales,*) all the way from astronomically long to microphysically short.

In this overview we will take a series of steps through a large variety of known time scales in the physical sciences, from the furthest reaches of the universe down to sub-atomic particles. We begin with the longest time scales known in the cosmos, and descend - via the time scales of human life - to the shortest, ending at the very frontier of the unknown.

As human beings, we occupy - more or less by chance - a kind of middle ground in the spectrum of cosmic time scales. The times with which we are familiar, span approximately the range from our reaction time (a tenth of a second) to the highest age we can reasonably expect to reach, say a century (three thousand million seconds). But the greater part of our conscious life is packed in a much smaller range, between a few hours (going shopping) and a few days (until that romantic dinner next Saturday). Due to this rather limited variety of experience, our intuition has not learned to encompass very slow or very fast phenomena. Therefore we must use an artifice : compressing a span of time by using a *logarithmic* scale, and comparing time scales by analogy with everyday events.

Powers of Ten

How can we abbreviate those astronomical numbers, which represent enormously small or large time scales, without resorting to endless trains of zeroes? Suppose we write down the time it takes a light ray to travel from the Sun to us, about 500 seconds. The same light ray would take 5000 seconds to reach the planet Saturn. What is the difference? Both numbers can be written as five times something : five times 100, and five times 1000. This allows us to use the following trick. We write $500 = 5 \times 100$ and $5000 = 5 \times 1000$. Then we abbreviate the number of zeroes by writing $100 = 10^2$ and $1000 = 10^3$. In this way, we compress all those 0000... in astronomical numbers in the form of a little reminder placed above the 10. This reminder is called an *exponent,* and the resulting number is called a *power* of *ten.* By using this trick we can write big and small numbers very compactly. Numbers smaller than one are written thus : $0.005 = 5/1000 = 5 \times 10^{-3}$. Thus, when writing numbers like $1/10^3$, we use the convention $1/10^3 = 10^{-3}$. Translated into words, $A \times 10^N$ means : multiply the number A with the number you get when you provide 1 with N zeroes. If N is negative, it means : *divide* A by the number one-with-N-zeroes.

We can take this one little step further. If we can write a hundred thousand million as 10^{11}, can we write 5×10^{11} in the form $10^{\text{something}}$? In other words, if we have a number T, is there a number x such that $T = 10^x$? Yes there is : we call x the *logarithm* of T. On a logarithmic scale, life is much more compact.

Powers of Ten in Time

By writing a very large number, such as 150,000,000, as something smaller times a clean multiple of 10, we can write big numbers very compactly. In this example, $150{,}000{,}000 = 1.5 \times 10^8$.

The time for a light ray to reach the Oort Cloud of comets is about 5×10^6 seconds (five million). The time it takes light to travel from the top of the Nieuwe Kerk to Berlage's Stock Exchange building is 5×10^{-6} seconds (five millionths). When plotted on a powers-of-ten scale, called a *logarithmic scale* by mathematicians, we can compare phenomena which take a very, very long time, such as the evolution of the universe, and those which last for an almost inconceivably short time, such as the process by which a neutron decays.

The idea of mapping our entire universe in steps of ten, or powers of 10, is due to the renowned Dutch schoolteacher Kees Boeke[1]. We will try to do the same in time, and compress the various time scales around us by means of a logarithmic Boeke telescope or expand them using a Boeke microscope.

Examples of time scales in nature

As viewed by a logarithmic Boeke telescope, we have (a) The age of Earth is 4500 million years, or $10^{17.2}$ seconds. (b) The time it takes for the Sun to complete one orbit around the galaxy is 250 million years, $10^{15.9}$ seconds. (c) Continental drift changes the landscape of Earth markedly in 25 million years, $10^{14.9}$ seconds. (d) The branch of evolution which led to *Homo Sapiens,* started to sprout about 4 million years ago: $10^{14.1}$ seconds. (e) The stars in the Orion Nebula have a characteristic evolution time of 400,000 years, or $10^{13.1}$ seconds On a logarithmic scale we therefore find the phenomena (a) through (e) at positions 17.2, 15.9, 14.9, 14.1, and 13.1.

The Time Scale of the Universe

It is appropriate to begin with the largest possible time scale. To put things in perspective, it is well to realize how insignificant human time scales are when seen in the context of cosmic Deep Time.

If the age of the universe were compressed into a week, starting at zero hours on Monday morning, the universe would be a seething, opaque mass of light and incandescent particles for the first thirty seconds of that day. At a minute past midnight, the universe would have cooled and expanded

$$150\ 000\ 000 =$$

$$1.5 \times \underbrace{10 \times 10 \times 10 \times 10 \times 10 \times 10 \times 10 \times 10}_{\text{8 factoren 10}}$$

$$150\ 000\ 000 = 1.5 \times 10^8$$

1

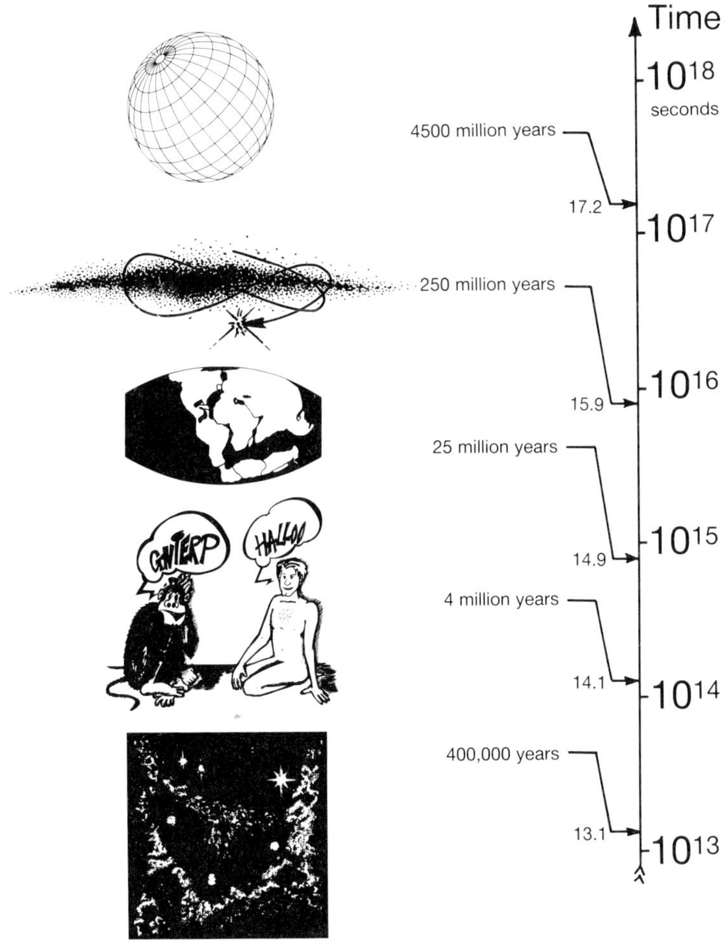

2

1. Boeke. For an updated version, see Morrison, Morrison; and Eames.

enough so that it became transparent. At a quarter to two that early Monday, the first galaxies would begin to form, lighting up with dazzling brightness as quasars at half past nine on Tuesday morning. Most of the mid-week would be spent making more galaxies, and all the stars in them. By Thursday, a thousand generations of heavy stars would have born and exploded, seeding the voids between the stars with the chemical elements from which you and I were eventually made. The Solar System, and Earth with it, would form around coffee time on Friday. The first appearance of living nucleated cells would be a good item for the late night news that Friday. Humans as a species would not appear until two minutes before Sunday midnight. The first human tool-makers made their appearance three seconds before the cosmic clock finished striking twelve.

How can we deduce the age of the universe? Indeed, how can we know that there is, in fact, such an age ; hasn't the universe always existed? It has not ; and we know this because the universe is expanding. From observations, astronomers have found that the average distances between the galaxies are getting larger, and this is happening everywhere at once and in the same way. More detailed observations show the remarkable result that *the universe is, on the average, the same everywhere,* no matter in what direction we look. Astronomers say that the universe is *isotropic.* Now we know, since Copernicus, that we are not at the centre of the Solar System. We know, since Shapley and Oort, that we are not at the centre of the Milky Way galaxy. Therefore, we assume that this isotropy does not mean that we are at the centre of the universe ; rather, we presume that we observe what we see because the universe is isotropic *everywhere.*

That means, that the motion of our universe can be described in a remarkably simple way. The isotropy of the universe implies that there are no special points in it, where the motion of the galaxies is much faster or much slower than elsewhere. Suppose that we were to draw a blueprint for the construction of the universe today. Then the blueprint for tomorrow will be *exactly the same,* except for one thing : in the corner of the drawing, where it says Scale 1 to ...', the scale factor is bigger. Likewise, the blueprint for yesterday's universe is the same as today, except that the scale is smaller. On the average, the change of the appearance of the universe as a whole is simply described by an overall enlargement in scale.

It can be seen straight away that this means that the galaxies move away from each other with a velocity which increases directly with the distance.

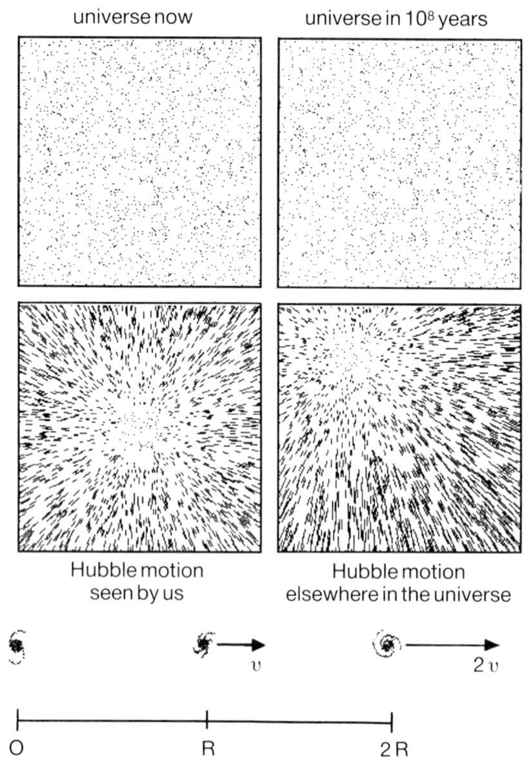

3

Hubble's Law and the age of the universe

If you compare the positions of the galaxies now and a little later, everything remains practically the same, except for an overall increase in scale. By putting these two pictures on top of each other, we can see the universal expansion quite clearly. Also, careful inspection shows that $v = H R$, *Hubble's Law.* The Hubble expansion gives the universe an absolute time scale. Since time=distance/velocity, we have $t = R/v$. Because we also have $v = H R$, we find immediately that $t = 1/H$: it was $1/H$ seconds ago that the scale factor of the universe was zero.

If R is the distance between two galaxies, and v is the velocity difference between them, then $v = H R$. This is called *Hubble's Law.* The Hubble expansion means, that the universe has a finite age, because we can calculate very simply how much time has gone by since the distances between the objects in the universe were zero. From daily experience we know that the time t taken to cover a certain distance R, is equal to that distance divided by the velocity v with which we travel: time=distance/velocity, $t = R/v$. Because Hubble's Law says that $v = H R$, we find immediately that $t = 1/H$: it was $1/H$ seconds ago that the scale factor of the universe was zero. Astronomers can observe the value of the Hubble parameter' H, and have found that $1/H$ is approximately 20 000 million years.

Thus, astronomy has shown that *there is a finite amount of time in our past.* Some people find this profoundly disturbing because, like so many astronomical discoveries, this upsets some of their philosophical tenets. But that's the way things are : a mere twenty thousand million years separate us from the beginning of time and space.

Shorter Astronomical Times

In what follows we will indicate times always as powers of ten in seconds, e.g. 20,000 million years is $10^{17.8}$ seconds. From this largest of all time scales we descend to smaller ones. The time scales of our galaxy (the Milky Way)and the Sun are comparable with that of the universe : observations indicate that the age of the galaxy is about 10 000 million years, fully half the age of the universe. From geological investigations, and from calculations of the behaviour of the Sun, we know that the Solar System is about half as old as the galaxy : 4,500 million years.

Almost all the 100,000 million stars in our galaxy have an age comparable with that of the Sun. The galaxy is held together by the mutual gravitational attraction of all its stars. Individual stars move in enormous, looping orbits in the galaxy. Each orbit takes about 250 million years. During the total existence of the Milky Way, our Sun - and all the planets attending it - has completed some twenty of these orbits.

The birth of a star from an interstellar dust- and gas cloud takes a few million years. Big stars evolve rapidly, and a mere ten million years after their birth they already explode in prodigiously violent events called supernovae. The remainder of the star stays behind as a pulsar, which is the best clock in the universe. So hot are supernovae, that radioactive elements like uranium and thorium are formed in the one-thousandth of a second of the initial outburst. These elements are scattered into space, eventually to become part of new planets like Earth. The radioactivity of these elements heats the innermost layers of Earth and makes it plastic. This generates convection motions, which we see on the surface as continental drift. The characteristic time scale of this remarkable phenomenon, which is responsible for the building of mountains, the opening of oceans, and earthquakes, is some 25 million years.

Human time scales : right in the middle

Where does our human sense of time come in? We really do not feel the continents shift under us, but somewhat shorter time scales are beginning to appear relevant. Somewhat less than a factor ten down we get to a scale of a few million years. That was about the time which was necessary for our distant ancestor *Australopithecus* to evolve into the genus *Homo.* It took a final 100,000 years or so for our species, *H. sapiens*, to become established. Not that we actually remember that, of course, but it is certainly a time scale relevant for humans.

Some astronomical times are shorter than this biological time scale ; the stars and the gas in the Orion Nebula (which we can see in the winter sky on the Northern Hemisphere), have a characteristic time scale of 400,000 years. If we were to come back in 4000 centuries, we would be hard pressed to recognize this nebula, which is so glorious today.

As we descend to shorter times, we enter the realm of the relevant. For example, the last Ice Age occurred a mere 40,000 years ago. The soil in some places on Earth is still slowly rising, after having recently lost its gigantic ice burden. Astronomically speaking, the latest Ice Age was but yesterday : we saw that the universe is half a million times older! Written history has existed for 6000 years. As soon as we get this time scale within our horizon, we perceive a world that is fairly modern, or at least recognizable. The Renaissance occurred 600 years ago. These are time scales with which almost everyone is familiar.

The ten time scales which lie between centuries and seconds are of crucial importance to us. The canonical life expectancy of threescore-and-ten years clocks in at $10^{9.3}$ seconds ; my daughter is 16 years old, $10^{8.7}$ and counting ; most people who work are paid once every $10^{6.4}$ seconds ; which gives us a positive bank account for 100,000 seconds, perhaps. An evening lecture about Time Scales' will occupy 7000 seconds of one's life. Don't hold your breath, even if the lecture is very good, because a reasonably well trained person cannot do that for more than 100 seconds. If you do not like the lecture, you may sprint away and cover the hundred meters to the bus stop in 10 seconds ; and your heartbeat, once calmed down, takes one second. The shortest time scale of everyday concern is our reaction time, just under a tenth of a second.

Some of the longer time scales in the universe

Brought within range by using the logarithmic Boeke telescope', these time scales stretch from $10^{17.8}$ seconds, the age of the universe, to 10^{-1} seconds, the human reaction time. One of the longest of sub-atomic particle processes, the decay of the free neutron, shows up at $925 = 10^{2.97}$ seconds.

It is interesting to see how many time scales our own experience encompasses: a dozen or so powers of ten between our reaction time and the last glaciation. Apparently we are capable of bridging and understanding quite a few powers of ten. The time scales with which we deal in our daily lives are not fundamentally different from those in astronomy and physics. All we have to do, is to allow our mind to extend our experience along the logarithmic scale.

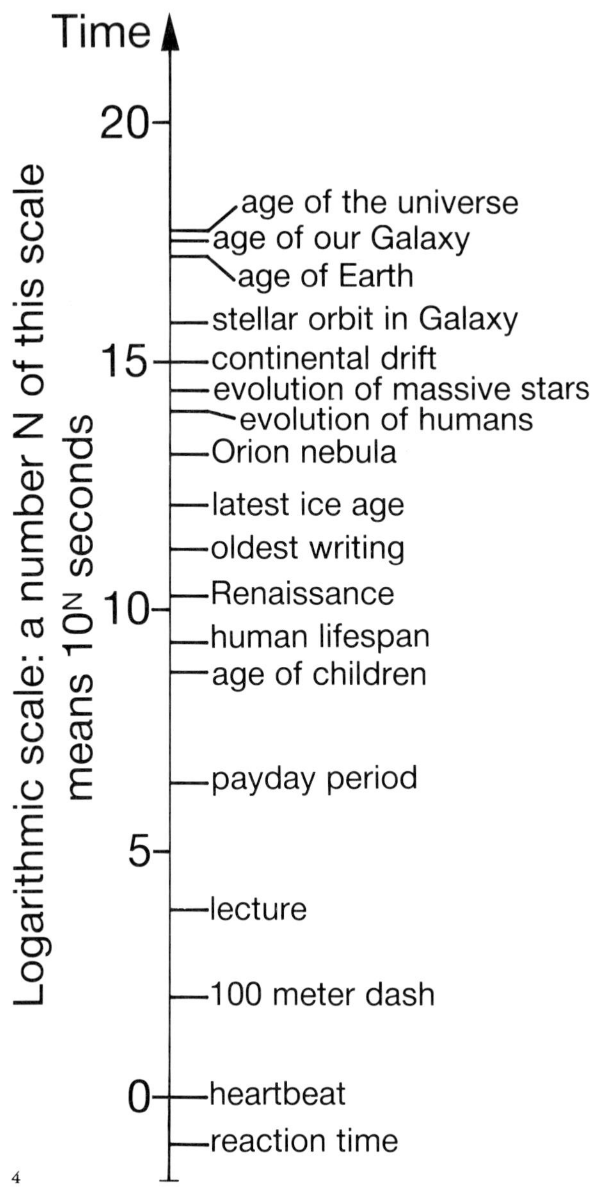

4

Radioactive Clocks

Let me remind you of a few of the key things which we know about the construction of the sub-atomic world. The processes which we will consider as examples of time scales are, from now on, mostly variations on what was generally called radioactivity' around the turn of the century. That is to say, we will look at the splitting of particles into other pieces, possibly as a consequence of a violent collision.

For example, consider a helium atom. On a length scale of 10^{-11} meters we see an atomic nucleus surrounded by two electrons. A ray of light would take only a very small amount of time to cross such a system: 10^{-19} seconds only. The helium nucleus, in olden days called an *alpha particle,* consists of two electrically charged particles, called protons, and two neutral ones, the neutrons. These two types of particles are fossils from the time our universe was very young: they formed when only one ten-millionth of a second had passed since the Big Bang.

From Atom to Quark

We are made of atoms. Each atom is built from a nucleus, attended by electrons. The nucleus is made of smaller particles: neutrons and protons. For example, the helium atom shown here has a nucleus made of two neutrons and two protons. These particles, in their turn, are made of yet smaller ones, called quarks. These quarks are bound in triples to form a proton or a neutron. In this way, we descend to smaller and smaller sizes, where shorter and shorter time scales are found.

They formed by a kind of fusion process, in which yet smaller particles, the quarks, bound together in triples. The neutron consists of two *d*-quarks and one *u*-quark, whereas the proton is a bound system of two *u*'s and one *d*.

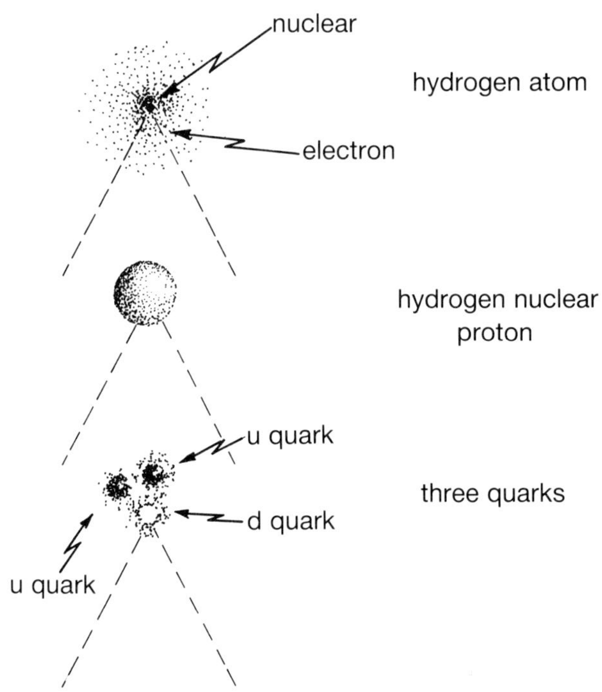

5

To give an example of an atomic radioactivity time scale, let us pry a neutron loose from the helium nucleus, and let us follow the life and times of its constituent quarks. For reasons which are not completely understood, the various types of quarks can change identity by a process of radioactive decay' : for instance, a *d*-quark can decay spontaneously into a *u*-quark and another particle, the W^--boson. The W^--may be absorbed by a nearby *u* quark, which then turns into a *d*-type. To use an analogy, if Sheriff Bloggs takes off his sheriff's star, he turns into plain Mr.Bloggs ; if the star is then taken over by Mrs. Murgatroyd, she becomes Sheriff Murgatroyd. Thus the exchange of *W* particles mediates an interaction (called the weak force) between quarks. However, every now and then it happens that the *W* itself decays before it can be caught by a quark. Usually, the W^- turns into two particles, an electron and an antineutrino (called $\grave{v}$). In that case, the three quarks in the neutron change from the triple *udd* into *udd* : the neutron has become a proton!

Decay of the Neutron : a Slow Sub-atomic Process

The internal structure of a neutron can change spontaneously, when a quark changes from one type to another. This causes radioactive decay : neutron - proton + electron + antineutrino. Of a cloud of free neutrons, about half have decayed in 925 seconds.

Thus, we get the reaction $n \text{-} p + e + \grave{v}$. Of a cloud of free neutrons, only half remain after 925 seconds ; the rest have decayed. In the next 925 seconds, half of those which are left will decay ; and so forth. Atomic time scales are often called half-life' because of this gradual decay.

In the type of radioactivity just described, an electron is produced. Because formerly electrons were called beta particles, this type of process is called beta decay. It is an important process in atomic nuclei. For example, the nucleus of phosphorus containing 15 protons and 19 neutrons is beta-unstable : one of these neutrons falls apart on a time scale of 12.4 seconds, which turns the phosphorus into a stable sulphur nucleus with 16 protons and 18 neutrons.

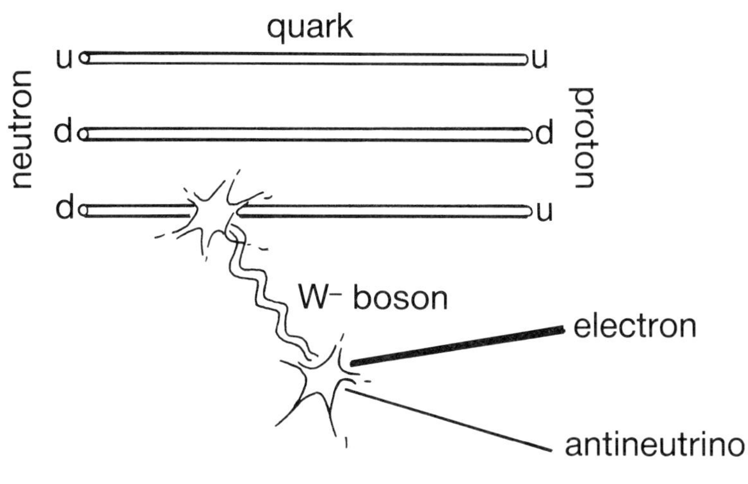

$n \rightarrow p + e^- + \bar{v}$ in 925 seconds

6

Thus, radioactive decay can act as a clock. A famous example of this is the carbon-14 clock. The element carbon normally has six protons and six neutrons in its nucleus. But there is a type of carbon with eight neutrons (and therefore 14 particles in the nucleus ; hence the name Carbon-14). One of these neutrons can turn into a proton, thereby turning the carbon into a nucleus with seven protons and seven neutrons. This is a nucleus of the nitrogen atom. The time scale for this process is 5730 years : the carbon-14 clock ticks slowly for us, but quickly enough to be extremely useful for the dating of objects of historical interest.

Radioactive Clocks

Radioactive phosphorus has a half-life of 12.4 seconds ($10^{1.1}$ seconds) ; this short time makes it useful for biomedical applications. Carbon-fourteen has a time scale of 5730 years ($10^{11.3}$ seconds) for turning into nitrogen ; potassium decays to argon in 1300 million years ($10^{16.6}$ seconds) ; and uranium eventually becomes lead, in a half-life of 4500 years ($10^{17.2}$ seconds).

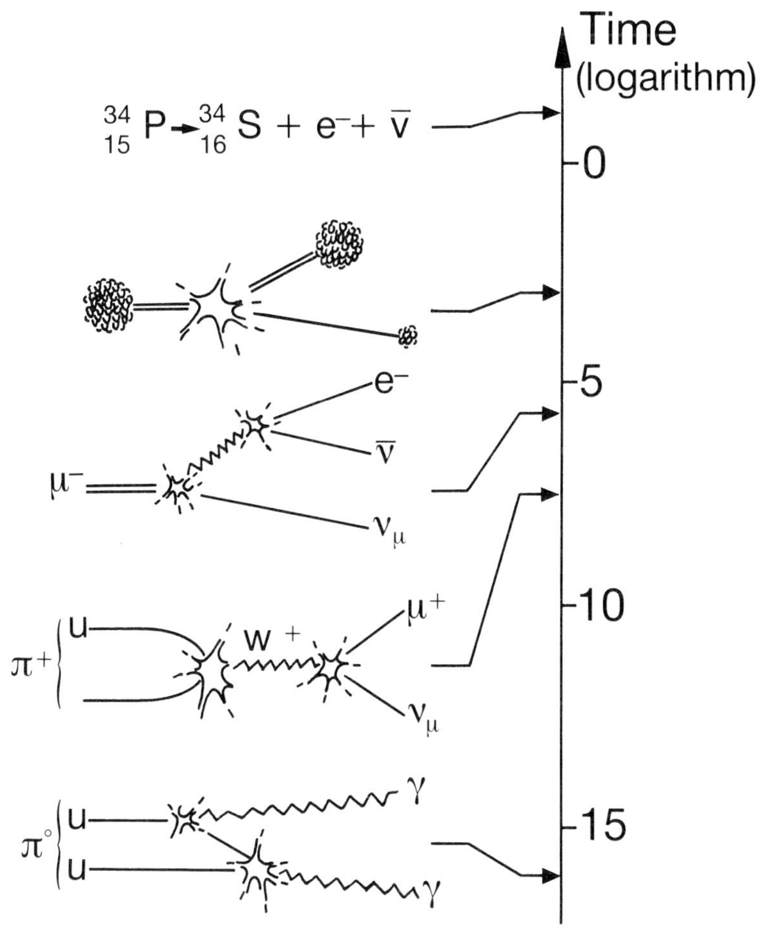

7

As seen in our large scale world, time scales for radioactive decay are typically not so short. For example, there is a type of nuclear decay in which a whole helium nucleus (alpha particle) is expelled. The shortest time scale for such alpha decay is about a thousandth of a second, but some radioactive materials have characteristic time scales up to many thousands of millions of years. Such forms of radioactivity are excellent geological clocks. For example, uranium turns into lead on a time scale of 4500 million years ; potassium turns into argon in 1300 million years.

Thus, of a given quantity of radioactive potassium, half has become argon after 1300 million years ; three-quarters after 2600 million ; seven-eights after 3900 million years ; and so forth. By using various radioactive clocks, geologists have determined the age of Earth : about 4500 million years.

The Smallest Time Scales

As the examples of the half-lives show, not all atomic processes have short time scales. As a rule, however, the times involved in the physics of subatomic particles are very short, and become shorter as we descend to smaller particles. Because the speed of light cannot be exceeded, we find that in general shorter times correspond to smaller distances. After all, the shorter the time of duration of an event, the shorter the distance which can be covered, even if one flies with the speed of light. Therefore, in order to find shorter time scales, we will have to visit the realm of the very small.

For instance, light crossing an atom is through in a flash, and typically it takes the atom a very long time to react to such a crossing. To give you an impression about atomic processes : a light ray races through an atom in 10^{-19} seconds. The atomic electrons typically take 4×10^{-10} seconds to return to normal. Returning to the old state takes an atom 4000 million times as long as the original perturbation. For comparison : 4000 million years is almost the age of Earth. If we build a house in a year, then it would - to be as stable as an atom, comparatively speaking - have to remain standing for the entire duration of Earth's life!

One example of small sub-atomic time scales we have encountered above, namely the decay of a *W*-particle into an electron and an antineutrino. This takes a modest amount of time : two millionths of a second. We will consider three more such examples.

The first involves a particle called a pi-plus meson. It is made of two quarks (instead of three, as in the case of the protons and neutrons) : a *u*- and a *d*-quark. These quarks can coalesce to form a *W*-particle, which then itself falls apart. The resulting time scale for the pi-plus particle is just under three hundred-millionths of a second ($10^{-7.6}$ seconds).

The second is a related particle, the neutral pi-meson. It consists either of a *u*- and an anti-*u* quark, or of a *d* and an anti-*d*. These quarks can annihilate one another and thereby turn into two photons (gamma rays, extremely energetic particles of light). A free neutral-pi particle thus has a fifty-fifty chance of turning into a double flash of gamma light in a very brief time, three hundred million times shorter than the decay time of the pi-plus : about 83 millionth-millionth-millionth seconds ($10^{-16.1}$ seconds). Even though the pi-plus and the neutral pi are members of the same family, their time scales are enormously different. To give you an impression how much longer a pi-plus lives than its unfortunate neutral brother : three hundred million times the duration of one heartbeat takes ten years of a human life. It is as if I (the neutral pi) were to succumb in ten seconds, whereas my brother (the pi-plus) will live a full century!

The third example is an exceedingly brief process : the collision between an electron and an antineutrino (we encountered these above, as the decay products of the *W* particle). This process is all over in $10^{-26.2}$ seconds. Compared with this, it takes an eternity to cross an atom, flying with the speed of light : this takes 10^{-19} seconds, fourteen million times longer.

The Bottom of the Ladder

When we plot the time scales on a powers-of-ten ladder, we find the longest time scale at the top : the age of the universe, $10^{17.8}$ seconds. But is there a shortest time scale? No, but there is a lowest rung on the powers-of-ten ladder, namely the time where physics as we know it ceases to apply. The absolutely shortest time scale with which we can hope to deal until our knowledge of fundamental physics improves, can be estimated as follows.

The Bottom of the Ladder of Time

Between the time scale of a human heartbeat ($10^0 = 1$ second) and the shortest experimentally known time, $10^{-26.2}$ seconds, lies a range which is even larger than the range from heartbeat to the age of the universe : half again as many rungs on the time ladder separate us from the subatomic world than are between us and the cosmos at large. Still, human beings occupy a kind of middle position on the ladder. The lowermost rung on the ladder of time is found at the Planck time, $10^{-43.3}$ seconds. Times shorter than this cannot be understood with present day physics.

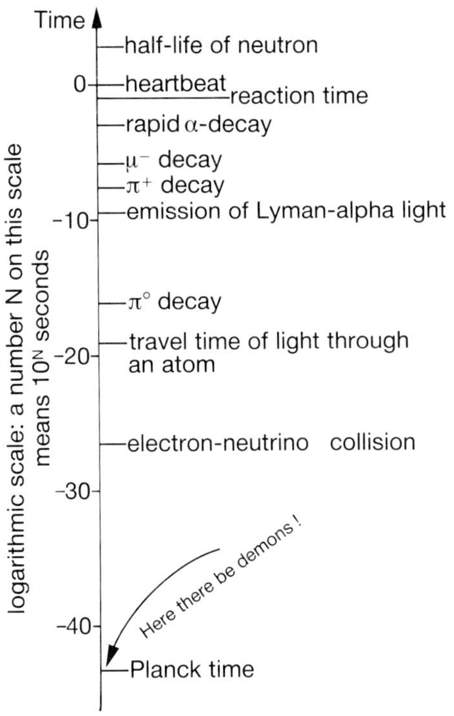

8

The pillars on which contemporary physics rests are three : relativity, gravitation, and quantum mechanics. Each of these three has a characteristic constant associated with it. For relativity, that is the speed of light, $c = 3 \times 10^8$ meters per second. For gravitation, the characteristic number is the constant of gravity, *$G = 6.7 \times 10^{-11}$ cubic metres per square second and per kilogram. The constant for quantum mechanics is Planck's constant,* $h = 10^{-34}$ square meter-kilograms per second. A physical phenomenon which is relativistic will involve the constant $c = 10^8$ m/s ; anything gravitational involves $G = 6.7 \times 10^{-11}$ m^3/kg ^{2} ; and quantum mechanical processes are described with formulae involving $h = 10^{-34}$ kg m^2/s.

Well, any process which involves all three, must contain a combination of the constants c, G, and h. What combination of these constants gives a time scale? The three constants all involve units of length, mass, and time. Thus, we can use a combination of two constants in order to remove the dependency on mass and length, leaving only time. Some tinkering shows that this combination is unique, namely $t_P = \sqrt{Gh/c^c}$. This time scale is called the *Planck time.*

The Planck time is phenomenally short: $t_P = 5 \times 10_{-44}$ seconds, or $10^{-43.3}$. This is the time scale of relativistic quantum gravity. If we want to describe nature on time scales shorter than this, we need a relativistic theory gravitational of quantum mechanics. But such a theory does not exist! Not yet, anyway. Until we make new basic discoveries, we will not be able to descend the ladder of time any further.

A.J. Turner

The Origins of Modern Time

The ability to measure time accurately over both very long periods and very short intervals has been and remains crucial to the development of modern science[1] ability to maintain regular precise measurements of time and to distribute them throughout the world, is crucial to the functioning of modern society[2]. Both developments have taken place in a relatively brief period, and both are the result of the ability to measure a phenomen to which a substantial body of philosophical writing would deny existence and which, for the greater part of human history, has been conceived in terms relative to individual societies. That philosophical disagreement about time continues is a matter of no great importance for philosophy, like poetry, makes nothing happen ; but that a unified, though irrational, system of time now extends throughout the world is the result of sustained investigations into time-measurement during the last seven centuries, combining with a desire to use measured time for social and scientific purposes.

Measuring time, an abstract and intangible aspect of the phenomenal world, is neither a natural nor an obvious thing to do, but from early periods of his history man has found it necessary. Time-measurement responded to a need and this need was to erect markers, sign-posts in the undifferentiated duration which, if we accept that time can have an existence outside of its measured self, makes up abstract time. Noting the alteration of day and night, the phases of the Moon or, more sophisticatedly, the return of Sun or star to a particular place, represent the first marking out of time, and gave rise to the first calendars. Inevitably calendars differed from place to place, from society to society. The needs of the farmer are not those of the urban dweller, calendars based on the Moon differ from those based on the Sun, while the demands of agriculture, religion or civic life all add their own differing, elements. Calendars moreover quickly acquired political importance as symbols of authority and legitimacy. The proliferation of different calendars was thus encouraged for each represented the identity and the independence of a particular city, state or empire. Only in periods when a dominant group exerted political or cultural hegemony over a wide area might the number of calendars be somewhat reduced.

The variety of calendars among early societies was then in part a function of political and cultural fragmentation. It also illustrates the lack of any dominant view of time in such societies. Time was not yet a single entity sweeping all in its course. Time in early societies was fragmented. The way in which time was conceived depended upon the role that it was to play. A calendar intended for agricultural purposes could differ quite markedly from one destined to mark ritual dates and religious festivals. Because both patterned out the year, which would inevitably repeat itself, both appear to be cyclical and much has been written on the cyclical nature of time in early societies[3]. While much of this may hold true for ritual practices, we should nonetheless beware of extrapolating too far. How the consumers of religion thought of their world is relatively little known, and a year in itself is a

1. Icke, *passim* and the relevant chapters in Whitrow (1 & 2).
2. For the development of time services and a world time see Howse, Dekker (2).
3. In particular by Eliade (2).

linear period which repeats simply as a linear period.

Repetition need not be cyclical a fact which, if disguised in agricultural and ritual activities, emerges much more clearly in politics, where although rulers succeed, patterns are not the same and where the break in continuity is under-lined by the habit of beginning a new year count for the new ruler. That there are cyclical patterns in time and that in some societies, most notably those of India[4], these were developed on a cosmic scale, cannot be denied, but we should not extend these instances too far. The evidence we have for early societies tends to be slanted. More materials survive concerning the ideas of priests, philosophers, and poets, than do for administrators, politicians, farmers, merchants and labourers. What ideas such people may have held about time we do not know, but the variety which reveals itself when ideas about time in Greek writers are examined closely[5], suggests that they were many.

Time as it is lived and used, and time as it is discussed by philosophers and theologians often seem to have rather little to do with each other. In part this is the result of a preoccupation with different time-scales, in part because different kinds of time are being discussed. The time in which daily life is lived is felt to be qualitatively different from the time of the gods, and this again is different from the ritual time in which man attempts to approach his gods, or the time into which he divides up and orders his past. Ritual time and divine time are not primary concerns of this essay, but they should always be born in mind as lying behind and giving meaning to the way time was used in everyday life.

A need to use time has been experienced by man virtually since his origin. Ability to predict new moons can be useful to hunters, and a day/night or lunation count in simple societies may be the only way of expressing distance. Counting regular phenomenon therefore is the first measure of time, but it is an essentially passive, recording activity. A naturally recurring phenomenon is used as a yard-stick and its occurrences are counted. Man has no control over his unit of measure. He cannot adapt it, cannot intervene if some catastrophe upsets the regularity. Unfortunately the regular phenomena such as lunar phases, annual movement of the sun, alternation of day and night, rising of a particular star, available to early man were relatively few, and for some purposes provided a unit, day month, year too large to be practical. From an early stage therefore, man had to intervene in the business of counting time-markers and play an active part. Anomalies had to be overcome, smaller units had to be imagined, and ways to determine them devised. Such active time-measurement involved the development of tools and machines. It also involved a change of relationship with time. Gradually time became manipulable. Time as a given element of the world gave way to a time which was created by the machines which measured it. In the process man became increasingly independent of nature. Whether this be seen as a liberation or denaturisation, it was an important consequence of the development of time-measuring devices and is apparently irreversible.

In the development of modern time three decisive changes ocurred before 1800. All of them may be associated with important technical innovations.

The first change is the move from a day which is not divided up at all, to one which is divided into hours. This change took place in Antiquity at different times in different places and even at different times in different sections of society. Broadly however it can be seen to have been effected in two stages. The first was the marking of noon as a fixed point in the daylight period previously undivided. The second was the addition of sub-divisions which marked the day out into 12 parts or hours. Hours, in the remote period with which we are now concerned, were of two kinds. The more natural and the most widely used were based on the total daylight period itself which was divided into 12 equal parts. The period of daylight however, thanks to the seasonal variation of the sun's declination, is not constant throughout the year, and the hours so obtained were therefore not constant but became increasingly long during the half year from the winter to the summer solstice, becoming correspondingly shorter again in the opposite direction. Except at the equinoxes, they were also not the same length as the hours of the night so unequal to each other day by day, and unequal to the hours of the night they are now, naturally, known as unequal hours. The second kind of hour was that based upon the period of the *nychthemeron.* This was divided up into 24 equal parts which were thus equal to each other both throughout a day and night, and throughout the year.

These equal hours, the hours which we use today (with the exception that they are differently determined) were however never commonly used for the ordering of social life before the 14th century ; they were restricted to astronomers

4. Eliade (1), but the extraordinarily long periods used mean that cyclical effects are rarely experienced in everyday life.
5. Momigliano.

who found them convenient and to those physicians, astrologers and calendar-makers, who had need of the astronomer's work.

The dividing up of the day into 12 parts took place at different times in different societies. Here, for want of space, both this and the problem of its origins will be ignored[6], attention being confined to the cases of Greece and Rome. In Greece the first mention of the twelve hour day occurs in that famous passage of Herodotus (5th century BC) - surely one of the earliest texts of the history of science - where he speaks of the origins of sun-dials, the 12-hour day, and land-surveying.

From this [the need to re-measure land in Egypt each year because of the Nile inundation] to my thinking, the Greeks learnt the art of measuring land ; the polos and the gnomon, and the twelve divisions of the day, came to Hellas not from Egypt but from Babylonia[7].

This passage has been frequently interpreted and re-interpreted. For reasons which will be set out fully elsewhere[8], I do not believe that either *polos,* or *gnomon* refer to true sun-dials (defined as instruments showing the hour of the day by a shadow cast by an indicator onto a scale permanently associated with it), and that reference to 'the twelve divisions of the day' refers rather to their use in learned circles in towns than in society at large in the Peloponesian peninsula. The earliest datable surviving Greek sun-dial comes from the 3rd century[9] BC and this seems to be the period (possibly the extreme end of the 4th century BC) when the instrument came into general use. That it did so was perhaps a response to the appearance in the 4th century BC of hour-indicating water-clocks running for 24 hours. These were developed in what must have seemed a startling manner to indicate time both visually and aurally as well as to drive automation side-shows, by Ctesibios of Alexandria (*c.* 280-230 BC)[10]. It was perhaps the appearance of water-clocks in the 4th century BC which stimulated the development of the true sun-dial out of the calendrical gnomon known in the 5th century BC, and probably earlier. More important than this speculation however is that such a development represents the move from an unstructured day, without time divisions, in which the nature of an activity determined the time spent upon it limited only by the time-markers of biology, hunger and the need of bodily relief, to one which is artificially divided up. Task-discipline rather than time-discipline structured Greek life as it was to continue to structure agrarian life for many centuries. With the appearance of water-clocks and sun-dials in civic life however a new structure appeared competing with the old. Although the time measured by these instruments was based on natural phenomena - the movement of the sun, the fall of water - the time divisions drawn from them were an abstraction. The time shown by clock or sun-dial was not directly experienced as was the time control which derived from the completing of a particular task or the constraints of the body. A first, if small step in the liberation of man from nature, or - depending on one's point of view - in his denaturisation, had been accomplished.

The innovation in time-measurement made in Greece in the 4th/3rd centuries, were paralleled, although later, at Rome. Details of the innovations were recorded by Pliny[11]. In the mid-5th century the day was undivided except by noon. It was not until 293 BC that the first sun-dial was set up. When a second was brought from Catania in 263 BC this was so little understood that it was set up unadjusted for the difference of latitude between the two places, and thus marked time incorrectly for 99 years until in 164 BC a dial specially made for Rome was erected by Q. Marcus Phillipus. Five years later Scipio Nasicus set up a water-clock. That the Romans, as Pliny remarks, should have 'lived such a long time without distinguishing the hours of the day', helps to underline a point implicit in what has been said earlier. Measuring time was not, even in the advanced cities of Greece or at Rome (where the use for 99 years of an incorrect dial confirms the suggestion), indispensable to social or private life. Whether this reflects a lack of need, given the small size of the communities involved, or that an innate sense of time was more highly developed in societies innocent of instruments to measure it, may provide matter for discussion, but the fact of the difference is important.

If the invention of the sun-dial, perhaps stimulated by the water-clock, thus marks a first stage in the development of the idea of time as an abstract, measurable, manipulable commodity, the abstraction was still close to the natural standard that gave rise to it and is unlikely to have been of much concern to men of the time. Not so however the clash between biological time and the artificial time of the sun-dial when this was adopted in social life. That the tiresomeness of waiting for the sun-dial to give one leave to eat furnished matter for the comic poets Plautus (3rd century BC) and

6. For a theory concerning these see Neugebauer (I), 84- 6 ; Krupp 192-4
7. *Histories,* II. 109. 3. Translation by A.D. Godley.
8. Turner (5) ; Turner (6), Ch 1.
9. Gibbs, 123.
10. Armstrong & MckCamp and *cf.* Drachmann.
11. *Historia Naturalis* VII. I have used the Budé edition by Robert Schilling (Paris, 1977) where the appropria te passage is found on p. 122.

Alciphron (1st century AD) is suggestive of a more widely spread reaction.

But with the spread of Christianity, and in particular with the development of monasticism, waiting for the time-measurer to give one leave to eat became far more familiar. For already by the 5th century, monasteries were time-regulated communities. True service to God required punctuality, the power of prayer was increased when the whole community offered it together, and the discipline of all time being communal time, the destruction of private time in the lives of the monks helped the monk in the struggle to overcome self and loose himself in god. Monasteries therefore were completely time-organised. Prayer was to be said at set times and no others. At night these times were determined by stellar observations, during the day by the position of the sun. But observations of the heavens can always be upset by cloud and bad weather, and so instruments were also developed. Sun-dials, water-clocks, and water-alarms were used in monasteries from early times, and it was this need to time, this familiarity with its manipulation, that makes a monastic origin for the weight-driven clock a highly probable supposition[12].

It was the appearance of this device that marks our second main stage of change in the consciousness of time. For the weight-driven clock, like the sand-glass but unlike the water-clock, can only count equal periods of time, it could thus only be made to indicate equal hours. The time counted out by the weight-driven time-keeper thus had, as it were, interchangeable parts. Each minute, each hour could replace any other in the sequence, something that could not be done in an unequal hour system. This characteristic of the weight-driven time-piece ought to have appeared as a disadvantage. The new machine was less flexible than the old water-clock which could be made to measure hours in either system relatively easily. The weight-driven clock was therefore of direct use only to that minute proportion of society which comprised the astronomers - a fact which argues in favour of its having originated in an astronomical context. One would have expected therefore that even if the weight-driven clock was not rejected entirely by European society as it was by Islamic ones, it would at least have been brought only slowly into use. In fact quite the contrary occurred. The weight-driven clock seized the imagination of Europe and spread with remarkable rapidity from city to city across the continent. Even more remarkably, instead of the machine being adapted in some way to indicate unequal hours following the norm prevailing in society, the machine itself forced a change in social life. In the course of the later 14th century, equal hours displaced unequal hours as the usual division of the day and night[13].

It thus rendered most existing sun-dials irrelevant and provoked the production of sun-dials which could indicate equal hours. Knowledge of how to do this was - at least in Islam - already available, but no reliable evidence of the existence of an equal-hour measuring dial is available in Europe from before the mid-15th century. However a good deal more investigatnce of an equal-hour measuring dial is available in Europe from before the mid-15th century. However a good deal more investigatquences of the weight-driven clock were thus dramatic, the move from unequal to equal hours entailed a drastic step away from natural time-measurement to use of an abstract scale. Not only was the perception of time now detached from the natural world, but it now also became anonymous, something received passively from a machine rather than actively experienced or determined by observation. With the invention of the spring drive this abstraction of time and its measurement from nature became even more marked thanks to the complete autonomy of spring-driven time-pieces. Nonetheless, remote though time now seemed to be getting from the natural world, a little reflection could quickly reconstruct the chain linking the two, and the movement of the sun remained always the ultimate standard, If sun-dials developed in accuracy as well as in the number of types available from the 15th century onwards, this was as much because they were needed to find the time in order to check weight- or spring- driven clocks as for any other reason. A man might not need a sun-dial for himself, but he did need it for his unnatural time-piece. The development of weight- and spring-driven clocks and watches was a major stimulus for the development of sun-dials.

The stimulus thus exerted on the production of dials by clocks and watches was one which continued to operate from the 14th to the 19th centuries. It disappeared only when telegraphy and the spread of standard time made the sun-dial redundant. Before this however further fundamental developments had taken place. Between 1656, when Huygens and Salomon Coster realised the first pendulum clock of Huygens' design, and 1675, by which time fitted with the

12. For times in monasteries see McCluskey, Biarne, d'Haennens, Landes and for some of the devices used numbers 165 to 176 in the catalogue below.

13. Robertson, 38-40 ; Clanchy 236.

anchor escapement it was in general use, the new clock moved from the drawing board to the astronomers's observatory, thence to the cabinets of the wealthy, and finally to every well-built domestic clock with pretensions to quality and to precision. Here is not the place to follow the historical and historiographical debates that were and are aroused by these events[14]. It is sufficient to note the remarkable increase in accuracy provided by the new machine whose average daily error of 10-15 seconds contrasts strikingly with the 10 - 15 minutes of its predecessor with verge escapement. This remarkable increase in accuracy brought into prominence a new problem, that of the equation of time. For the one-second-beating pendulum clock was the first commonly used weight-driven time-keeper of sufficient accuracy to reveal the irregular behaviour of the sun to the common man. This meant that the sun-dial took on a new significance as the representative of true time, ultimate arbiter of the time marked by clocks. A sun-dial motto sums up the popular understanding of the situation:

Je marque le temps vrai
l'Horloge marque le temps moyen[15].

But the strength of the feeling that true time was the natural time of the Sun, and not that of any machine is shown by the many attempts in France and Britain to build clocks which would indeed 'go with the Sun'. That this enterprise appealed to the better clockmakers as a way of displaying their skills is certainly true, for equation clocks were a challenge to technical virtuosity, but something of the underlying attitude emerges in the *Traité générale d'Horlogerie* of Jacques Allexandre where instructions for making equation clocks join those for making ordinary meantime clocks and sun-dials.

Toute la plus grande perfection he tells us *que M. Huygens ait donné aux Pendules étoit de leur faire suivre le moyen mouvement du soleil ; ici je donne le moyen de leur faire suivre le mouvement apparent & de marque toujours le veritable heure du Soleil*[16].

With the appearance of pendulum clocks with anchor escapement the idea of time and its measurement was turned topsy-turvy. From being an abstraction, time now became arbitrary. Two times were available, clock time and solar time, either of which could be used. No longer was the time derived from the natural world the standard to which all others aspired. Although solar time remained the standard of *comparison* for another 300 years, the time now used in everyday life was in effect the artificial mean time of the machine. *Temps vécu* was now literally *temps conçu*. The flurry of publications to explain the equation of time, the tables made to be pasted to the inside of clock-case doors, the making of equation dials and clocks, are all evidence of the confusion felt and sometimes expressed at this reversal of natural order. Even so, in the same way that weight-driven clocks had caused equal hours to be adopted throughout Europe in the 14th century, so now the pendulum clock imposed use of the machine-made mean time.

If the pendulum clock thus changed the nature of the time by which men lived, it did not change the standard of comparison. No clock was yet reliable enough to act as a standard for others nor was there any practical way in which its time could have been transmitted to more than a few other clocks. Clocks therefore had still to be regulated against the Sun or stars, and for the majority of clock-owners this meant by a sun-dial now used in conjunction with an equation of time table. Better clock-makers such as Thomas Tompion or Daniel Quare printed their own equation tables, supplying them with their clocks, and for exceptional pieces might even supply a sun-dial. That the equation table should be engraved on the dial itself, rather than on a perishable piece of paper was an obvious idea quickly adopted. More important were efforts made to refine the accuracy of dials themselves both in their construction and in the way they were read. In France, Jean Picard and Philippe de la Hire devised and described new methods for drawing dials which offered greater accuracy, while in the workshops of sun-dial makers, the hour scales began to be divided more carefully and more finely to read to minutes. The search for precision also produced new forms of dials of which the most important - overtly showing the influence of horology in its use of gears and a watch-type dial with revolving index - was the mechanical equinoctial dial. This according to its size and the care with which it was made, could give reliable readings to at least 1 minute. A degree of mechanisation was thus imported even into the determination of true solar time.

With the pendulum and its effects this brief presentation of three turning points in the measurement of time and in the way that man thought about it may be brought to a close. It is not the end of the story, for the further realisation that time

14. The account by Robertson may be supplemented by that in Landes 116-21 and the important study by Leopold. For the problem of the anchor escapement see Beeson, Robinson & Ward, *passim* and Turner (1).
15. Eden & Lloyd 303 no 545.
16. Allexandre, 142.

may be considered as arbitrary, and the freeing of its measurement from all dependence upon the natural world was still to follow. Nonetheless, the three stages described represent the transformation of the time of primitive man into the modern time of urban communities. As described here, the stages are to a degree symbolic, chosen to represent the thousands of individual changes in different times and different places which make up the transformation in technique and thinking.

Even as late as 1797, when Berthoud and Lalande argued against the practice before the Institute, the public clocks of Paris were still being set to true solar time, not mean-time[17]. In the long perspective used here no one time-measuring instrument stands out as fundamental to the development described, rather it was the continuing response by horologists to specific social and technical problems which accumulated to transform not only time-measuring instruments, but the nature of time itself. From being a product of nature, time became a product of the machines that measured it and thus itself an independent measure of the natural world and the universe.

17. De la Lande, 789.

Elly Dekker

From a World Ruling Time to a Time-Ruled World

Time has been measured for ages by the daily rising and setting of the Sun. At 12 o'clock local noon the Sun culminates and reaches its highest point above the horizon. The true solar day is defined as the time between two subsequent (upper) culminations. By dividing the solar day into twenty four parts the hours are fixed, by dividing the hours into sixty parts the minutes, and so on. These definitions satisfied mankind for quite some time. Only when mechanical clocks became sufficiently accurate that their role in time measurement increased considerably, were changes introduced. The true or apparent solar day is not constant during the year. Depending on the season the Sun is quicker or slower in returning to the same point in the south each day. Unless especially corrected therefore, mechanical clocks indicate a time that differs from the apparent solar time. Special instruments were designed to put the clocks right daily at noon. Another solution to the problem was to introduce a definition of time more suitable for mechanical clocks. This artificial time is called the mean solar time and can be thought of as the time indicated by an artificial Sun that is moving across the heavens regularly so that the length of a day remains the same throughout the whole year. The introduction of an artificial time was the first step in modern society to manipulate time to suit our purposes. More steps would follow, leading in the end to a system of time reckoning in which science, technology and the needs of society are intimately intertwined.

Time services

Modern time services respond to a variety of needs. The first time services worked primarily for navigation. Knowledge of time is essential to determine the longitude at sea. Using his chronometer a navigator knows the time in his home port and by comparing this with the local time as measured with his sextant, he can work out the difference in longitude between his sea position and that of his home port. But what if the chronometer would not run regularly? Then his position at sea would be at fault. In the 19th century very few chronometers maintained a daily rate of less than a second. If chronometers were to be relied upon, they had to be checked regularly. This was very necessary but it was a very elaborate thing to do. Each ship in the harbour had to send an officer ashore, equipped with a watch, a sextant and an artificial horizon in order to find the right local time. Back on board this time was compared with the times indicated by the ship's chronometers. If instead signals were to be provided at preset times, visible to all ships in a harbour, officers could verify their clocks without going ashore and without making a lengthy calculation. For instance, a big ball mounted on a device visible to all ships and dropping at known times would serve the turn. If such time services were operated in harbours all over the world, captains everywhere could be assured of an up to date knowledge of the performance of their chronometers. The first proposal of this kind was put forward in 1824 by a captain in the British Royal Navy,

Robert Wauchope, at a time when the use of chronometers at sea was becoming the rule rather than the exception[1].

The reliability of time signals was of great importance since a single error could cause errors to chronometers in all the ships lying in the roads. The institutions designated to run and control such time services were astronomical observatories for these were best equipped with instruments and know-how to work out the precise time. In England the time service was run from 1833 onwards by the Royal Observatory at Greenwich. A time ball was installed on top of the observatory and initially this was dropped by a manually operated lever system. In 1852 a much more sophisticated system was installed using newly invented telegraph technology. Following a plan devised by the Astronomer Royal George Airy in cooperation with the London clock-maker Charles Shepherd, the dropping of the time ball was automatically triggered by a master clock. This clock also served to send time signals to a secondary clock mounted in London Bridge station whence the precise time was available to all in need of it. In 1860 time balls at Greenwich, the Strand, Cornhill, and in Deal and Liverpool were dropped by signals from Greenwich and soon every part of Britain (if connected by telegraph with London) could obtain Greenwich time[2].

The installation at Greenwich was technologically the most advanced of its day. The naval interests of Great-Britain in the 19th century justified important investments in this field. In most other countries as in Holland money was not available for the construction of such elaborate automatic time devices. Moreover, Holland never had an astronomical observatory especially devoted to navigational problems. However, at the initiative of a few naval officers, a time ball was installed in Djakarta (then Batavia) in 1839. In the Netherlands itself, time balls were being dropped from the masts of guard-ships of the Dutch Royal Navy from 1845. Their introduction was severely criticized for the lack of proper astronomical facilities to determine time. This particular problem was only solved in 1859 when a regular time service was implemented by the then director of the Leiden Observatory, Frederik Kaiser, who had been officially charged with the care of Dutch naval instruments. Naval Harbours were provided with an astronomical clock which was regularly verified by time signals sent by telegraph directly from Leiden observatory. The time indicated by these clocks served to operate a manual time signalling apparatus. The latter was designed by Kaiser and thought by him to be more effective than the more common time balls. As in Britain, other institutions like the *Rijkstelegraaf* and clock-makers like A. Hohwü, also used the time signals sent out by this observatory[3].

In the United States of America also an early start with a national time service was made. In 1845 the Naval Observatory at Washington was instituted and in 1865 a public time service was started by sending time signals to interested parties like the 'fire-alarm and police stations of Washington' and 'the Western Union Telegraph Company for communication to the offices of the company's lines throughout the United States'[4]. In 1845 a manually operated time ball was erected at the naval Observatory and in 1880 time signals were sent automatically instead of by hand. This was done by a device invented by an employee of the observatory, the instrument maker William F. Gardner[5].

Clocks and Electricity

The first time services depended heavily on the telegraph. In his *History of Electric Clocks* Alexander Bain wrote about it thus : *The instantaneous transmission of intelligence by the Electric Telegraph to distant places and its influences upon society, are as yet only partially developed ; but enough has already been accomplished to show how speedily the whole world may be brought together by its mysterious and centralizing agency*[6]. The advantages of telegraphy made it possible indeed to have a centralized system of time reckoning on all sorts of scales. All that was needed was one precise clock, some wires to send time signals along, and a series of electrical devices to convert the signals into movements of the hands of a clock. Bain was among the first to develop impulse dials, as the electrical devices to keep count of the received pulses were called. His design seems to have failed, not because of the dial movements but because his transmitter was a pendulum which made contact every second. Through lack of precision in the make and break of these frequent contacts the duration of the pulse could not be controlled. This led to irregular behaviour of the dial movements. These difficulties could be reduced somewhat by using a pulse every minute instead of every second as was done in France from 1855 on. These and other attempts to make impulse systems work led to a variety of more complicated devices : *It drove us to the synchronization of ordinary*

1. Bartky and Dick (1), 155-8.
2. Howse, 79, 89-105.
3. Dekker, 29-35. A detailed description of the Dutch time service is given by Kaiser (2).
4. Weber, 34.
5. Bartky and Dick (2) give further details on the first time balls at the Naval Observatory ; see also Weber, 34-6 and Bartky, 3.
6. Bain, 3.

clocks by electro-mechanical devices; it drove Paris to a pneumatic system and produced complicated methods in Germany and Switzerland in which the current was reversed at each impulse[7]. This latter solution has become the common practice in Europe. The famous Swiss clock-maker Hipp seems to have been the first to design a polarized dial movement. By automatically inverting the polarity of his battery leads, the impulse was alternativy sent in one and the other direction and this caused an armature to move alternately to the left and to the right. Any irregular behaviour of the contacts were thereby more or less compensated by the continual change in polarity. The first dials made by Hipp received an impulse every minute which further improved the control over the dial movements.

Uni-directional impulse dials seemed for a long time to be a dead end in the development of impulse systems. In 1895 however Hope-Jones invented the Synchronome switch by which a perfectly defined signal is transmitted and the step by step movement of impulse dials is not hampered by irregularities occurring in the contacts. It opened the way to a new development of uni-directional pulse dials and the first patent was registered in 1897 by Hope-Jones himself. Nowadays most impulse systems in England use uni-directional impulse dials but the system never received much attention on the continent, whereas polarized dials never became popular in England[8].

Although not unimportant, the development described here was not the major goal in the many efforts made by clock-makers and electricians to improve horology by electrical means. All kinds of horological problems were attacked. The new source of power offered a wide perspective that attracted many inventors. But it was only dimly realized that the real problems in applying electricity to timekeeping still had to be discovered. The difficulty of making good contacts is just one example of the many technical problems that had to be solved. To keep pendulums going by means of electricity is another example. For centuries clock-makers have tried to invent mechanical devices to keep pendulums in constant motion. A pendulum slows down or its movement becomes irregular for many reasons, of which friction and temperature variation are the best known. Just to keep it going, energy must be supplied continually, either at a constant rate or in pulses. In mechanical clocks this is done either by gravity, using a falling weight or by a spring. Both exert force on a gear work, the motion of which is controlled by a clever device called the escapement. This causes a small portion of energy to be administered to the pendulum at every swing. At the same time the number of swings made by the pendulum is counted and transformed by the same gear work into the movement of the hands of the clock. The main attraction of horology for electricians seems to have been the replacement of the cumbersome system of weight and escapement by some electrical device that prevents the pendulum from slowing down. For that purpose one has to close a contact at a precise moment in order to exert a suitable force on the pendulum generated by electrical means. The various designs of electrically-driven pendulum clocks illustrate the great practical difficulties that were encountered in making such contacts and in finding the right sort of force. Many attempts were made from 1841 when Bain registered the first patent in England for such a clock, but very few were successful[9].

The history of the electric clock is much too complex to mention even all the successful inventions. The few achievements we shall bring here to the fore are selected primarily for their relevance to the story of time dissemination. Design of an electrically driven clock that could compete in accuracy and reliability with existing mechanical precision clocks was very difficult. In the 1880's the Swiss clock-maker M. Hipp invented an electrically-driven pendulum clock with a contact device that very cleverly worked only when needed, namely when the pendulum had slowed down below a certain level. This device was a mechanical rather than an electrical ingenuity. Hipp's precision regulators could maintain a daily rate of 0.03 seconds and for decades functioned successfully as master clocks in time services all over Europe[10].

Another example of the successful use of electricity in horology is the invention of the Shortt free pendulum clock. In 1921 Shortt 'hit upon a means of making two pendulums swing together in *precise* sympathy'[11]. The Shortt free pendulum is a highlight of horological wisdom. It employs the synchronous oscillations of two pendulums and pays tribute to the principle that the main pendulum - used to measure the time - must be exempt from all other tasks but swinging. For these tasks, counting the number of swings and making contact for the release of gravity levers, consumes energy and interferes with the free oscillation process. These tasks are performed by a second, slave, pendulum. The function of the gravity lever, a mechanical device, is to give both pendulums

7. Hope-Jones, 97.
8. *Ibid.,* 92-107.

9. For the patents of Bain see the introduction by Hackmann in Bain, x-xiii.
10. Erbrich, 22-3; Hope-Jones, 64-6.
11. Hope-Jones, 166.

an impulse once every half minute. The synchronization and the release of the gravity lever are electrically controlled. The care taken in this design to free the master pendulum from as many jobs as possible resulted in the most precise mechanical clock ever : the Shortts were accurate to 1 second in three months[12]. The Shortt free pendulums were widely used at astronomical observatories which formed the core of the time services in many nations[13].

High precision time-keepers have always been outside the domain of domestic time-keeping because their high price and the care they needed stand in no proportion to the requirements of daily life. For domestic purposes less elaborate electric clocks were marketed by firms like Bulle, Brillié and Siemens & Halske. Many new contact devices and synchronization mechanisms were developed by these firms[14]. A different line has been to use electricity simply as a power source to wind up a traditional weight-driven clock. All these developments can be seen to be still in line with the existing clock-maker's tradition. However, a real break-through came with the invention of the synchronous motor clock. Through it time could be brought into the house of anyone attached to the mains simply by connecting a clock to them. The synchronous motor clock is totally different from all clocks we have discussed so far because the hands of this clock are driven through suitable gearing by a very small electric motor the speed of which is entirely controlled by the frequency of the external power supply.

Although the use of alternating current for domestic purposes and its application to time measurement had been envisaged already before 1900, it was only in 1918 that the first synchronous motor clock was realized (by the American engineer H.E. Warren). Synchronous motor clocks are very simple in construction and are therefore very much cheaper than mechanical clocks. Their successful application depends however very much on the quality of the electricity supply. In the last two decades of the 19th century the 'Battle of the Systems', a struggle between supporters and opponents of alternative current was decided in favour of domestic A.C. supply. Next, the electricity companies had to decide about the use of a standard frequency which after many debates was ultimately settled at 50 cycles per second in Europe, and 60 cycles per second in the USA. Now, the accuracy of a synchronous motor clock depends very much on the stability of the frequency of the mains. A clock designed to work on a 50 cycle main supply will lose nearly half an hour per day if the frequency dropps to 49 cycles. The precision of the synchronous motor clock thus stands or falls with the stability of the mains frequency. For a power station, on the other hand, the frequency is only of secondary importance. It would require great efforts and much money to maintain the frequency on a level that would turn a synchronous motor clock into a precision time-keeper. And as one author remarked in 1942 : 'as long as the factory does not make a profit by keeping the frequency constant, one can not count on it'[15]. This holds even today and explains why synchronous motor clocks can serve only simple domestic purposes[16].

Standard Time

The rapid industrial expansion of the 19th century left its mark on the development of time-keeping and in particular on the dissemination of time in various different ways. Railway companies were among the first to show an interest in the standardization of time. The local times used in cities and towns for civil purposes were not practical for use in railway timetables. As a result railway companies started to adopt a so-called 'railway' time. In England the local time of Greenwich was adopted for this purpose, in Holland the local time of Amsterdam, in France that of Paris. The practice thus led in many nations to the introduction of a single standard time. In small countries like Holland, where the difference between this railway time and the local time of the most eastern or the most western towns is a matter of minutes, this had no consequences. In larger countries, especially the United States of America and Russia, the use of one national standard time was impossible : the difference between the local times of New York and San Francisco or between Moscow and Vladivostok amount to many hours. Therefore in 1870 the American Charles Ferdinand Dowd proposed *A System of National Time for Railroads.* The North American continent should be divided into a set of zones, each of which would have its own standard time. This system was introduced in 1883[17].

Railway men now had only to change their watches by one hour when passing into another zone. The trends described here eventually led to the adoption of the international time zone system. This system is in a way an extension of the American system but the zones are counted from the Greenwich meridian which by international agreement was

13. Hope-Jones, 166-181 ; Erbrich, 25-27.
14. See for details Hope-Jones, 21-2 ; Hoogeveen.
15. Van Herk, 173. The stability of the mains frequency is primarily fixed by the need to calculate the customer's consumption accurately.
16. Hope-Jones, 226-242.
17. For more details see Howse, 120-6.

adopted in 1884 as the prime meridian of geographical longitude. At the same conference a new concept of time was defined, the universal day 'for all purposes for which it may be found convenient, and which shall not interfere with the use of local or other standard time where desirable'[18]. This universal day was 'to begin for all world at the moment of mean midnight of the initial meridian'[19]. Originally the universal day or universal time was introduced to serve science and telegraphy but in due course universal time indirectly came into use all over the world. By gauging the international time zone system on the prime meridian, international time standards are gauged to universal time.

In the zone system the world is divided into 24 zones. The zero zone is centred on the Greenwich meridian. The standard time is Universal Time (UT) which equals Greenwich Mean Time (GMT) when the latter is counted from midnight instead of noon. The time zones to the east have standard time later and those to the west earlier than the standard time at Greenwich - one hour difference for each zone[20].

The introduction of standard times, first on a national and then at international level was not always accepted without protest. Changing from local time was resisted. Local histories of time reckoning can therefore deviate considerably from the story told so far. In Holland, for instance, telegraph time, being the local time on the meridian of the Westertoren in Amsterdam, was first used as a standard time since it was indicated by the clocks in telegraph offices all over the country. From 1859 time signals sent out from Leiden Observatory were used to verify these clocks. After 1866 the Dutch railroad companies also started to use telegraph time. However, in 1892 these companies introduced Greenwich Mean Time in order to facilitate international traffic. Since telegraph time had been closely connected with railway time, the telegraph offices followed suit and GMT became the new standard time. However, for political reasons the local time of the Westertoren in Amsterdam was again adopted, in a new law dated 1 May 1909, as the Dutch standard time. This remained until in 1940 the Germans invaded Holland and introduced for their own convenience a new time standard, the Middle European Time (MET). This standard has not been changed since. The occasional introduction and repeal of daylight saving time, or summer and winter time, since 1916 has complicated the picture further, but the Dutch seem to have suffered without grumbling[21].

New Technologies in Time Measurement

In the last two decades of the 19th century the German Heinrich Hertz carried out experiments which opened up a whole new field of communication : radiotelegraphy. The man who actually paved the way for practical applications was Guglielmo Marconi, an Italian engineer, working in London. In 1899 he succeeded in detecting a signal sent across the English Channel. In 1901 he successfully transmitted the first wireless signal across the Atlantic Ocean. These exciting events were followed by rapid development of radio technology in which transmitters and receivers were produced with increasing power and sensitivity. Pretty soon too radio signals were used for the dissemination of time. The first time signal by radio was sent on 9 Augustus 1905 from a station at the Navy Yard at Boston, USA and in the next year the Naval Observatory started regular transmission of time signals by radio[22].

The regular transmission of time signals was naturally of great interest for navigators for it made the use of the chronometer in principle superfluous. By listening to their radio, captains could obtain the time in their home ports directly. In practice, the chronometers remained aboard for use if the radio transmission failed. Even so, it was now possible to verify their rates daily during the voyage.

The first low power transmitters did not reach very far. In order to cover a reasonable area many transmitter stations had to be used. In 1909 the number of stations operating in the USA was nineteen, each having a range of about one hundred miles. Many of these stations were to disappear when in the twenties high-power transmitters were built with ranges up to 3000 miles. The first radio transmissions in Europe were experimentally carried out by Nordreich Radio in Germany in 1907. In 1910, a regular service of radio signals from there began. In that year also France started such a service, broadcast from the Eiffel Tower in cooperation with Paris Observatory[23]. With these various radio services in operation it was soon found that the time signals transmitted from different stations differed from each other. Clearly the times worked out at the various national astronomical observatories did not agree as they should. This worried the astronomers. In 1912 international coordination was called

18. Howse, 144.
19. *Ibid.,* 145.
20. *Ibid.,* 521-62.
21. Schell, 465 ; Draaisma en Kaarls, 125.
22. Weber, 35-6.
23. Bigourdan, B75.

for 'to study ways and means of effecting practical unification of radio time signals, and to prepare plans for an international time service to suit the needs of all'[24]. An international organisation, the Commission Internationale de l'Heure, and an executive office, the Bureau International de l'Heure (BIH) were established 'with the task of coordinating results from observatories and deducing the most exact time - *l'heure définitive*'[25]. Until 1988 the BIH has been the central international agency carrying out this task.

The technology developed for radiotelegraphy also opened up totally new methods for the measurement of time. The discovery in 1904 of the diode by Fleming and the subsequent development of the thermionic valve by Lee de Forest and the use of his triode for generating oscillations by the feedback mechanism discovered by Franklin in the first decade of the twentieth century, are cornerstones in the dramatic development of the electrotechnical and telecommunications sciences. With them came a demand for accurate measuring instruments, a.o. to measure the frequency of the radio wave. In order to interpret the measurements reliable standards had to be developed. The successful development of high frequency standards resulted in a very precise new type of clock, for frequency and time are directly related to each other. Frequency simply means the number of times a thing happens in a second. By measuring the one the other is also fixed.

For instance, in an accurate 1-second pendulum clock time is in fact measured by counting the number of oscillations of the pendulum (that is determining the frequency) and subsequently multiplying that number with the duration of one oscillation (1 second). Radio waves are electrical oscillations with in practice frequencies of at least 100,000 cycles per second and usually much higher. Therefore, counting the oscillations of a radio wave generated at a very stable frequency yields in principle an highly accurate measure of time. As Hope-Jones put it: *The scientists who produced the frequency standard only regarded the time element as a means of fixing their frequencies but their product was also a clock - the quartz crystal clock - and since their standards demanded the highest accuracy, the clock is of the same quality*[26].

The essential element of a quartz crystal clock is the quartz oscillator, a thin lense-shaped wafer cut in a special way from a quartz single crystal. It has the important property that mechanical stress creates an electrical potential over it and vice versa. When placed in an electronic oscillating circuit which applies an alternating voltage to the crystal its mechanical response is appreciable only when the frequency of that voltage is close to the natural or resonance frequency of the wafer, usually a few times 100,000 per second. In this way a standard high-frequency generator is constructed which can be turned into a readable clock by reducing high frequency by frequency dividers to the much lower frequencies needed to drive a small motor. The latter can be applied to turn a set of gears that drive the hands of the clock.

The first quartz clock proper was described by Horton and Marrison in 1928. In the early forties quartz clocks started to replace the mechanical precision time pieces that for about 300 years had served astronomers and the public. Subsequent technological developments increased the precision of the quartz clock such that by 1950 for the first time in the history of mankind the rotation of the Earth as a perfect time-keeper could be verified and rejected! Clearly, the Earth was no longer suitable to rule time as the variation of its rotation rate proved to be about 1 millisecond per day. Clearly too, the central role of astronomical observatories in finding true time came to an end. Time from now on was in the hands of the physicists who with their newly acquired knowledge of the structure of the atom would soon add further to the scope of measuring time independent of the Earth's movement[27].

The crystal clocks used in laboratories and observatories in the early fifties were still very elaborate structures. Their application seemed to be restricted to institutions that did research or were responsible for providing standards of time or frequency. This and the typical nature of the quartz crystal clock - a high frequency device - convinced many that there was 'no prospect of a popular domestic clock in a quartz crystal'[28]. As we know, this expectation of future developments would later be completely undone because of developments in solid state electronics, leading to the transistor and the integrated circuit. Nowadays quartz clocks and watches are at least as common as mechanical time-pieces were in the past, but they are much cheaper and are ten to hundred times more accurate.

New Physics

The development of the quartz crystal frequency generator has been followed by that of the even more accurate 'atomic

24. Howse, 165. See also Bigourdan, B104.
25. Howse, 165.
26. Hope-Jones, 253. See also Dunsheath, 266-282.
27. Howse, 172-8.
28. Hope-Jones, 262.

clock'. At the end of the 19th century atoms had been shown to exist of a nucleus surrounded by a cloud of electrons. Material properties could be directly correlated with the properties of the individual atoms or their combinations called molecules. The emission or absorption of light at definite frequencies, the spectral lines, appeared to be closely connected with the structure of the electron cloud. In the first three decades of our century a whole new physical theory, quantum mechanics, was formulated to explain the structure of atoms and their interaction with radiation[29]. Atoms are systems that can exist only in definite states of energy. When an atom makes a transition from one state of energy to another, energy is absorbed or emitted, usually as light waves or radio waves of a definite frequency. This frequency is largely independent of whatever happens around the atom. The impact of this property of atoms on modern science has been far-reaching and has also put its stamp on the measurement of time. Through it an accuracy and stability of frequency standards has been achieved that surpasses that of the crystal clock by a factor of 10,000 or more.

An atomic standard is a complicated device which is often misleadingly called an 'atomic clock', for the core of each atomic frequency standard is still the quartz crystal oscillator discussed above. Owing to various influences such oscillators, when left to themselves, are accurate and stable only up to a limited degree. If the performance of the quartz crystal resonator is to be improved upon such a system should not be left to itself. In an atomic standard a feedback device is coupled to the quartz crystal oscillator that acts as a kind of frequency stabilizer. The way this works is as follow. The feedback device includes a complicated equipment in which atoms of a special kind, for instance of hydrogen or of caesium, are manipulated in such a way that their spectral properties are applied as a stabilizing agency. For that purpose the atoms themselves are first brought into a state in which transitions between two definite states of energy are favoured above all other transitions. Then actual transitions are triggered in a resonant cavity by interaction between the atoms and a radio signal, the frequency of which is in its turn controlled by the quartz crystal oscillator. The closer the frequency of the radio signal to the frequency of the atomic transition the greater is the number of atoms that respond. Next a selection takes place which separates the atoms that actually responded from those that did not, and the former are counted by a detector. This latter device translates the number of detected atoms into a signal, which is maximal when the frequency of the radio signal is precisely the same as that of the atomic transition. This feedback signal is then used to control the frequency of the crystal oscillator. A servo-system serves to sense the effects that small variations in the oscillator frequency have on the output of the detector. Through it the oscillator stays automatically tuned at the frequency for which the detector signal is the highest. Once the system is operating at the atomic frequency, it is maintained there forever. The whole process is taking place continually and compares well with the automatic tuning of a radio till a signal is heard loud and clear : the radio receiver remains exactly at the frequency at which the signal is emitted by the radio station.

Because of the necessary feedback mechanism involved in atomic frequency standards the relevant frequencies are usually chosen in the region of ultra high frequency radio waves (micro-waves). This explains why historically the most important steps in the development of atomic standards were closely connected with spectroscopic experiments made at the time in the micro-wave range. The radar technology developed during the second world war stimulated such experiments greatly. The first 'atomic clock' was built in 1948 by Harold Lyons based on a micro-wave transition in the molecule ammonia. Further progress in atomic beam spectroscopy and maser technology resulted in 1955 in the caesium beam frequency standard which nowadays is in common use[30].

New Times Scales

For the time being the atomic clock is the last station on the long road of time measurement. The long term accuracy and stability of atomic standards has drastically improved the accuracy of time reckoning. Nowadays scientists have atomic standards at their disposal that take 300,000 years to gain or lose a second. Compared with such accuracy our Earth is but a poor time-keeper. Consequently, astronomers in the early fifties were not longer satisfied with the existing definition of time in terms of the mean solar day. In retrospect it might have been wiser if the astronomers had then and there switched over to a more physical definition of time connected with the best performing clock then in existence, the quartz

29. See for instance Heilbron, 40.

30. Kartaschoff, 56-9.

clock. But the new standard of time proposed by astronomers remained coupled to astronomical phenomena. Their choice followed the existing traditions. Astronomers, more than physicists, needed a unit to be trusted upon in the long run and the new quartz clocks did not seem able to provide that reliability. Instead of deriving time by the Earth's rotation, time was now considered to be derived from observations of the movements of the bodies of the whole solar system and primarily of the motion of the Earth in its orbit around the Sun. This new time scale, the Ephemeris Time (ET), is a complicated concept based on fictitious mean motions along fictitious mean equators and with respect to fictitious mean meridians. In order to fully understand its origin one has to go through the whole nightmare of astronomical constants involving time which is beyond our present purpose. Let it be sufficient that Ephemeris Time is measured from 1900 January 0^d 12^h (ET). The Ephemeris second was elaborately defined in 1958 as 1/315569259747 of the tropical year for 1900 January 0^d 12^h (ET). For a short period, from 1956 till 1967, the international unit of time, the second, was agreed upon to be equal to this ephemeris second[31].

The new astronomical time unit, although useful to astronomers and still in use for astronomical almanacs, could not serve the needs of physicists and telecommunication engineers. Taking into account the increasing prospects of the newly developed caesium-beam resonators, it was obviously only a matter of time before astronomical methods of defining time were abandoned and replaced by physical methods. In 1967 at the General Conference of Weights and Measures in Paris the ephemeris second was put aside and the atomic second adopted as the fundamental unit of time in the International System of Units (SI system) : the second is the duration of 9,192,631, 770 periods of the radiation corresponding to the transition between the two hyperfine levels of the ground state of the Cs^{133} atom. By continuous counting of such SI-seconds a new atomic time scale is obtained the origin of which was set on 1 January 1958 on $0^h0^m0^s$ (UT2, see below).

Today our basic time scale is the *International Atomic time Scale* (TAI). From a physicist's point of view this is the ideal time scale. In dozens of physics laboratories all over the world the number of oscillations of their atomic clocks, the primary standards, are being continually counted. The results of these individual counts are sent to Paris. There all data are collected and compared by an international organisation. It is the task of this organisation to establish an average which is hoped to be more uniform than any individual result and which constitutes the *International Atomic Time Scale.* In 1988 this task was transferred from the Bureau International de l'Heure to the Bureau International des Poids et Mesures (BIPM) at Sèvres, near Paris. This seems a natural move for by its new definition the standard of time has become part of the overall packet of standard physical units like the metre, the kilogram, the volt that is taken care of by the BIPM. However, time remains distinct from such other standards in the sense that next to the fundamental unit, the course of time (the time scale) has to be watched and to do this with sufficient accuracy continual counting of seconds in a number of institutes all over the world is still needed[32].

Next to the *International Atomic Time Scale* we keep several others because time is associated with the night and the day, the summer and the winter, and one's position on Earth, and that feeling cannot easily be abandoned. The most important of these is the *Coordinated Universal Time.* Its origin goes back to the universal day defined at the 1884 conference[33].

In spite of the fact that Universal time is affected by irregularities in the Earth's rotation it generates a time scale that is very suitable for civil life and the demands of navigation. Consequently, when atomic time scales were introduced universal time did not disappear but continued to be the time scale from which time services derive their signals. The increasing precision in time however gave rise to more refined definitions distinguishing between the various irregularities observed in the Earth's rotation. From 1956 three sorts of universal time have been defined[34].

1 UTO is the mean solar time of the prime meridian obtained by direct astronomical observation.

2 UT1 and UT0 corrected for the effects of the wanderings of the position of the Earth's pole, the polar variations.

3 UT2 is UT1 corrected for seasonal fluctuations in the rotation rate of the Earth.

Of these UT1 is the time scale that on the average most truly reflects the rotation of the Earth and therefore is of greatest importance to navigation. It also appeals most directly to our daily sense of time. For that reason life on Earth is now by international agreement regulated by a new time scale, called the *Coordinated Universal Time (UTC),* that is kept artificially as closely as possible to UT1, the astro-

31. Ibid., 100-103. For ephemeris time see Smart, 139-41.

32. Howse, 156-160.
33. Kartschoff, 100.
34. Howse, 183-7.

nomical time scale, although it runs as fast as TAI, the atomic time scale. This operates as follows. In 1972 UTC was set 10 seconds slow on TAI. After that UTC and UT1 are regularly compared with each other and as soon as the two time scales start to deviate by more than 0.9 s a *'leap second'* is added to UTC. Because of the regular addition of leap seconds UTC always differs from TAI by an integral number of seconds. Since 1972 15 seconds have been added, indicating that since then the rotation of the Earth has been slowed down by an averaged amount of 0.83 seconds a year[35].

The Bureau International des Poids et Mesures (BIPM) in France is the central office where all activities on time are coordinated and the decision to add leap seconds is taken. From there, reports are distributed to national bureau of standards like the Van Swinden Laboratorium in the Netherlands, with data on the performance of atomic clocks, the leap seconds. Apart from running the atomic clocks and reporting the measurements back to the BIPM, these national bureau, take care of the adjustment of their own clocks to UTC in case a leap second must be added and of the dissemination of UTC on a national scale. Local standard times (MET in Holland) are of course directly obtained by adding the required number of hours to UTC. Clearly, worldwide all clocks are set to the instructions of the Paris office. In civil life one is hardly aware of all these efforts to keep precise time. We often have our own watches and clocks and all we need is some reference time to compare it with. We can pick up the telephone and dial for an indication of time or we can listen to radio peeps to set our watches. Navigators (air and sea) use a very complicated and sensitive system of radio signals. soon to be fully regulated by satellites. Some of these service has existed for some time, telephone time service 1905, of radio signals for navigation for almost a century. In the course of time technology has changed tremendously and with it the precision attainable in the measurement of time[36].

35. This number of leap seconds was kindly provided by G. de Jong of the Van Swinden Laboratorium, Delft. It applies at the time of writing, that is for 1990.

36. Howse 181 ; Kartaschoff, 192-202 Zweedijk, 40.

H.F. Bienfait

Time in Nature

Thanks to traces of the past in the human world and in human memories man has evolved a concept of time. This allows him to develop ways of behaving which take account of time. But mankind is not alone in this ability. Micro-organisms, plants and animals also display behaviour which takes account of the fact that the world changes, and when such changes can be predicted, they are frequently anticipated by the organism concerned.

The alternation of day and night, outgoing and incoming tides, winter and summer are all traceable in the rythmical behaviour patterns of ourselves and of other living organisms.

This chapter and the section of the exhibition 'Times of the Natural World' examine both the traces of time in lifeless matter and the ways in which time is assimilated in the behaviour of living organisms.

A succession of events can be detected clearly if they are homogeneous and leave behind comparable traces, such as forming sediments on the ocean floor. The sequence of sediments illustrates the changing conditions in the place where sedimentation occurs. At certain times in the past, the sea withdrew completely so that only the wind could transport matter, at other times the water was so turbulent that only coarse particles of sand could sink, then again, complete calm prevailed (in deep water or an inlet) so that finely powdered sand sank together with other tiny particles such as dead algae and the calciferous skeletons of plankton.

Rivers carried grit from the mountains which was never allowed to settle because of the constant movement of the fast-flowing waters, but as soon as the water discharged into a broad sea delta it flowed more slowly and the grit was able to sink to the sea-bed. Some sediments have only accumulated millimetre by millimetre over the centuries, others have accumulated so fast that each tidal changeover has formed a layer that is more than a centimetre thick. This process can be seen clearly in mountains where old sediments have been pushed up and torn apart through folding to reveal a series of different layers.

When drilling operations are carried out, for example by a company searching for oil (formed out of sedimented remains of dead algae) or gas, the hollow inside of the drill reveals a cross section of the earth's floor that has pierced. Nevertheless, it is no easy task to piece together the Earth's complete history from the successive layers of sedimentation. When the ocean floor has been pushed up into a mountain, a period of erosion followed in which the history of the sedimentation was erased, like pages torn from a book. The pattern of sedimentary deposits can be likened to the diary of a somewhat chaotic person : periods in which an extensive account is given of all events taking place and the associated feelings are followed by days or even weeks with significant events which the author had neither the energy nor inclination to record. And all this is interspersed with more or less clear evidence of torn out pages.

The stone foundation of the 'Nieuwe Kerk' is the product

of geological sediments and is supported by poles on a firm layer of sand. The upper layer, 12 to 20 metres under the church, was deposited from over land, by the wind, during the latest Ice Age, between 100,000 and 10,000 years ago. From 21 metres downwards the layers have been formed by sendimentation from a sea. The deeper layers have increasingly hardened as a result of the pressure from above ; at a depth of 1 kilometre they give way to layers of rock which are considerably older.

At that point a number of pages are missing which should have been written between 70 and 150 million years ago. The deepest level to be struck while drilling (near Haarlem) reached as far as 3 km down and dates from the Carboniferous Period, about 300 million years ago (to reach molten rock, or magma, one would have to drill another 30 km deeper).

The stones used for the building of the 'Nieuwe Kerk' have been taken from those layers which are absent here (torn out pages) or lie hundreds of metres or more deep, but which have surfaced in Belgium or Germany.

How a sequence of different layers is formed through sedimentation is demonstrated in the Sedimentator which has been specially constructed for this exhibition. A tube filled with water is connected at the top to two barrels. One barrel contains water with white sand, the other with red sand. The white sand is continuously agitated by a stirrer with a rotor so that at regular intervals a sand grain lands in the tube via an opening and there sinks to the bottom. In this way a gradually accumulating sediment is formed in the tube. In the other barrel, the red sand is only agitated for a couple of seconds when a disturbing factor (the passing visitor) enters the vicinity. Then a couple of red grains land in the tube and red sediment forms amongst the white sediment. The sequence of red and white sediment discloses the past of the sedimentation. Should a stirrer break down, or somebody by accident take the plug out of the socket, the result would be a disturbance in the rhythm of the layers.

One sedimentator accumulates sediment at a rate of 1 mm per hour ; this one was set up on the 5 November 1990, at the start of the construction of the Great Time Exhibition and will reflect its complete history from start to finish. The other sedimentator accumulates sediment at a rate of approximately 1 mm per minute and will provide a recording of the daily history.

One sediment formation in which life forms determined the development of a rythmical stratification can be found in the limestone interplates in a water duct near Cologne. This, for the most part underground, water mains was built by the Romans in the first century A.D. to transport water from the Eifel mountains and it has operated for five centuries. In the gutters through which the water flowed, a rapidly accumulating layer of calcium formed which had regularly to be chipped away. The calcium deposit was built up of thin, brown layers, rich in iron, and thicker, mainly calcareous layers. The origin of these calcium deposits was, it is now thought, to be found in the source of the water supply. The rainwater that falls on the slopes of the Eifel trickles through the soil into the heavily calcareous ground. In the soil, carbon dioxide is continually released by the respiratory activity of plants and micro-organisms, which is unable to escape into the open air in wet weather and is transported by the rainwater in which it eventually dissolves. When this water, rich in carbon dioxide, reaches the limestone stratum, the following reaction takes place :

$$CO_2 - H_2O - CaCO_3 \quad Ca_2{-}{-}\; 2\, HCO_3{-}$$

so that the insoluble $CaCO_3$ is broken down into the soluble Ca_2– - and HCO_3– - ions which therefore promptly dissolve in the water and are ultimately transported to the source of a stream or river. Once exposed to the open air, hence, in the mains water canals, the carbon dioxide in the springwater escaped from the solution so that the reverse reaction took place and the insoluble $CaCo_3$ was formed and precipitated on the floor of the aqueduct. At those places where most access shafts to the waterworks were built for maintenance or cleaning activities, and fresh air could transport carbon dioxide the fastest, the thickest layers of calcium deposit were found.

In wintertime, when vegetation is leafless, the respiratory activity of the plant roots in the ground is much lower, and the same is true for micro-organisms in the ground which are dependent for their respiration on materials that come from the roots. This means that less carbon dioxide dissolves in the rainwater and that therefore the water carries less calcium with it from the calcareous rocks.

The dissolving of iron in the groundwater is mainly dependent on iron solvent compounds that seep out of rotting plant material ; in this process seasonal changes probably play a less important role. These iron solvent compounds eventually lose their iron to hydroxyl-ions in the water, resulting in the reddish brown $Fe(OH)_3$ deposits. In this way, thin calcium layers, rich in iron and hence dark brown, were formed from the wintertime water, while from the summertime water, thicker layers were formed in which the iron merged with the enormous amounts of white $CaCO_3$.

Upon sawing the chipped out limestone slabs, fine laminations were obtained, since the different layers were not all sedimented in a horizontal plane. The effect resembled marble and indeed, the slabs were also used as imitation marble ('Marmorersatz'). A number of these slabs have been used in the St. Lebuinuskerk in Deventer.

A small piece of 'kalksinter' ('chalk sinter') with approximately 30 'year layers' is displayed.

Nature not only influences the forming of alternating layers of dead matter, but also produces them herself: trees form their wood in a rhythm that is determined by the seasons.

Situated between the bark and the wood is a layer of cells (cambium) which by division discharge cells inwardly (wood) and outwardly (bark). The function of the wood is to transport water upwards; in the bark, carbohydrates are transported to the roots. In springtime, the cambium cells build up a thick layer of large wood vessels whose task it is to carry the rising flow of water to the new leaves. Later in the season, narrower vessels are constructed in which the density of the wood is much higher. In wintertime, the division activity of the cambium layer comes to a standstill. In a sawn off tree trunk, this rhythmical activity can be seen in the growth rings as alternating soft springtime wood and hard summertime wood. In a summer in which the tree has suffered from drought and a short supply of nutrition, the tree ring will be thin, whilst a good summer will betray itself as a thick ring. A succession of good and bad seasons is therefore apparent in the pattern of thick and thin growth rings. This pattern will be identical for all trees of the same type in the same area. If a tree of 100 years old is felled, the annual ring pattern for the past 100 years is obtained. Another tree, felled in the same area 80 years ago and then used to make a roof-beam, must show the same growth pattern in its last 20 annual rings as is apparent in the first 20 rings of the 100 year old tree that has just been felled.

In this way we can look back to the youth of the second tree, and so on; by studying the annual ring patterns of oak trees in Germany, it has been possible to observe the history of the region as far back as 9000 BC.

The timber used in buildings and picture frames can also be dated on the basis of the annual growth ring pattern: dendrochronology. Oak is particularly suitable for this type of dating method: oak was used extensively as timber for beams in roof constructions and unlike other tree types, never missed a year. (Another possible complication can be that on a hillside, the effects of a period of drought can be far more serious than in low lying areas with the result that trees growing on hillsides display an annual growth ring pattern that is different from those of the same type growing in lower lying regions.)

Living nature does not only react passively to the changing seasons or to other changes in circumstance, but can also develop activities in order to respond more appropriately to those situational variations which, by their cyclical nature, are predictable. Fungi, plants and animals have a system that for their organism functions like a clock. People naturally become sleepy in the evening, even if it is still light in the place where they are at that moment, when experiencing Jet lag. Many plants flower, not when they have reached a particular size or age, but when the length of the days and nights indicates that the appropriate season has arrived. The mechanism of this biological timekeeping system is not yet clearly understood, but it is, of course, the subject of ongoing research. How such a clock might work is demonstrated with a chemical model, the reaction of Bray-Liebhafsky, in the version of Rauscher and Briggs.

The principle of this model is that a chemical chain of reactions progresses in such a way that one particular component is involved in more than one reaction. If the concentration is high, then the one reaction progresses quickly, the other slowly, and vice versa. In the case of a favourable composition of the different components, the reaction develops in such a way that one component, which can be made visible, regularly increases and decreases in concentration for the duration of the reaction. In the demonstrated chain reaction, the substance used is Iodine, which with starch colours blue, and of which the rhythmical change in concentration is observable for several minutes after the onset of the reaction.

Similar cyclical variations have also been encountered in plants and animals, such as in the concentrations of certain compounds and the activities of enzymes. It is becoming increasingly clear that there are several systems which display clocklike properties which together or separately constitute the phenomenon that is popularly known as the 'Biological Clock'.

One variety of the bread fungus *Neurospora crassa* is known that clearly demonstrates the operation of an internal clock in the form of its reproductive organs, the conidia.

The effect of the clock can be made visible by inoculating the fungus on a substrate in a tube so that it can only grow in one direction ('race tube'). At regular intervals corresponding with the growth in a twenty-four hour period, pads of conidia can be seen. In the cells of the fungus, the clock has to be synchronized once : this is done with a light pulse (see below). The clock continues to operate for a time, dictating the rhythm of conidia formation. As the action of the clock can be displayed so easily using this method, and since disturbances in the rhythm become immediately visible in the irregular distances between the conidia pads, this fungus is very popular in the research into the mechanism of the 'biological clock'.

As night falls, many plants close their flowers or fold their leaves together. A limited period for opening the flowers can be of importance for minimizing the loss of water through evaporation and the loss of honey to non-pollinating insects. The pollinating insects appear to adapt their flight programme to the opening time of the flowers ; if one variety of flower is past its flowering stage and another variety with a different opening time starts flowering, then the activity pattern of the bee changes accordingly.

More than 200 years ago, the Swedish botanist Linnaeus constructed a 'flower clock' consisting of a flowerbed with plants that 'kept time' by the opening and closing of different flowers at different times during the day. The Morningstar *(Tragopogon pratensis)* and the Night campion *(Silene nutans)* are of course to be found in his clock, but also the Hawkbit *(Leontodon hispidus),* Pimpernel *(Anagallis arvensis),* Evening Primrose *(Oenothora lamarckiana),* and many others.

Some plants let their biological clock control the closing of their leaves ; one of these is the French bean. The bending of the leaves is brought about by the swelling and shrinking of special groups of cells in the stalk joining the leaf-blade to the stem (petiole). The change in size of the cells themselves is caused by changes in the flow of potassium in and out of the cells. It is not yet known what controls the potassium pumps in the cell membranes that are responsible for this change, and therefore it is also a mystery how the clock influences this process.

It is known, however, that in a variety of algae and plants a disturbance in the calcium metabolism, for example, by withholding calcium from the growth substrate or by administering compounds that influence the calcium absorption in the cell, can completely disrupt the plant's 24-hour rhythms. It is also known that calcium influences the potassium pumps in the cell membranes. By now there is sufficient information available about cellular calcium metabolism to formulate plausible and verifiable models of a rhythmical variation in the available calcium concentration in the cell. One particular compound (NADP) which is also essential in the light-dependent production of carbohydrates in the chloroplast plays an important role in this process. Thus, a link can be made between light and a calcium-dependent rhythm in the plant. In the bean-plan, this Ca-dependent rhythm is thought to lead to rhythmical changes in the position of leaves, through changes in the potassium pumps in the specialized cells in the leaf stalks. If a night were to be skipped, then the independently continuing calcium rhythm would ensure that, even so, in the hours that it should have been night, the leaves would bend downwards and a few hours later, upwards again.

Is the biological clock therefore a type of cyclical change in calcium ions, similar to that of the iodine ions in the Rauscher-Briggs reaction?

It is of course, not that simple. An example from zoology :

The fruit-fly *Drosophila pseudo-obscura* crawls out of the pupa around sunrise. At that time the moisture in the air is at its maximum which is important as it enables the straightening of the wings which come out of the pupa still folded up. The moment of emergence is determined by a system that fluctuates with a 24-hour rhythm. Flies have been found with genetically determined abnormalities whose pupae opened too early, too late or at an unpredictable moment. In almost all cases the defect is found at a place in the chromosomes that is called 'per' (from periodicity). This 'per' gene appears to be responsible for the production of a protein that displays all the characteristics of a cell membrane protein. In addition, it appears that during the larval period, the abnormal fruit-flies have a disturbance in the transport system between cells in the tissue of the salivary gland. Evidently the transport of one or more substances between the cells is essential for the operation of the relevant 24-hour rhythm. For the previously mentioned calcium dependent 24-hour rhythm, no between cell transport is necessary. This is also true for the single-celled alga Euglena.

Humans also have a biological clock system consisting of a main clock in the brain and a number of subsidiary clocks, that work more or less independently, in the other tissues. Someone who travels from Amsterdam to the United States becomes sleepy at the moment that it is evening in Amsterdam, even though it is still broad daylight in New York or San Francisco. The main clock, a 24-hour oscillator, operates in the brain close to the place where the optic nerves cross over (supra-chiasmatic nucleus). The mechanism of the oscillator is unknown. It is, however, known that the 'synchronization' takes place by means of a signal from the optic nerve, that is, through light. The amount of light necessary for an effective synchronization signal appears to be relatively large, far more than was supposed until recently. In wintertime, getting up and going to work in the dark and subsequently spending the day in an office, it is possible for this minimal light intensity not to be reached. The main clock and the subsidiary clocks must then continue working on their own. If synchronization could not take place for a long period, the clocks would lose time with respect to the actual time as the individual cycle of the clock is not exactly 24-hours, but mostly somewhat longer. In such cases, psychological symptoms can develop that are very similar to those associated with jet-lag and which manifest themselves as depression. In recent years, experiments have been carried out with light therapies consisting of supplementary light baths in the early morning.

In a light booth different light intensities can be obtained, among others, that of the minimal intensity of light necessary for the synchronization of our main clock (2500 lux), where one can (for a moment) experience a light bath.

The mechanism by which the main clock is synchronized by light is unknown, except that the signal is communicated by the optic nerve. There is, however, an additional centre in the brain, the intergeniculate complex, that effects the suprachiasmatic nucleus. This centre is thought to be activated not by light, but rather by darkness, thereby transmitting a 'dark signal' to the suprachiasmatic nucleus. From the suprachiasmatic nucleus, the position of the clock is passed to the pineal gland (epiphysis), probably via neuronal tracts. In the pineal gland, a chemical signal (hormone) is produced and excreted into the blood whereby it passes to the rest of the body and synchronizes the other clocks that operate in the tissues. The signal is melatonin and it means: night. Melatonin production is suppressed if a person sees more than 2500 lux of light. By administering melatonin at the right moment, the effect of jet-lag on sleep can be considerably reduced.

Those who do not have a melatonin remedy readily available can best combat jet-lag by surrendering completely to the new day and night rhythm, and especially by not going to sleep at the moment night falls at their previous location as this only further delays the synchronization of the clock. Other methods have been devised as well, such as following special diets before, during and after the flight but in all likelihood any possible curative effect lies mainly in the fact that one takes the trouble to do something about jet-lag and wants to invest energy in a fight against it. So called 'early birds' and 'night owls' probably differ in the cycle of their biological clock: there are indications that morning people have a shorter cycle than night people. They take longer to recover from the symptoms of jet-lag.

The existence of a biological day-night rhythm is clearly apparent in the difficulties experienced by night shift workers: sleep disorders, problems with concentration, and a variety of health complaints. Working through the night once in a while is easily remedied by catching up on sleep the following day. But working every night of the week and sleeping (or trying to) in the daytime causes a type of jet-lag, and is actually worse: in the case of jet-lag, the new light regime at least stimulates the adaptation of the biological clock, whereas those on night shift receive, both at home and on the way to and from work, completely the wrong light signals. In addition, sleeping in the daytime is difficult. Not only because of the light and the noise of life going on elsewhere, but because the amount of time one sleeps is influenced by the biological clock. Going to sleep in the morning results in a couple of hours less sleep than having an early night, even in controlled laboratory conditions of dark and absolute silence.

It is not always possible to avoid night shifts and it is

therefore a matter of limiting the negative effects as far as possible. Current research aims to discover whether this could be done by changing the sequence of consecutive shifts (morning-day-night shift or morning-night-day shift) and the optimal duration of each shift.

Theoretically speaking, a permanent night shift would be the best solution to this problem as our biological clock would be able to adjust completely, but this too would cause problems (free time, social activities, family). As a rule, therefore, it is a matter of finding an alternating schedule that disturbs the 24-hour cycle as little as possible ; in this, the type of work that has to be done is also a determining factor.

The fact that certain symptoms tend to develop at a particular moment of the day or night does not necessarily imply that they are connected with the biological clock system. They can, of course, easily be linked to something else that varies in accordance with the 24-hour cycle in daily life. For instance, certain forms of depression and asthma mainly occur in the early hours of the morning (around 4 a.m.). The question is, what causes such a distinct 24-hour rhythm. Histamine, epinephrine (adrenalin), and cyclical AMP display their highest or lowest concentration in the blood at exactly 4 o'clock in the morning and are perhaps partly responsible for causing the problems. But also familiar, is the nightly worrying around the same hour which prevents everyone, once in a while, from getting a good night's sleep. The most common explanation for this is that problems trouble us most at the time that we can do the least about solving them, the possibility of escaping them in work or social contacts is at a minimum, and the deepest sleep is already over.

In animal husbandry, extensive use had been made of the advances in the biology of timekeeping or chronobiology. By administering melatonin to sheep, it is possible to deceive them (or rather their clock) into believing that the nights have become longer so that their fertile period, which normally takes place in the autumn, commences. In this way, the birth of lambs can be planned exactly so that lamb can be put on the market when the price is at its highest. Melatonin administration is also used to ensure that furbearing animals produce their winter coat at the desired moment. (Artificial fur can, without the administering of hormones, be produced in all possible varieties).

The 'biological clock' as it has been described so far concerns systems that show a cycle of (approximately) 24-hours. In fact, there are also oscillations with shorter or longer cycles.

Field mice display a rhythm in their activities that covers a period of about 2 to 3 hours : 1/2 hour walking around outside searching for food, 2 hours sleeping in the nest. What, we may ask, is the purpose of this type of rhythmical behaviour. It may be related to the digestion time of the food in the gastrointestinal tract. If the mouse is full, there is little sense in walking around any more, and moreover, the risk of being eaten by predators dictates that is wiser to stay inside. If, driven by hunger, the mice venture outside, it is better that they all go out at the same time so that the risk of being caught is evenly shared. They also save energy by sitting together in their nest and keeping each other warm.

Just as the 24-hour rhythm in humans, the rhythm of mice is not exact and constant : individual mice fall quickly out of step with each other. In a group, the mice continually synchronize each other's clocks : they leave together and return together.

In humans, ultradian rhythms have also been demonstrated, apart from the obvious heart rate (1 per second), and respiration (20 per minute). A slower rhythm of approximately 1 per minute is measurable in the depth of the breathing and the blood flow in the skin.

There are also cycles with a periodicity longer than 24 hours : the most well-known is the menstrual cycle in women (28 days). A great deal is known about what happens during this cycle, particularly with respect to the sequence of the hormones in the blood, and the role played by the different glands and cerebral centres. However, exactly how the precise timing of this whole complicated interplay hangs together is still not fully understood.

Various types of bamboo have a flowering cycle of as many as 20 to 60 years, the Chinese Bamboo *(Phyllostachus bambusoides)* beats them all with 120-year cycles. The seeds that are then produced can assume gigantic proportions and are made in vast amounts ; the mother plant invests all her reserves and then perishes. The underlying strategy is probably that if the seeds, which are very popular with rodents and birds, fall off all the plants at the same time, the population of animals present becomes completely glutted. To be sure, the creatures quickly propagate themselves, but in the meantime some seeds have germinated and survive. Seeds that are produced a year later land up in a hungry population of rats and birds which make quick work of them. The plants therefore have to wait for a number of years before flowering until their attackers and their offspring have been reduced to their normal quota.

Cycles of this sort are, of course, difficult to study, but as humans have spread this plant over the entire world and the plants maintain their rhythm, the effect of different environments on the expression of the rhythm can be examined. In this way it has been observed that around the equator, bamboo has difficulty in maintaining its cycle : this suggests a year counting system on the basis of the seasonally-dependent alterations in the length of the day.

That measuring the length of day and night is not just a matter of registering the position of a 24-hour rhythm appears, among others, from an investigation into the way in which many plants decide whether they will flower or not.

The Red Goosefoot *(Chenopodium rubrum)* can be induced to flower by subjecting it to a day-night rhythm in which the nights are longer than 12 hours. Grown in a lighting regime in which the nights are shorter, it fails to flower. For one particular variety of this species one single night of 16 hours, in a life of otherwise continuous light, is sufficient to start flowering anyway. Such a plant is ideal for this kind of research : with just that one flower-inducing night, all kinds of wonderful experiments can be carried out.

In this investigation it also appeared that a much longer night, of 46 or 76 hours, can induce the plants to flower. However, neither a 30-hour night, nor a 60-hour night produces the same result.

In this case, the 'biological clock' seems to consist of the capacity to switch on, at the onset of darkness, a fluctuation in the sensitivity to the light going on again, with a periodicity of 30 hours.

In this field too, the business community has recognized the value of chronobiology. Chrysanthemums, which normally flower as the nights grow longer, are now not only available in the autumn, but throughout the whole year. It is simply a matter of turning off the lighting in the green-house on time and screening off daylight. A somewhat uneasy feeling at seeing a bouquet of Chrysanthemums in springtime is equally experienced by morning and evening people.

We shall attempt, throughout the exhibition, to indicate the date by the progressive flowering taking place in a long row of plants. When the last plant flowers, the Sedimentator will be full and the Great Time Exhibition over.

Willem Dijkhuis

A Hinge in Time

Introduction

Mass culture of the 20th century in work, entertainment, sports, leisure, health, tourism, education and politics, is a direct descendant of the elitist scientific and technical civilisation of the 19th century. The mastery of time and space achieved then, set the clocks for our current temporal awareness.

This temporal awareness of western cultures and societies has shifted and changed dramatically during the last quarter of a millenium. Now at the end of the mechanical industrial era, we may ask if it is legitimate to speak of an appreciation of time that is characteristic of that period. Can we talk about a temporal awareness typical of the first industrial age? In what way is that awareness based on previous stages of social evolution? And what does this imply for our temporal awareness now and in the future?

Quantification implies precision. Manmade objects like the megalithic monuments of Stonehenge and Avebury still testify to this long quest for quantification and precision in coping with time and space : a pedigree much longer than that of weightdriven clocks or that of the unequivocal measurement of space. Many cultures do not know this preoccupation. The Hopi language, for instance, has no words for time and space. Its verbs know no tenses. Western languages abound in time dependent words. For many millenia Western cultures and societies have felt the urge to quantify their realities and environments in temporal aspects.

In Europe an explicit, ever growing social fascination with time and space started to manifest itself. A slow but steady harvest of measuring devices resulted therefrom : astronomical instruments, clocks and methods to record quantified measurements.

The Middle Ages knew a great number of boundaries in its view of the world : its earth was finite, and man was limited in his possibilities of transportation and communication. When a medieval monarch offered a kingdom for a horse he was talking good sense in his day. His great need was for the highest speed available for purposes of survival and the communication of information. His Majesty put his hopes on what may be the most tenacious, functional upperlimit culture has ever known. The speed of a fine horse carrying as fine a horseman was for thousands of years the unquestioned, finite, barrier to the celerity with which man could put his mark on the face of the globe. Napoleon's *Grande Armée* swept through Europe hardly faster than Alexander the Great's through Asia, more than two thousand years earlier.

At the end of the first half of the eighteenth century the fascination with time grew into a fullyfledged obsession. During its second half, society started to capitalize on the accomplishments of the age of Pascal, Huygens, Leibniz and Newton, a century before. When the eighteenth century craftsman James Watt experimented with a new kind of power, clocks with a precision better than one second per day already existed. The obsession with space and time and with its offspring, speed, escalated during Queen Victoria's reign in the century after Watt ; particularly in anglophone and urbanized parts of the globe.

The size of the earth has contracted and man now gives earth its time. Here some of the 19th century developments which created modern time are discussed and the barrier of time is hinged open.

Listening to Time

Almost in spite of its associations with the Enlightenment and its hopes for the final triumph of rationality over things nasty, brutish and short, the second half of the eighteenth century puts a spell on us, who live two centuries later. The first cast-iron bridge, the invention and hey-day of the noble savage, Jean-Jacques Rousseau throwing his watch away, the tinkering with steam, electricity and magnetism of craftsmen and aristocrats seem to obey one single pattern.

There is a lot we cannot see. No past can be completely reconstructed or re-enacted. Individual objects of art and craft of that period can give our present eye, professor among our senses, only a very narrow vision of the cradle of modern industry. What we can still see from that period is, as always, the result of history's elimination. What was too fragile, what was deemed irrelevant, or considered a hindrance, was put out of sight by practicality and pragmatism.Rather than putting all our trust in our eyes, as we are wont to do, we should trust our hearing, the policeman among our senses, a bit better. We should try to listen to the era in which Mozart was born, and died, 1756-1791.What we hear is the sound of ticking clocks and the beginnings of mechanical, automated regularity. This is the time that the foot regiments of armies started to march in step, that human aggression took on a clockwork rhythm. It is the totally controlled rattle of the stop-watch, originally invented, before the eighteenth century's first half came to its end, to measure the performance of race-horses. There is another sound. It is the inanimate metre of the steam engine, levelling off into a majestical, mechanical regularity, beyond the particularities of the individual horse.When that age had come to its grinding halt in 1815 at Waterloo, all the conditions were fulfilled for the emergence of another new sound : that of Maelzel's metronome, defining, once and for all, the pace and tempo of musical performance, the most abstract of all arts. One of the greatest innovators in the European musical tradition, Beethoven, was characteristically the first to prescribe a Maelzelmetre : for his Opus 106, 'Sonate für das Hammerklavier', that musical instrument as new to the period as the metronome of his friend Maelzel. During the first two decades of the nineteenth century the first wave of the publishing of musical scores manifested itself, allowing music from earlier days to be played, re-enacted, on a large scale.

Bach's Saint Matthew's Passion had not been performed for a century when Felix Mendelssohn presented it once again in 1829 to a musical audience that knew its contemporary composers, but few others. Mendelssohn, who is said to be the first conductor to turn his back to his audience - Mozart still faced them - embodies, literally and metaphorically, a new temporal awareness of culture. His musical experiment was quite unprecedented, - and a phenomenal success. In order to discover the new in the old a reversal was necessary.

A New Kind of Time

After the Renaissance, the development of historical time, and of time in general, was largely held to be a purely linear affair : a subsequent phase of a socio-cultural phenomenon depended, almost organically, on its previous stage.

During the seventeenth century a new political world emerged, where the word 'revolution' implied the return to an old, often fabled, ideal set of conditions. Originally, it stood for a cyclical behaviour of heavenly bodies : it was an astronomical term, as in the title of Copernicus' work *De Revolutionibus orbium coelestium,* published in 1543. Only during the drama following the fall of the Bastille, and because of it, did the word 'revolution' gain its present sense : a radical break with the past. By the same token, the French revolutionaries were only doing what comes naturally when they ordained a brand-new calendar, starting with year one of the Revolution ; it lasted some ten years.Yet, time would never be the same again. During the nineteenth century, the single line of Enlightened Progress as designed during the Ancien Régime was transformed into a texture of a higher complexity. To this age, the perennial dilemma of life - that it must be lived forwardly, but that it can only be understood by looking back on it, to use Kierkegaard's words - presented itself as a vast web of existential choices, rather than as one rational thread. It is an age desperately searching for the one style, the style of old. It wouldn't find it ; it couldn't find it anymore, for the old, simple, consistency and homogeneity had gone.

For the first time in history, eclecticism in art, -architecture, literature and in history itself - became a direction on the compass of western cultures. Anachronism, the well-considered and artfully produced clash between two temporal domains, warp and weft contrasting with one another, is very characteristic of the nineteenth century. It fed its mind *à la carte,* chose from a menu which became more and more elaborate, where its predecessors had been satisfied with potluck.

While that age did not develop a style of its own, it invented a new concept : the neo-style. The invention presumed a wider and deeper view of times past and of the consequences of time passing.

The Past of Tradition Refuted

Once upon a time, Europe had a well-kept, clearly surveyable past. Basing himself on the appropriate data of the Old Testament and a suitable method of calculation the German astronomer Hevelius, for instance, ascertained, in his *Prodomus Astronomiae* (1690), that the world was created at 6 o'clock p.m., 24 October 3963 BC. Dozens of these determinations of the moment of the 'big bang', Old Style, circulated, reflecting commendable precision but between them disagreement. However they all agreed that the world was less than seven thousand years old.

During the fateful journey of HMS *Beagle* its most famous passenger must have had a miserable time. Sea-sickness claimed Charles Darwin for its victim. Back in England, in 1836, it took him twenty-three years to recover from his dizzying experience : in 1859 *On the Origin of Species* was published. The delay reminds one of the long and brooding years Copernicus needed to come to terms with the ideas expressed in his *De Revolutionibus.*

It seems as if the sheer notion of the evolution of life along Darwinian lines - a notion that would expand the past's magnitude beyond the reach of tradition and conventional wisdom - caused a commensurately long period of gestation of the book on behalf of its author. The past of biological life ceased to be a well-defined conceptual space.

Biological life was to Darwin what geological stones were to Charles Lyell. In the three volumes of his *Principles of Geology,* published in 1830, 1832 and 1833, this son of a botanist finally discredited the catastrophic, traditional school of geologists. Lyell's work had inspired Darwin's venture, just like inanimate matter precedes and makes possible biological life. The two men supported one another in the refutation of the ultimate temporal assessments - and limits - of our culture in that period.

When Lyell published his book *The Antiquity of Man* in 1863, four years after the 1850 copies of Darwin's initial print-run was sold out, the cooperation had come to full fruition : the first seeds of a new temporal awareness in Western society were sown. The past of the entire earth had ceased to be a well-defined conceptual space, controlled by an apparent god and an obvious command.

There is another book. Twelve years after Lyell's work, Heinrich Schliemann published, in 1875, his *Troy and its Remains.* Homer's fabled heroes - Achilles, the first true individual in all of literature, among them - joined the historicity of stone and of life. The past of mankind's recording of itself had ceased to be a well-defined conceptual space.

The third quarter of the nineteenth century changed our view of the landscape of time's past as thoroughly as, given the grim prospect of a world-wide ecological disaster, the fourth quarter of the twentieth century is changing that of its future.

Because of the new temporal perspective, and only because of it, the assessment of non-Western cultures by the dominating political and military forces, those of the West, could start to change. From tribes of savages - noble as in the early Romantic view, or exploitable as in the colonizing one - these cultures could become, slowly, our 'contemporary ancestors'; the word 'anthropology' was a neologism when Tylor used it in the title of his seminal text-book in 1881.When the industrial era was in its prime and in full vigour it created, perhaps by its nature, a new awareness of time lacking all precedents.

New Ways to Judge the Progeny of Time

Edward Gibbon, the author of *The Decline and Fall of the Roman Empire,* that unique example of eighteenth century temporal awareness of a distant past, and an enduring bestseller ever since, died in 1794. The next year, Frau von Ranke gave birth, in Germany, to her son Leopold, destined to become the founding father of history as a pursuit of the university, rather than as a calling in the life of a man of letters as in Gibbon's age, which was also the age of Voltaire.

Subtracting Gibbon's birth-year, 1737, from that of von Ranke's death, 1886, results in 149 years, some five human generations. During that elapse of time, the concept of history in the West was completely changed.

The established tradition of the historian as the worthy descendant of the incomparable amateur Tacitus, unrestricted by academic discipline and training at universities, was slow to die. Thomas Carlyle, born in the same year as von Ranke, and who lived only five years less, still represents the old tradition of history as literature, and vice-versa, while a different breed of historian was raised at the German universities of the period, basking in the sun of 'Bildung' as von Humboldt made it shine.

As the study of history - until then the work of *literateurs* - became universitized, slowly but surely, a new wave of scholarly specialization manifested itself. The 'philosopher' of the eighteenth century could study the classics, Locke, Newton, Montesquieu, Leibniz and Descartes as one whole, one entity. Now, two hundred years later, we have separated their work into physics, mathematics, political science, history, and philosophy as an academic pursuit. The contemporary of Diderot and d'Alembert still employed one, and only one, concept of time: largely an unmathematized one, taken for granted by him and by his cultural envelope. By specialization, perhaps the ultimate cultural characteristic of the Industrial Age, this monolithic, homogeneous concept of time and its epistemology exploded into a galaxy of times, all of them quite, and sometimes alarmingly different, while most of them demanded long years of exclusive and dedicated scholarship. The technological and scientific achievements of the nineteenth century are the true agents of new ways to judge time's faces.

The Moment Eternalized; the Recurrence of Events

The recording of visual images and of sounds, Nièpce and Louis Daguerre's 'pencil of light', photography, and Thomas Alva Edison's phonograph are the two key inventions of nineteenth century. Daguerre and Nièpce formed a partnership for the development of their photographic inventions in 1829; phonography became a reality in its second half, in 1877.

Both inventions were, from their beginnings, linked with a fascination with the passing of time, the urge to capture evanescence and by intimations of immortality. Photography derived much of its early and rapid social and commercial success from the entrepreneurial suggestion that making photographs of your loved ones would keep them from oblivion in all possible futures; while Kurt Vonnegut, in his youth an employee of the Public Information Department of General Electric, successor to the Edison Company, has attested that, when asked by a journal to what purpose the phonograph might be put Edison, after a few seconds, suggested, 'Last Wills and Testaments'.

When in 1895, Auguste and Louis Lumière invented cinematography and a motion-picture camera - the same year that Wilhelm Röntgen discovered x-rays- the foundations were laid for an entirely new temporal awareness. The impulse behind these discoveries was not cultural change nor the need for it: it was just scientific curiosity of the typical Victorian brand. Movement in nature often takes place either too slowly or too quickly for the unaided eye, so time-lapse and high-speed photography makes them accessible to our vision. Muybridge settled with his photographs the long-standing dispute as to whether all four of a horse's hooves were ever off the ground together (they were).

These inventors were not consciously seeking to find new forms of mass entertainment and art ; they were fascinated by motion and time, and tried to master them.

Time conspires with space in the enigma that is motion. The photograph, the visual still, eternalizes one fleeting image : 'detemporates it' would be a better expression.

Cinematography and phonography reenact sequential events in time and space, 'retemporates' them. Slow-motion rendering of a recording dilates time to its witnesses, allowing them to analyse, at leisure, visual complexities too rich for technology-independent reality.

Because of the technical achievements of the nineteenth century, Western culture gained immortality, - admittedly one of a totally artificial kind. Any effort to attain permanency in human affairs must be intrinsically paradoxical, while human reality lies beyond repetition.When Albert Einstein published his paper on the special relativity of time in 1905, time had already become largely materialized in Western culture, less and less an event, more and more a thing.

The Mystique of Speed

It is important to note that the first precise timing of motion was connected with horse-races. In comparison, the Greek athletes and audiences do not seem to have cared how fast the race was run. Of course they lacked sufficiently accurate chronometers, but they must have been perfectly capable to measure accurately in space how far they jumped, how far they threw the javelin. But as far as we know they did not do it. The achievement of athletes in Classical Antiquity and their reputation consisted of lists : of the events they had won, and of the adversaries they had beaten. Precise measurement of distance, time and of weight and shapes (of the discs and javelins) - and their standardization were matters of no consequence. Classical philologists even maintain that the Greek language lacks expressions for 'setting a record', - or for breaking one. Our current noun 'record', an abbreviation of 'fastest recorded time', as it can be differentiated from the verb 'to record' is a neologism dating back to the 1880s.

When the Olympics were revived in 1896, on the initiative of the Baron de Coubertin, an American observer, Rufus B. Richardson complained in *'Scribner's Magazine'* of September of that year that gymnastics were not very popular, since they couldn't be measured precisely : that is mechanically, not depending on the judgement of informed individuals.

The decades before this event are replete with the fascination of speed : Pitman's 'stenographic shorthand' dates from 1837 and coincides almost with the discovery that the new contraption, the typewriter, could command speeds which out performed the writing human hand. During the same time, Charles Babbage, professor of mathematics at Cambridge, and founding father of computer technology, published his book *On the Economy of Machinery and Manufacture,* 1831. Babbage quotes here the data of the Frenchman Perronet, who in 1760 timed by the clock, and computed, the cost of each operation in the manufacture of twelve thousand needles.

It should be noted that both the Industrial Revolution and the concept of modern sports hail from the same heartland : Great Britain. The coincidence is by no means accidental. Sports and industrialized capitalism are birds of a feather.

By 1912, following the time studies that began in Frederick Winslow Taylor's machine shop of the Midvale Steel Company in 1881, when Frank Gilbreth realized, assisted by the motion-picture camera, the first precise recording of movement, 'scientific management' of time had already become a reality. By 1915 Henry Ford could reap the harvest of Taylor's toil at his plant at Highland Park : the conceptual results gained by the dogged perseverance of an innovator had become common knowledge, accessible to the self-taught man that was Henry Ford. A new belief had been born : that skill in trades, in all forms of athletics, even in professions such as surgery, was based on a common set of fundamental principles. In 1936 the nature of speed and efficiency was mythologized, once and for all, in Charlie Chaplin's 'Modern Times' ; in the medium of cinematography, that had, curiously, been instrumental in the taming of time and of its exploitation.

Life Lengthened

The virtual absence of any kind of anaesthesia before the invention of chloroform in 1831 - simultaneously by the

American Samuel Guthrie and the German Justus von Liebig - (ether was first used by Morton, in 1846) had accustomed surgeons to the necessity to work at incredible speed. Dominique Jean Larrey, surgeon to Napoleon's Imperial Guard, personally performed two hundred amputations in 24 hours after the battle of Borodino : not as an efficient butcher, but with incredible deftness and consideration for his patients and with excellent results. The feat passed into medical legend. No surgeon of our time could possibly amputate a limb with the speed Larrey commanded. He would be at best four times slower, even with today's most sophisticated equipment.

In its quest for permanency and its battle against time, the nineteenth century even managed to circumvent human physical pain by anaesthetics, at least for some time. It is a story of a great social gain. When Queen Victoria used chloroform, while giving birth to her seventh child in 1853, twenty-two years after the chemical was invented, she bestowed regal blessings on an artful new way to manipulate the human experience of duration.

The delay of decay and death is one of the grand themes of the industrial age : in the middle of the eighteenth century life-expectancy in England was less than 35 years ; at the dawn of the Industrial Revolution a fifty-year old was indeed an old man. In 250 years life-expectancy has been doubled in our part of the world, mainly due to the great inventions of the century preceding ours : names like Pasteur, Jenner, Semmelweis come to mind immediately, but the list is almost endless. Take Nathaniel Ward (1791-1868).

He was a practising London doctor and an expert amateur naturalist. Due to the propensity of gardeners to improvise, perhaps, but certainly armed with a sound dose of natural curiosity and the type of inquisitiveness common in his age, he took up the study of the growth of plants in glass containers, under practically air-tight conditions. He constructed a case to become known as the Wardian case : in it, moisture-loving plants could survive desiccated air and polluting fumes.

This simple invention became quite fashionable in the smoky and overheated drawing-rooms of the period, but its enormous significance lies elsewhere. Ward's invention made the introduction of the tea plant to India possible, of the banana to other countries than China, and of quinine-producing plants to places outside of the American continent.In a number of ways this Victorian achievement is typical for the period. It is directly associated with the extension of biological life and the postponing or abolition of decay - Ward's case is really a time-capsule facilitating vegetal survival - with far-reaching, long-lasting socio-economic consequences ; now largely forgotten outside the small circle of historians of gardening ; and its inventor remained almost unrewarded.

Robert Fortune, that giant among plant-collectors of the period, was the first to use the Wardian case in the grand manner during his first visit to China, 1843-1845. For the first time in the history of the world living plants could securely and predictably survive sea air and sudden changes in temperature. Once more the industrial age had increased its command over time and space.

The Paradox that is Time

Exactly one hundred years ago Robert Routledge published a book entitled *Discoveries and Inventions of the Nineteenth Century.* One year ago Bracken Books published a complete reprint in facsimile. It is a typical *fin de siècle* work. In 681 pages it sings the praises of a world at last securely encompassed by science and technology on a basis of permanency.

A quotation from page 478 might serve as a beginning to a conclusion for the present essay :

'By furnishing the means of almost instantaneous communication between distant places, the electric telegraph has enabled feats to be performed which appear strangely paradoxical where expressed in ordinary language. When it is mentioned as a sober fact that intelligence of an event may actually reach a place before the time of its occurrence, a very extraordinary and startling statement appears to be made, on account of the ambiguous sense of the word *time.* (author's Italics). Thus it appears very marvellous that details of events which may happen in England in 1876 can be known in America in 1875, but is certainly true ; for, on account of the difference of longitude between London and New York, the hour of the day at the latter place is about six hours behind the time at the former. It might, therefore, well happen that an event occurring in London on the morning of the first of January, 1876, might be discussed in New York on the night of the 31st of December, 1875'.

This was written well before world-wide telephone, radio

and satellite systems had become social realities. The inheritors of the nineteenth century could not easily come to grips with this brand-new paradox. As Whitrow remarked 'One of the main reasons for the catastrophic failure of diplomacy to prevent the outbreak of the First World War in August 1914 was the inability of diplomats to cope with the enormous volume and unprecedented speed of telegraphic communications in the last days of July. The rate at which messages could be sent from one capital to another necessitated rapid and often ill-considered response'.

Routledge wrote his rather endearing observation years after Jules Verne published Phileas Fogg's adventures in *Round the World in 80 Days*: that was in 1872. H.G. Wells would see his *Time Machine* printed in 1895. The roots of science fiction all lie in the nineteenth century.

In 1889 a watch-maker in Munich, Riefler, invented an escapement with a pendulum that was almost free from interference. Its error per day was one hundredth of a second. In daily affairs today there is still no need for greater precision. Only scientists have everyday use for high precision. But the cry of anguished Lear 'O, reason not the need', is truly universal and time-independent. High precision for contemporary man is not a need but an aspiration, and the precise time of which he can dispose sets him on the threshold of new truth. Old wisdom here meets modern knowledge for as Bacon formulated an older saying, 'truth is the daughter of time'.

Catalogue

Time which antiquates Antiquities,
and hath an art to make dust of all things,
hath yet spared these minor *Monuments.*

Thomas Browne, *Hydriotaphia* (1658), V.

Origin and the Course of Time

Neither in Time, nor yet before the same,
but in the instant when Time first became.
Guillaume de Saluste, Sieur du Bartas,
Divine Weeks and Works (1578),
translated by Joshua Sylvester (1605)
I.1.37-8.

The Beginning of Time

In the beginning of the universe, a lot happened in an almost inconceivably brief time. As the universe expanded, things became quieter and went more slowly. To accommodate this dramatic change of pace, we take steps in time which become progressively longer. We start with the time before which we are ignorant of any physics at all: 10^{-43} seconds. The time when physics-as-we-know-it started was 10^{-18} seconds (one million-million-millionth) after the Big Bang. We end with the longest time for which some sensible physical speculation exists, a staggering 10^{34} years into the future.

6 Time and the World Begin Together
The 'Big Bang' video presentation showing a summary of the events which led from the beginning of our universe to the world which we observe today.

7 The Big Bang Compressed Into a Week's Time
When you look up in the sky, you realize how insignificant we are as compared with the vastness of the Deep Space of the universe. When you look at your watch, you should realize that we are equally insignificant in the vastness of universal Deep Time.
If the age of the universe were compressed into a week, starting at zero hours on Monday morning, the universe would be a seething, opaque mass of light and incandescent particles for the first thirty seconds of that day. At a minute past midnight, the universe would have cooled and expanded enough so that it became transparent. At a quarter to two that early Monday, the first galaxies would begin to form, lighting up with dazzling brightness as quasars at half past nine on Tuesday morning. Most of the mid-week would be spent making more galaxies, and all the stars in them. By Thursday, a thousand generations of heavy stars would have formed and exploded, seeding the voids between the stars with the chemical elements from which you and I were eventually made. The Solar System, and Earth with it, would form around coffeetime on Friday morning. The first appearance of living nucleated cells would be a good item for the late night news that Friday. Humans as a species would not appear until two minutes before Sunday midnight. The first human toolmakers made their appearance three seconds before the cosmic clock finished striking twelve.
The universe expands. Today, the distant galaxies are farther away from us than they were yesterday; tomorrow, they'll be further still. Thus, the progress of time is reflected in the expansion of the cosmos as a whole. We may think of this cosmic expansion as a constant creation of space. In the past, there was less space in the universe; we say that matter and energy were denser at earlier times. If we follow the universe back in time, back to its origins, there comes a time when it was infinitely dense in mass and in energy. It was infinitely compact and infinitely hot.
This 'infinity' point marks the beginning of time and space. We call it the Big Bang. There's no such thing as 'before' the Big Bang, because nothing existed, not even Time itself.
Astronomers have partly unravelled the early history of the universe. We now know something about the way in which the universe expands, and how it has expanded in the past. We can prove that it must always have been expanding, more rapidly in the past than today. We know its age: the universe started expanding some 20,000 million years ago. Thus, we know that there is a finite amount of time in the past. One of the great surprises of astronomy is that the universe does not have an infinite age.
We can represent the motion of the universe as a nested sequence of expanding spheres. We, and all that is in the universe, live in such a sphere. Actually, the spheres are three-dimensional, and they expand in the direction of the fourth dimension: the direction of Time. As the universe expands, the universal sphere moves outward in Time. Thus, the future universe surrounds the universe which is past.

8 Cosmic Time
We can draw the motion of the universe as a nested sequence of expanding spheres. We live in such a sphere. Actually, the spheres are three-dimensional, and they expand in the direction of the fourth dimension: the direction of Time. But that is too difficult to draw, so this figure pretends that Space is two-dimensional. The future universe surrounds the universe which is past.

'What is outside the universe?' is one of the most frequently asked questions. By thinking of Time as something real, as real as distances, we see the answer before us.

9 What Is Outside the Universe?
Outside our universe lies our universe in the future. Inside lies the universe which is past. As we follow the motion of our world through time, it always expands. If we trace

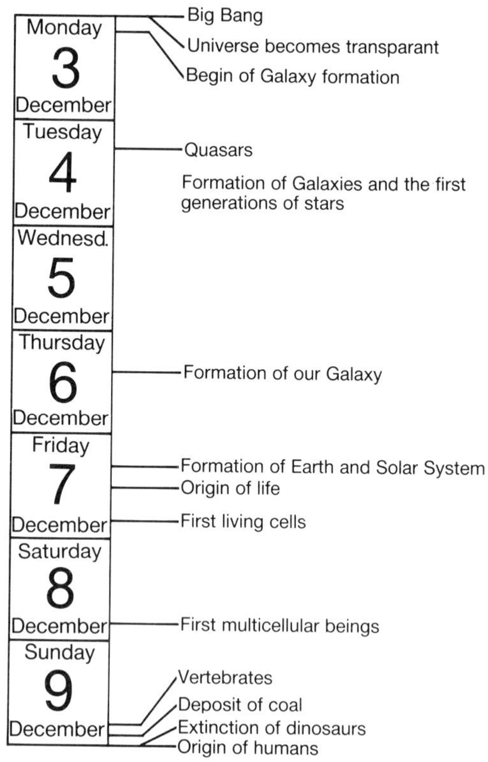

7

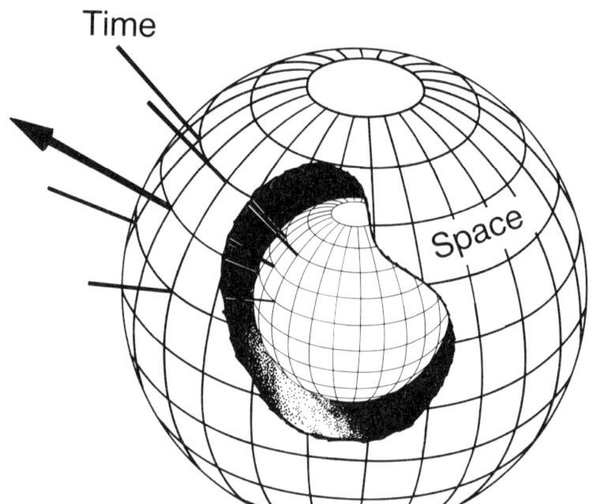

8

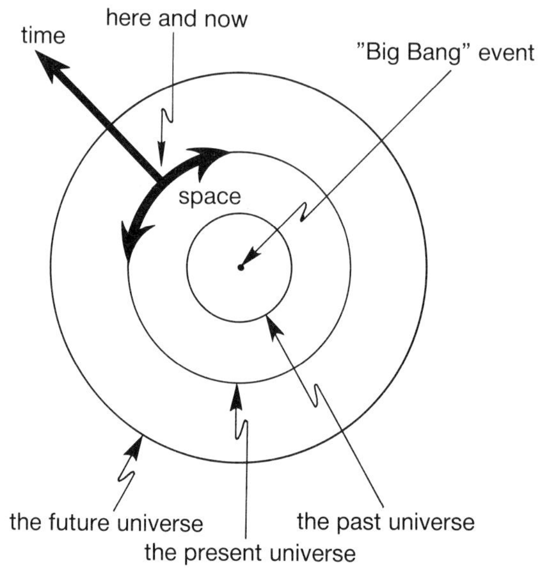

9

it back in time, it shrinks; ultimately, it vanishes to an exact point in the past. This is the point where Time and Space began. You cannot shrink something to a size smaller than zero; thus, there is no 'before' this beginning. The Big Bang is the true Alpha of our universe.

Let us now follow the universe in its expansion, and see when its various contents arose. Because in the beginning everything went so enormously quickly, we will take extremely small steps in time at first. Later events were slower, and we will adjust our time steps accordingly.

10 Steps in Time from the Big Bang Until Today

In the beginning, a lot happened in an almost inconceivably brief time. As the universe expanded, things became quieter and went more slowly. To accommodate this dramatic change of pace, we place some signposts about key events in the physical history of the universe on a *logarithmic scale*. On such a scale, each number is written as a power of the number 10. For example, when the universe was one ten-millionth of a second old, hydrogen nuclei – protons – were formed. Because 0.000 000 1=10^{-7}, we find the birth of protons at number -7 on this scale. Helium nuclei were formed when the universe was about 100 seconds old; 100= 10^{2}, and so the marker indicating the synthesis of cosmic helium points at number 2 on the logarithmic scale.

Looking Back in Time

Can we see anything of this majestic spectacle? Yes: due to a special property of nature, we can actually see the unfolding of the universe before our very eyes.
By following the path of the light that reaches us from the universe back to its

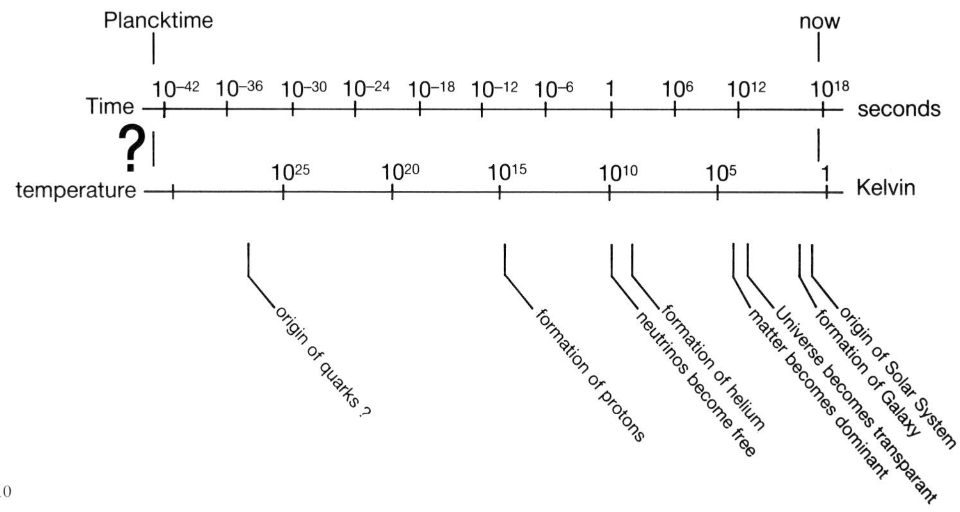

10

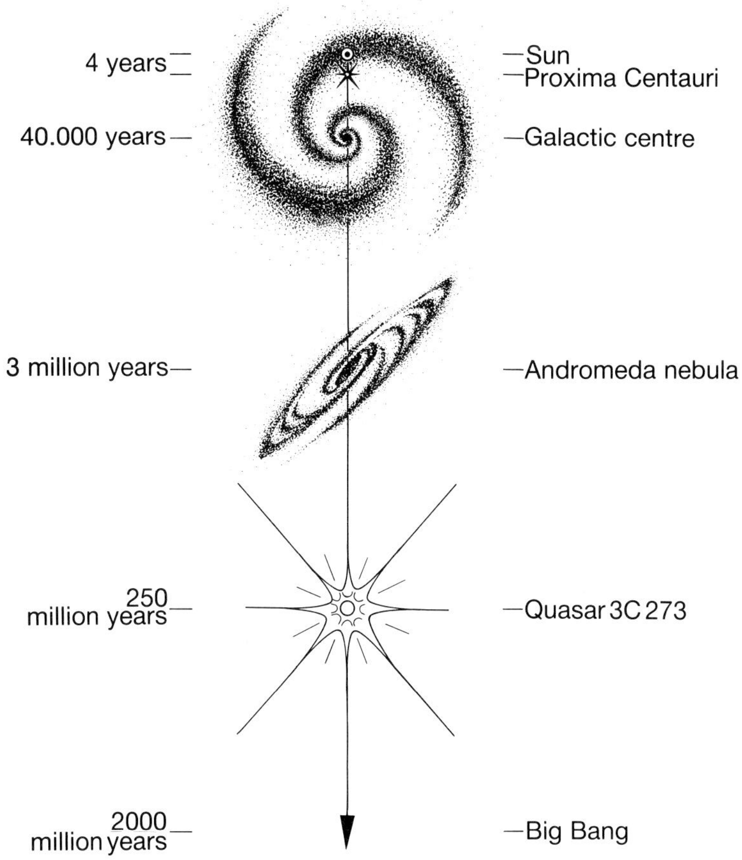

11

the universe was one ten-thousand-millionth of a second old.

21 Time Fossils: Hydrogen
In the apparatus shown here, you see water being split up, by means of electricity, into hydrogen and oxygen. The nucleus of the hydrogen atom is a proton. Each such proton is a time fossil, which formed when the universe was one ten-thousand-millionth of a second old. Hydrogen reacts chemically very quickly, and therefore it is mostly captured in chemical compounds.

But we can liberate the hydrogen by using electricity. When you push the button of this electrolysis experiment, the bubbles which rise in the tube on the left are hydrogen gas. Each hydrogen molecule contains two hydrogen atoms. Each atom has a nucleus which is a single proton. Each such proton formed just after the dawn of Time, when our universe was 0.000.0001 seconds young.

22 Time Fossils: Helium
The universe continued to expand after the protons and neutrons were made. Soon, the temperature became low enough (about 20 million degrees centigrade) for these particles to stick together in their turn. This process is called nuclear fusion. The most important fusion product is helium. In the glass vial is caught a little helium; this is a time fossil, which formed when the universe was two minutes old.

23 Quark Soup
When the universe was younger than one ten-thousand-millionth of a second it was so prodigiously hot, that not even subatomic particles like protons and neutrons could exist. Instead, the universe was a seething soup of particles called quarks.

24 A Proton is Born
As soon as the universe had expanded a little, the temperature had become low enough to allow the quarks to stick together, bound into tiny triples by means of intermediary particles called gluons. Thus, protons and neutrons formed. The time elapsed since the beginning of all time: one ten-millionth of a second.

25 Time Fossils: Oxygen
But what about the oxygen which is being made in the electrolysis tube? Is it as old as the hydrogen? No: the universe expanded so rapidly that there was not enough time to make any oxygen at all. And yet the oxygen is a time fossil too. It was made inside a star. The inner core of a star is so hot, that hydrogen nuclei – protons – fuse together and form heavier elements, including oxygen.
When you push the button of the electrolysis experiment, water is decomposed into hydrogen and oxygen. The hydrogen is in the left tube; the hydrogen nuclei are almost 20,000 million years old. In the right tube rise the bubbles of oxygen.

26 Formation of Nuclei in Stars
Most of the oxygen which you see bubbling up in the glass was made about seven to ten thousand million years ago, long before our Sun and the planets formed. In fact, all elements heavier than hydrogen are time fossils made in stars, by a process called *nuclear fusion.* The energy liberated in this process makes the stars shine (and eventually explode). Some of the atoms cooked in stars end up in planets, and in our bodies.

Ancient Myths of Creation

Before all Time, all Matter, Forme and Place God all in all, and all in God it was
Guillaume de Saluste, Sieur du Bartas, *The Divine Weeks and Works* (1578), translated by Josuah Sylvester (1605), I.1.4344.

27 Creation Myths: a Photomontage
The creation that modern man describes in terms of time, space, matter and energy was explained in ancient societies in terms of the will of gods.

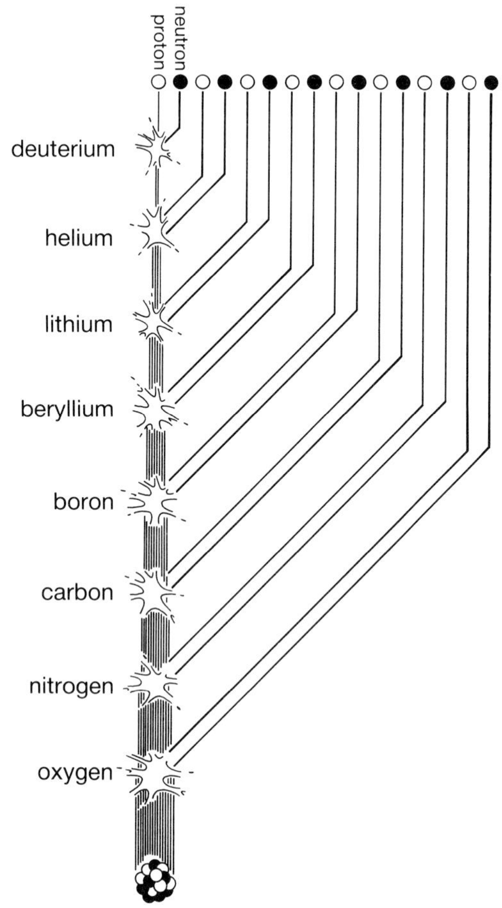

26

Time in Pre-history and the Ancient World

Time is an appearance which takes the form of day and night.
Sextus Empiricus.

Early Evidence for a Sense of Time

Time and space are the two basic dimensions of the phenomenal world.
We do not know when or how man first became aware of them. Recognition of repeating patterns in behaviour of the sun and moon, which gave rise to calendars, may have provoked the idea of periods, and so of time.

28 Incised Bones from the Upper Paleolithic Period
Photographs and schema.
a. From the rock shelter site of Blanchard (Dordogne)
Middle Aurignacian
b. From the rock shelter site of Lartel, Gorge d'Enfer (Dordogne)
Early Aurignacian
The bones shown here are typical of many from sites throughout Europe which have been interpreted by Alexander Marshak as lunar records. If this be the case then such records may have supplied the data for, may themselves have been, the first calendars. Marshak's findings however have been disputed. For a critique see d'Errico & Thevenin.
Photographs Musée Nationale des Antiquités, St. Germain-en-Laye.

But a similar recognition could have developed from understanding that after longer or shorter periods, which could be measured in terms of lunar or solar movements, men die.
Whichever perception prompted recognition of the phenomenon of time, the appearance of mortality rituals, ceremonies in which provision is made for the future needs of the dead, is evidence for a sense of before and after, and so for a rudimentary sense of time.
There is evidence for ritual burials from as early as 35,000 BC, just as there is for lunar observations.

29 An upper palaeolithic Burial
Photograph.
Burial place of a Neanderthal man, La Ferrassie, Dordogne
Photograph Musée de l'Homme, Paris.

30 A Model of Stonehenge
Stonehenge is perhaps the most famous of all megalithic monuments. Although much about it remains mysterious, its astronomical, and less certainly, calendrical functions have been sufficiently demonstrated for it to stand here as a symbol of the astronomical and calendrical activity of the Megalithic period as a whole. Alexander Thom, pioneer of modern study of this subject, and his son Archibald, from their extensive study of sites throughout Britain have discerned a possible megalithic calendar of sixteen months. Eleven of these have 23 days, four have 22 days and one has 24 days. An alternative offers a year of thirty two months of 11 or 12 days each.
Beiaardmuseum, Asten

Types of Time

Time is the experience of duration or succession. The regular succession of repeating phenomena measure duration.The movement of the earth around the Sun in 365 days 5 hours 48 minutes and 46 seconds is the measure of the mean solar year.The apparent motion of the heavens around the Earth is the measure of day and night.

31 'And God divided the light from the darkness' *Genesis* I.5.
Miniature from Pierre le Mangeur, *Bible Hystoriam ou les Hystoires Escolatres,* 14th century.
Photograph Bibliothèque Nationale, Paris.

The day and night period is determined either by the successive return of a particular star to the same position - sidereal time - or by the successive return of the Sun - solar time.

32 The Relation of Sun and Earth
Diagram by David Penney.

The sidereal day and night period 24 hours 3 minutes 56.5554 seconds is 3 minutes 56.5554 seconds longer than the mean solar day and night period.

33 A Modern Sidereal Clock
Designed by Allan Mills.
The illuminated area indicates that portion of the sky that can be seen from the latitude for which the clock is set.
Allan Mills, Leicester.

He appointed the moon for seasons.
Psalms CIV. 19.

Time may also be reckoned by the Moon which passes through the cycle of its phases in about 29.5 days.

34 The Phases of the Moon
Diagram by David Penney.

The attempt to divide up the solar year by these lunar months led to a twelvefold division of the year.The solar year, the solar day and the lunar month are naturally determined time periods. All other divisions of time are arbitrary.
Dividing the light and dark periods of the day each into 12 equal parts had already occurred in Egypt by early in the 2nd millennium BC, perhaps replacing an earlier division into 10. But in Babylon a division of each period into six kaspu prevailed.Except at the equinoxes, the lengths of the daylight and night periods are not equal to each other and neither are the hours which divide them. Such unequal hours change in length throughout the year.
If the whole period of day and night is divided into 24 equal parts, the equal hours which result are invariable.But before the 14th century such equal hours were only used by astronomers.
The point from which hours are counted is also arbitrary. Astronomers have generally counted from noon to noon. In ancient Egypt, as in modern Europe, the count is from midnight to midnight, but sunrise and sunset have also served as the starting point.
The experience of time is both linear and cyclical. Each day is an entity with a beginning and an end, but each day repeats. The counting of any series of time divisions gives a linear result and the starting of a new count of years from the beginning of each new reign stresses theruniqueness, and so the linearity, of the time of each ruler. But the repetition of the years and the seasons and of the religious rites associated with them independently of the ruler, suggests greater cycles of events in which there is a constant renewal, and so a denial of time in eternity.
The calendars and chronologies created in all societies served as records of time past, as devices for orienting ones self in the year, and as a means for remembering the religious/magical rites to be carried out now and in the future so as to ensure that the pattern established in the past, and lived in the present, would continue to determine the future.
This 'ritual perpetuation of the past' seeks to assure the future by making past acts of symbolic power eternally present. It lies at the heart of most religions.
This cyclical repetition of the past also brought reassurance to man disturbed by the change and dissolution that the action of time inevitably brought about. Linear or cyclical, eternal or finite, time is the envelope of man and all creation.

35 **Images of Gods and Time**
Photograph.

I saw Eternity the other night,
Like a great ring of pure and endless light,
All calm as it was bright;
and round beneath it, Time in hours, days, years,
Driv'n by the spheres
Like a vast shadow mov'd; in which the world
And all her train were hurl'd
Henry Vaugan, *The world.*

Calendars

Calendars may be solar, lunar, or a combination of the two (luni-solar).
Since the periods of the Sun and Moon are incommensurable, lunisolar calendars and lunar calendars that try to remain in step with the seasons are complex and difficult to use.
Although solar calendars are often therefore preferred for civil use, religious organisation retains use of the lunar calendar. The Babylonian calendar of the 3rd millennium BC and later was strictly lunar.
Each month began with the evening when the new lunar crescent became visible soon after sunset.

Nisan: Sacrifice
Aiaru: Blossom
Simanu: The fixed or appointed time
Du'uzu (Tammuz): Vegetation leaves the parched earth
Abu
Ululu
Tashritum
Arakhsamna: The eight moon
Kislimu
Tebitum
Shabat
Addaru

The beginning of the year was in the period of the spring equinox, and intercalation took place whenever the king noted a 'deficiency' in the year such that Nisan would have arrived more than a month before the equinox.
Before the 3rd dynasty of Ur (2300-2150 BC), each major city of Sumeria maintained its own calendar. During the 3rd dynasty however the calendar of Nippur was adopted for official use. Since the month names related to seasonal activities.

Month of the Dweller of the Sanctuary.
Month of the Leading Out of the Oxen.
Month for Brick Making.
.....
Month of Setting the Fire.
Month of a festival of 'Ishtar'.
Month of the Sacred Place Duku.
Month of Opening Irrigation Canals.
Month of Ploughing.
Month Name of Religious Festival.
Month of Emmer-grain.
Month of Corn Harvest.

An intercalation system was essential to keep the months in step with the activity described.

36 **Babylonian Almanac, probably 5th century BC**
Cast 128 mm x 97 mm.
The original, in the British Museum, bears twelve columns, one for each month. It is inscribed in cuneiform script and dates from the time of Ezra and Nehemiah. This cast shows the columns for the 7th to the 12th month (reading from right to left). Within each column (reading from left to right) the top line includes the name of the month. Each line, except the top line and where an entry runs on to two lines, begins with the number of a day in the month, from 1 to 30. The numerals, which may be seen clearly at the left of each column, are followed by a brief note mentioning, for instance, whether the day is favourable or otherwise, or whether certain actions, such as the purchase of a slave, may favourably be performed on that day.
Provenance: C.J.S. Thompson.
Museum of the History of Science, Oxford.

A correspondence was found between the Sumerian and the Babylonian calendar, and the latter also lies at the base of the Hebrew calendar.

Babylonian month :	Hebrew month :	European month :
Nisannu	Nisan	April
Airu	Iyar	May
Sivanu	Sivan	June
Duzu	Tamuz	July
Abu	Ab Elul	August
Ululu		Sept

Babylonian month :	Hebrew month :	European month :
Tasritu	Tisri	Oct.
Arah-samna	Narchesvan	November
Kislou	Kislev	Dec.
Tabitu	Tabit	Jan.
Sebatu	Sebat	Febr.
Addaru	Adar	March

And the evening and the morning were the first day
Genesis I.5.

As in the Babylonian system, the Hebrew day and so the month began at dusk.

33

36

37 **The Gezer Calendar, ?10th century BC.**
Photograph.
Discovered at Gezer in 1908 by the noted Irish archaeologist R.A.S. Macalister, this small limestone tablet (114 mm x 70 mm), one of the oldest surviving examples of early Hebrew writing, is an agricultural calendar. The text reads
'Two months of ingathering ; storage of grain and wine : September/October.
Two months of sowing : November/December
Two months of late [Spring] sowing : January/February
Month of pulling flax : March
Month of barley harvest : April
Month when everything [else] is harvested : May
Two months of pruning vines : June/July
Month of Summer fruit' : August
Bibliography: Mauchline, whose translation has been used.

The Gezer calendar illustrates the primordial position of agriculture in early Hebrew society. It is an entirely secular calender. Religious life however was also involved with time, and in later periods this led to the development of a number of instruments.

38 **Calendar for Religious Festivals, 1800 1850**
Dutch
Ms on vellum roll written in blue, red and black, 2 m 37 x 50 mm.
Joods Historisch Museum 526.

39 **A Roll-on Hebrew Calendar, late 19th century**
Wood and vellum, 380 mm x 270 mm x - 90 mm.
Joods Historisch Museum 492.

40 **A Sabbath Calendar, 19th century**
Wood with linen rolls, 360 mm x 250 mm- x 60 mm.
Joods Historisch Museum B307.

41 **A Watch with Hebrew Alphabetical Numerals,** *c.* 1880. Brass and steel with enamelled dial, diameter 48 mm.
Joods Historisch Museum 689.

42 **A Watch with Hebrew Alphabetical Numerals,** *c.* 1900
Signed 'Boîte artistique de Genève Médailles d'or Genève 1896 Paris 1900'.
Brass and steel with enamelled dial, silver-plated case diameter 55 mm.
Joods Historisch Museum 1371.

Time Measurement in Egypt : Calendars

The earliest calendar known from Egypt is lunar but regulated by a star observation. Introduced in the 5th or 4th millennium BC, its year began with the start of the Nile inundation which was linked to the first helical rising of Sirius equivalent with the goddess Sothis. An intercalation of one month was made every 2 or 3 years to keep the two events in relation.

The year

1st season
Rise flood fall of Nile
1 Thoth, 2 Phaopi, 3 Arthur, 4 Oholak

2nd season
Seeding tilling harvest
5 Tybi, 6 Mechir, 7 Phamenoth
8 Pharmuthi

3rd season
Low water
9 Pachars, 10 Payni, 11 Epipni, 12 Mesare

43 **An Artist Painting the Three Seasons, VIth Dynasty (c. 2345-2121 BC)**
The artist is, seated on a stool with a mixing bowl in front of him beside which his son, Khenu, waits to help. He is painting images of the three seasons, *akhit* (inundation), *perit* (emergence [of the land] and sowing), *shemu* (harvest).
Wall painting from the entrance of the tomb of Merruka.
Bibliography : Daressy ; Montet 243-4.

The Egyptian day began at dawn.

44 **The Days of the Lunar Month**
Table from Parker, 11-12.

...the only intelligent calendar that ever existed in human history
Neugebauer, 81.

39

42

At the beginning of the 3rd millennium BC a schematic lunar calendar was introduced for civil purposes. This was composed of 365 days made up of 3 seasons each of 4 months each of 3 weeks each of 10 days in which 5 extra days 'epagomenal' were added at the end of each year to make 365.
This calendar was invariable and therefore of the greatest convenience for civil life.
The old lunar calendar continued to govern religious life. Shortly however it became badly out of step with the civil calendar.
In about 2500 BC an intercalation rule was adopted which made the lunar calendar dependent upon the civil. The national invariable Egyptian civil calendar is of great historical importance. It influenced the calendars of Sassanid Persia, Armenia, Cappadocia, Hellenistic Greece, Rome, and Revolutionary France. In addition it was the standard astronomical reference calendar up to the time of Copernicus.

Time Measurement in Egypt: Star-clocks

Knowledge of early Egyptian time-measurement using the stars derives from the diagonal 'star-clocks', copies of which were drawn on the inside of coffinlids.

45 Coffinlid Star-Clock from the Tomb of Idy, end of 3rd millennium
Photograph.
Only eighteen of the thirty six decades are shown.
The principle of such star-clocks is shown in the following scheme which reads from right to left. The top line T gives the thirty six decades which divide up the year in ten day divisions. These are followed by four columns for the epagomenal days although only the last (column 40) is actually used, the other three merely listing the names of the thirty six stars found in the decades.
The columns are divided into two sets by a space containing representations of the sky goddess Nut and some stars. The horizontal inscription (R) dividing the hours contains invocations in favour of the deceased to various stellar deities. The principal of operation is that sets of stars are chosen that rise heliacally (i.e. at daybreak) every ten days. Because of the Earth's movement around the Sun, each day such a star will rise earlier. Positions are therefore plotted in each ten day group at which the chosen star rises one twelfth of the total night period earlier. In the course of 120 days therefore a star which rises at the twelfth hour gradually rises diagonally higher across the clock until, after having reached the first position, it drops out completely since it now rises after daybreak.

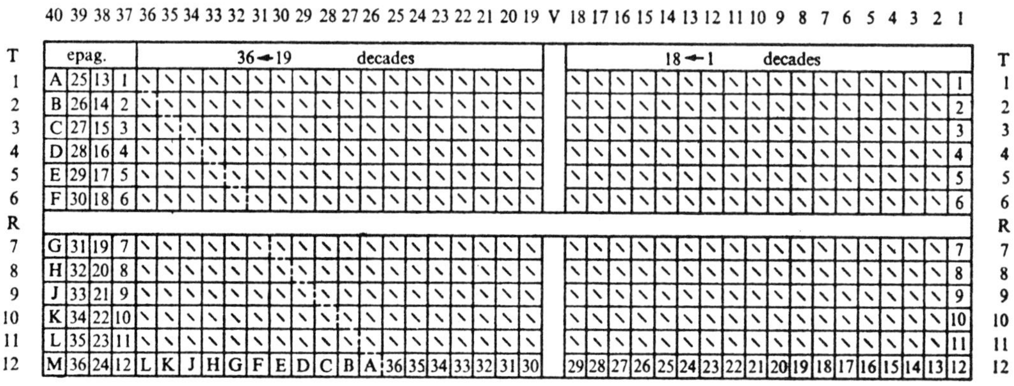

45

46 The New Star-Clock of the 2nd Millennium BC
Photograph of a diagram on the ceiling of the tomb of Ramses VII.
By about the twelfth dynasty, the star-clocks using helical rising had become deranged, and were replaced by a new form using transits. The times at which a number of selected stars crossed the meridian were noted and by extrapolation onto a grid other times, before and after, were determined. At the centre of the grid was a seated human figure, the meridian line passing through his heart. Other times were noted in the text by reference to this figure. At night if two men sat facing each other at a distance along a meridian line, or if the observer faced an artificial cutout figure, time could be found by noting which star was where in relation to the body of the target figure or to the grid.

47 An Indicator and Level for Transit Observations, c. 600 BC
Copy c. 1960 of the original presented to the Science Museum, London in 1929 by Dr Howard Carter. The inscription states that the instrument is the property of Bes, astronomer-priest of Horus, at Edfu (Upper Egypt).
Horizontal square-section bar inscribed on the side, with endblock carrying the plumbline and bob, and at the other end, a head of Horus.If properly aligned the plumbline of the instrument would mark a meridian and provide a clear sighting-line for an observer using a slit palm-leaf front sight.
Private collection
Bibliography : Ward 7, no 2 ; Zába 55-8.

48 A Y-shaped Front-Sight of Palm wood, c. 600 BC
Photograph.
Such an instrument may well have been used in conjunction with the merkhet.

Egyptian Time-Measurement : The Water-clock

Used from at least the 16th century BC in Babylon and Egypt, water-clocks were probably designed for use at night when cloud made time-finding by the stars impossible.

49 The Earliest Surviving Egyptian Clepsydra, c. 1400 BC
Plaster facsimile c. AD 1970.
Diameter of rim 495 mm, diameter of base 273 mm, height 356 mm.
Outflow clepsydra graduated on the inside and with a small hole in the wall just above the base. This hole was probably originally surmounted by a dog-headed ape, emblem of Thoth. The wall is inclined to the base at an angle of approximately 110° and is decorated on the outside in three bands. The lowest of these is divided into six compartments and shows King Amenhotep III with the twelve moon gods, two to each compartment. In the middle band are the circumpolar stars and images of the deities of the days of the week. Opposite the orifice is a large panel extending from the middle to the upper band in which the king is depicted protected by Thoth and offering sacrifice to the Sun-god Harmachis. The remainder of the upper band contains symbols of the planets, and the constellations through which the sun passes in the course of a year. The astronomical disposition of these bands corresponds with a date between 1630 and 1510 BC, not with that of the clepsydra c. 1400. It therefore probably indicates the period when clepsydrae of this form were first developed, a conclusion for which there is some textual evidence.The form of the clepsydra is determined by the need to maintain a constant pressure of water immediately above the outlet so that the discharge rate will be regular. On the inner wall twelve hour scales are marked one for each of the twelve months. The hour intervals marked on these scales are adjusted to correspond with the longer and shorter unequal hours throughout the
shorter unequal hours throughout the year. The original clepsydra, now in the Egyptian Museum Cairo, was discovered in fragments in a refuse pit during excavations at the Amon Temple, Karnak by G. Legrain in 1904. It is made of translucent alabaster decorated on the outside with carnaline and coloured glass incrustations.
Beiaardmuseum, Asten.
Bibliography : Borchardt 67 ; Sloley 44-5 ; Ward 8 no 4 ; Turner 46-7 ; Desroches-Noblecourt, 8.

50 Principle of the Outflow Clepsydra
Diagram by David Penney.

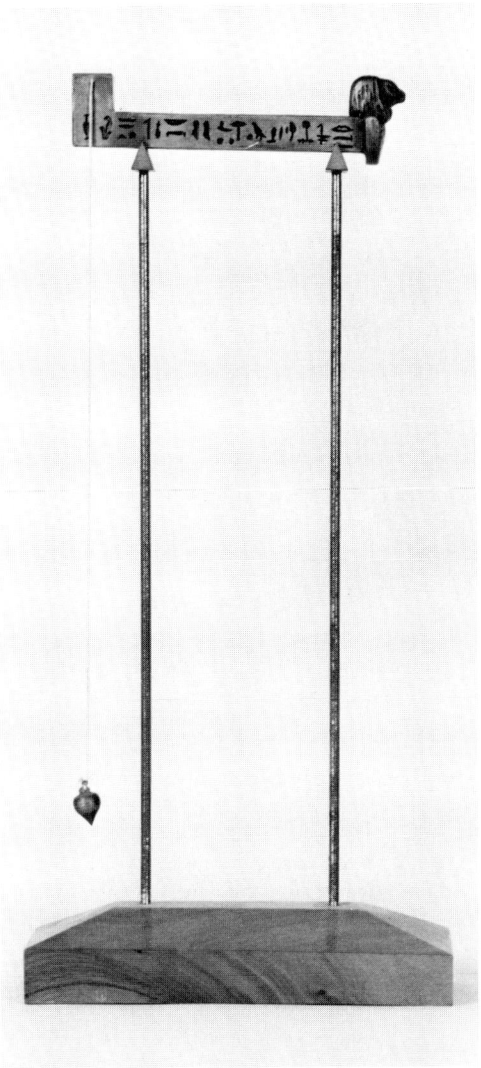

47

51 a

51 b

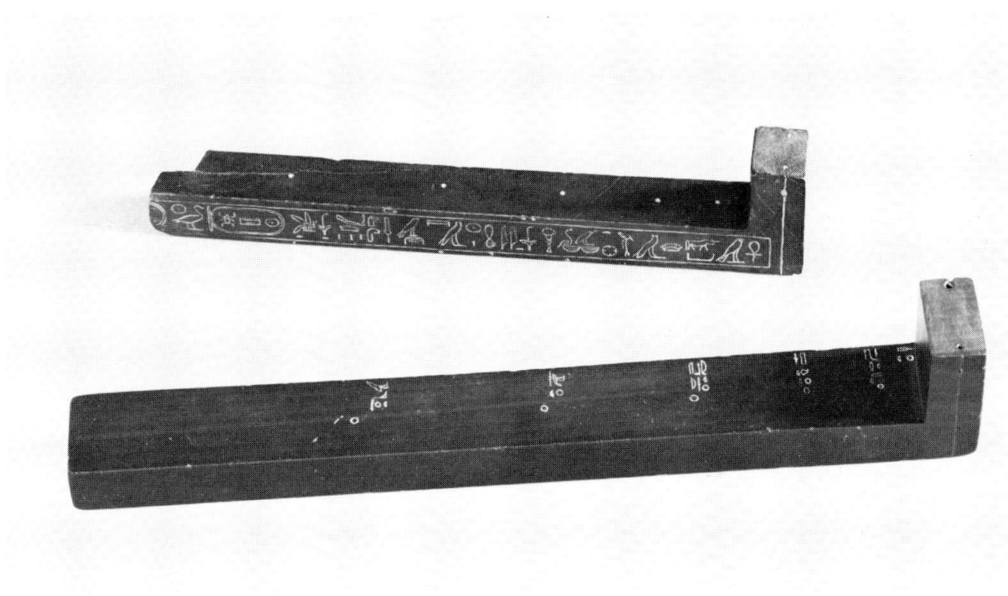

53

51 Fragment of a Ptolemaic Clepsydra
Inscribed '...to know the hours of the day when one cannot see Ra [the Sun] and to know the hours of the night...'
Granite, 112 mm x 156 mm x 43 mm.
The inscription confirms that clepsydra were used for time-finding at night, or in cloudy day-time conditions when sun-dials could not be used.
Musées Royaux d'Art et d'Histoire, Brussels.
Bibliography: Capart, 53; Berryer & Lébioles, 7.

Time-Measurement in Egypt: Sun-dials

Sun-dials simulate the apparent daily movement of the Sun across the sky by means of the shadow of an indicator falling on a specially constructed grid.
This grid carries divisions of the daylight period and is permanently attached to the shadow-caster (the gnomon).The earliest known sun-dials are Egyptian and date from the mid 2nd millennium BC. They are thus only slightly later than the earliest evidence for water-clocks.

52 The Oldest Illustration of a Sun-dial, 1318-1304 BC
Photograph of a drawing of a sun-dial with descriptive text from the Sarcophagus.
Room of the Cenotaph of Seti I at Abydos.
The sun-dial shown divides the day into only eight hours. Since a surviving sun-dial of the mid 15th century BC (see Borchardt, 34) already has a 12 hour division, this may be ascribed to the survival, in the highly conservative tradition of funerary decoration, of a dial which predates the introduction of the twelve hour day.
There is some evidence in the text that the writer was aware of the problem that the picture created.
Photograph Department of Egyptology, University of Leiden.
Bibliography: Frankfort.

53 An Early Egyptian Sun-dial, c. 1000 BC
Facsimile and photograph.
A horizontal bar marked with an hour scale 5 - 1 has a rectangular block at one end which carries a plumb-line and bob. In use the instrument was placed in a level position with the end-block towards the sun. The shadow it cast on the hour scale would then give an approximate indication of the time. At noon the instrument would lie on the meridian and no shadow would be cast. At sunrise and sunset, it would lie along the east-west line.
The similarity of this instrument with the Merkhet (no 46 above) will be noted.
Photograph and facsimile: Ägyptisches Museum, Staatliche Museen zu Berlin.

54 A Developed Form of Egyptian Sun-dial, early to mid 4th century
Faience, 65 mm x 75 mm x 27 mm.
An horizontal sun-dial of the same basic form as the precedent but with the hour-table inclined to the gnomon-block at an angle corresponding to that of the co-latitude of the place where the dial was to be used. The hour-table is divided vertically by six lines for the paired months, and the hour lines are drawn across them at an angle to

give a more accurate reading for the unequal hour throughout the year.
The present example, which includes the name of King Nepherites (c. 39993 BC) in the inscription was probably a votive instrument for there is no trace of an hour scale.
Musées Royaux d'Art et d'Histoire, Brussels (Inv E.73003).
Bibliography: Capart; Berryer & Lébioles, 5.

55 A South Facing Vertical Sun-dial of the Ptolemaic Period (304-30 BC)
Steatite.
Diameter 56 mm, height 92 mm.
A simple semi-circular dial in which the gnomon was set at right-angles to the dial surface at the centre of the diameter, and with the unnumbered hour lines radiating from it.
Dials of this type, of which several Hellenistic examples are known, were to have a long history. Probably because of the simplicity of their construction, they were adopted by early christian communities and spread to Central and Western Europe in the wake of monasticism.
Musées Royaux d'Art et d'Histoire, Brussels (Inv. E 7330).
Bibliography: Capart, 51-2 ; Berryer & Lébioles, 9 ; Gibbs 45-6.

Gnomons and Shadow Length Tables

A vertical gnomon an object specifically designed to cast a shadow permanently set up can be used to locate oneself in time either in the year or in the day. The earliest gnomons were probably calendrical, marking the dates of the seasons. But the change in the shadow length cast by the gnomon throughout the day when related to the season could indicate the hour.
Tables expressing this relationship are known from 7th century BC Assyria, but the tables that lie behind the long tradition of using the human body itself as a gnomon probably derive from the Greek scolar Philip of Opus in the 5th or 4th century BC.

56 Finding Time from a Gnomon Shadow Length
Diagram by David Penney.

57 Shadow Lengths from Palladius 'de Re Rustica', 5th Century AD
The values given in various chapters of Palladius' very popular agricultural manuals are here presented in a single table similar to many others used throughout Europe and around the Mediterranean basin during the Middle Ages and later.

58 Borneo Tribesmen (kenyahs) Measuring a Noon Shadow Length
Photograph from C. Hose & W. McDougall, The Pagan Tribes of Borneo, London, 1912, pl. 60.
The tribesmen are using a template to determine whether the shadow is at the length which indicates the season for planting padi. Although they are thus conducting a calendrical, timebased, operation, this has become a purely mechanical perhaps ritual act which requires no conscious understanding of the time element involved.
Photograph : The Bodleian Library, Oxford. RHS. 916.9.R.2.

Sun-dials first appeared in Greece in the 4th century BC. Whence the Greeks derived them is still a matter for debate.Very quickly, Greek scholars began to study the instrument mathematically, giving it much greater accuracy and many different forms. Greek and Roman craftsmen carved the dials in marble to make fine, monumental sculptures of them.

59 Graeco-Roman Roofed Spherical Sun-dial, 3rd - 4th century AD
Marble, height 64.5 cm; diameter 55 cm.
The dial is constructed from a single block of marble. It was probably intended to be mounted free-standing on a pedestal (see no 60), inclined at an angle appropriate for the latitude of the place where it was to be used.
The triangular block set at the bottom of the instrument was cut appropriately to supply this inclination. A square recess (approx 4 cm deep) in this block received a locating peg set in the upper surface of the stand. The exterior of the dial is relief carved with acorns and stylised oak leaves. At each side of the instrument, at about 90 degrees from the base is a decorative bracket holding an open cylinder. At the top the outside of the instrument is flattened to allow attachment of the gnomon plate. This plate is now missing leaving a roughly circular hole of approximately 5 cm diameter around which are sixteen small holes of which eleven contain the remains of (?) lead plugs. Only three of the holes however pierce the wall of the dial. Originally there would have been a much smaller hole or a metal diaphragm would have covered the space its perforation being aligned on the meridian (noon) line of the dial.

54 a

54 b

54 c

55

The interior of the dial is incised with eleven hour lines cut by seven transversal season lines. Reading from the outer edge inwards these are :

1 Line for the summer solstice and July marked ΛΙΚΛΟC ΘΕΡΙΝΟC ΤΡΟΠΙΚΟC ΠΡΟ–ΟΡΤΙΟΙ ΚΑΛΑΝΔΩΝ ΙΟΥΝΙΩΝ

2 June and August marked ΠΡ̃ Η̃ Κ̃ ΙΟΥΝΙΩΝ ΠΡ̃ Η̃ Κ̃ ΑΥΓΥCΤΩΝ

3 May and September marked ΠΡ̃ Η̃ Κ̃ ΠΑΙΩΝ ΠΡ̃ Η̃ Κ̃ CΕΝΤΕΜΒΡΙΩΝ

4 April and October marked ΠΡ̃ Η̃ Κ̃ [ ] ΠΡ̃ Η̃ Κ̃ ΟΚΤΩΟΡΙΩΝ

5 March and November marked [ ] ΜΑ [ ] ΠΡ̃ Η̃ Κ̃ [ ]

6 February and December marked [ ] [ ] ΔΕΚΕΜΒΡΙΩΝ

7 Winter solstice and January [ ] ΚΛΟ [ ] C ΙΛ— ΝΑ Ι— [ ] Ν

The letters or symbols that precede each month name, in so far as they can be read, are all the same. They are abbreviations. The line over each of them is the usual way abbreviations were indicated in both Latin and Greek. ΠΡ is probably short for ΠΡΟΟΡΤΙΟΙ the word in line 1, and which presumably is a misspelling of προεορτιοι which must mean the equivalent of our 'eve', as in Christmas Eve. Η̃ and Κ̃ are probably abbreviations respectively for the Greek equivalents of Ides and Kalends, for all that the usual Greek way of expressing Ides is Ειδοί. There was very little difference in pronunciation between η and ει. These days were known in Greek as εορρται -- feastday : K was a common way of abbreviating for 'Kalends', even by the Romans (information from Michael Vickers).

The surface of the dial is marked by 14 small holes similar to those on the outside around the gnomon plate hole. These are almost symmetrically placed on the hour line network but their purpose is not clear.

The dial has not previously been recorded and it differs from the other twenty-four known examples in having seven day curves. Of these specimens, one (Gibbs 7001G) has five day curves, all the others have only three.
Provenance : Found in Tunis and probably from a Roman villa in the region of Carthage.
Private Collection.

60 A Late Antique Sun-dial in a Garden, 6th century AD
Photograph of a miniature from the Vienna Genesis. The illustration shows Joseph in prison visited by his friends. Outside stands a free-standing concave sun-dial. On stylistic grounds the miniature has been ascribed to either Constantinople or the Syro-Palestine region. Whatever its source it shows clearly the way in which dials were set up as useful public ornaments in the late Roman period. That it had symbolic importance in the context of the present picture is also possible.
Photograph : Nationalbibliotek, Vienna. Cod Theol. gr 31.
Bibliography : Weitzmann 84-5.

58

61 **How Unequal Hours Change by Day and Throughout the Year**
A demonstration apparatus designed and constructed by Allan Mills.
The display shows how unequal hours are generated at latitude 52° 30' N. The arc apparently followed by the sun is shown at three points in the year to indicate its changing height. The tip of the shadow of the gnomon on the horizontal hour scale indicates the time.
Dr. A.A. Mills, Leicester.

62 **Types of Sun-dials in the Graeco-Roman World**
Photographs.
a. Spherical dial with eleven vertical hour lines and three horizontal season lines.
Found at the Citt Lavinia, 1891 and presented to the British Museum by Lord Savile.
British Museum London, Inv. no 2547.
b. Conical dial with eleven hour lines and two season lines on an elaborate, sculpted pedestal. A portion of the bronze gnomon remains.
c. Horizontal plane dial by M. Antistius Euporus, late 2nd or early 1st century BC. The inscriptions in the circle around the dial are a list of wind names.
d. A double-faced (east and west) vertical plane dial.
Museo Nationale, Rome.
e. A cylindrical dial with bull head support.
Archaeological Museum, Constanta, Rumania.
f. A multiple dial with cylindrical and spherical faces.
Found at the Villa Scipio, near Herculaneum in 1769.
Museo Nationale, Rome. Bibliography: Gibbs 1017G, 3053G, 4002G, 5024, 6002G, 7004G.

Such dials were public time-keepers intended to regulate urban life, or that of centralised farms. But they were not always looked upon with favour.
'The pointer doesn't mark the sixth hour yet, and I am in danger of withering away, so goaded am I by hunger. So now Lapedethambus [dish-crazy], it's time for you to produce a scheme, or better still a crowbar and a rope to hang ourselves with. For if we throw down the whole column which supports this hateful sun-dial, or bend the gnomon this way where it will be able to mark the hours sooner, that will be a scheme worthy of Palamedes [Nauplian inventor-name used proverbially]. As matters stand I'm parched if you please, and dried up from hunger. And Theochares doesn't take his place on his mattress until his boy runs up and tells him that it's six. Brought up by a severe frowning pedagogue, he has no youthful ideas, he is strict in his ways and doesn't let his stomach take it's fill until the dinner hour has arrived'.
Alciphron : *Letters tr.* Benner & Forbes, 148-51.
Dials and the time they measured could be important imperial, ideological symbols.

63 **The Sun-dial of Augustus Caesar, 13 BC**
Photographs and diagrams.
The largest time-measuring and calendrical instrument of the Ancient World, Augustus' monumental dial used an Egyptian obelisc 100 roman feet (29m42) high for its gnomon. Designed by the mathematician Novius Facundus, the dial was laid out in the Campus Martius at the same time as the Ara Pacis (Temple of Peace) was erected, in such a way that at the Autumn equinox - which was Augustus' birthday - the shadow of the obelisc travelled along the equinoctial line to end in the centre of the Ara Pacis which was placed axially on this line.
Symbolism was therefore complete. On the Emperor's birthday, and he was born slightly before sunrise, the shadow of the sphere surmounting the obelisc travelled 150 m along the straight line of the equinox to the very middle of the Ara Pacis, thus tracing a direct line from the birth of Caesar to Pax, peace. The idea that he was natus ad pacem, born for peace, was thus made visible to all, while the occasion of this peace, Augustus conquest of Egypt was aptly expressed by the obelisc and his world dominion by the sphere that surmounted it.
Bibliography : Buchner.
a. The Egyptian obelisc used as a gnomon for Augustus' dial as it is today in the Piazza Montecitorio, Rome.
b. Ground plan of the Campus Martius showing the obelisc, the dial grid and the Ara Pacis.
c. Perspective and ground level views of the Campus Martius with the dial and Ara Pacis.
Based on his calculations of the probable position of the lines of the Augustan sundial, the Munich archaeologist Edmund Buchner conducted a series of excavations during 1979-81. In 1980 bronze letters making up the Greek word Parthenos (Virgo) were found, and shortly afterwards parts of the seasonal lines with cross bars indicating days.
Although these finds confirmed Buchner's theory concerning the dial, the depth at which they were found did not correspond with the Augustan level. It therefore seems probable that the dial and calendar were relaid at a higher level at a later period - probably during the reign of Domitian.
d. Photographs of the bronze letters of the word Parthenos (Virgo).
e. General view showing the line for Virgo and the inscription Etesiai pauntai (the summer winds cease).
But that they could also be humbler pocket instruments implies that for a few men to be able to find the time at any moment was important.

64 **Roman Vertical Dial, c. AD 250**
Bronze, diameter 60 mm.
The dial consist of two discs. The smaller of these is countersunk in the other being held in place by a pin and wedge, and may be rotated. The gnomon and hour scale are mounted on the pin and may also be turned. In use the small plate is turned until a radial line drawn at right angles to a diameter indicates the latitude of the place of observation on a degree scale (30°-60°) marked on the edge of the larger disc. The gnomon is set for date against the calendar scale drawn in two fan-shaped scales joined by the diameter on the smaller disc. The dial is then suspended and turned until the shadow of the gnomon falls exactly on the hour scale. The tip of the shadow then indicates the time. On the back is a list of the names and latitudes of 30 places.
Museum of the History of Science, Oxford.
Bibliography: Stebbins, Price (IV) no 7; Field & Wright, 23-4.

Greek scholars made some fundamental improvements to water-clocks.
Development of the inflow clepsydra in which time is measured by the period taken to fill a recipient, is associated with Ctesibios (c. 300 BC - c. 230 BC) and others working at Alexandria.

65 **Remains of a Greek Outflow Clepsydra, 4th century BC**
Stone, depth of tank 1.93 m.
a. General view from west, and b. detail of tank from south west of the clepsydra in the sanctuary of Amphiaros at Oropos, north east Attica. The square central tank is surrounded by a service area to which access is provided by ten steps on the west side. A bronze outlet pipe near the bottom of the north wall of the tank of 90 mm diameter ends in a hemisphere pierced with a hole of about 2.6 mm. The walls of the tank slope inwards towards the bottom to compensate for pressure change with the decrease in the height of the column of water.
Bibliography : Armstrong & McKCamp.

66 **Inflow Clepsydra with Overflow Tank**
Principle of operation.
Diagram by David Penney.

67 **A Romano-Egyptian Clepsydra Vessel, AD 100**
Facsimile, height 381 mm, interior diameter 168 mm.
The original of this white limestone vessel (now in the Museum of Egyptian Antiquities, Cairo) was found at Edfu, a capital of

Upper Egypt, during excavations in 1901. It is commonly considered to be the receiver of an inflow clepsydra. In use it would have been fitted with a float carrying an index which would have marked time on a scale set behind the cylinder or, if the index itself was graduated, against the rim of the cylinder. Some scholars however have had doubts about this interpretation and at present there is no consensus.
Bibliography : Borchardt 22-5; Sloley 48-9; Turner 50-52.

Some water-clocks were arranged to indicate time aurally using whistles or the sound of falling balls.
Others gave visual indications.

68 A Seventeenth Century Interpretation of a Romano-Greek Water-Clock
Photograph from Claude Perrault, *Les Dix Livres d'Architecture de Vitruve, corrigez et traduits nouvellement en Francais...*, 2nd edition, Paris, 1684, pl. 56, p. 289.
A cylindrical drum is mounted vertically on a pedestal which contains the water-machinery. The drum is divided vertically into columns representing the zodiacal signs, which are marked round the base.
Crossing these columns horizontally are twenty-four curved hour lines. Also carried on the pedestal are two miniature figures, one (a satyr) holding a pointer, the other (a cupid) crying. The illustration shown here encapsulates much of the history of the inflow clepsydra. It shows the reconstruction by Claude Perrault of a clepsydra of Ctesibios of Alexandria, according to the description given by Vitruvius. The model has, therefore, a very limited resemblance to Ctesibios' original conception, the hydraulic arrangements of which are likely to have been simpler. The interior mechanism of Perraults' machine is as follows. Beneath the figure of the crying cupid is a pipe which leads water from an external source into its body and causes its tears. These drop into a basin at his feet, whence they are led through an horizontal pipe in the top of the pedestal into a vertical cylinder beneath the pointing figure. Inside this cylinder is a float to which the pointing figure is connected. As the water level rises, therefore, the figure is gradually lifted higher up the cylinder, against which it indicates the hour with its pointer. At the end of twenty-four hours, when the float tube is full and the figure at its highest point, water from the float tube overflows into a siphon that operates to release the water into a compartmented water-wheel with six divisions and which thus makes one complete revolution in six days. The water-wheel has a six-leaved pinion which turns a wheel of sixty, which in turn, through a pinion of ten leaves, turns a wheel of sixty-one. This is attached to the

65 a

65 b

67

69

central shaft of the hour column, which thus makes one complete revolution in 366 days. By so doing, the appropriate part of the columns of zodiacal signs is presented to the pointer, and the curved hour lines enable the unequal hours to be shown.

Water mechanism could be used to drive a wide variety of devices. The anaphoric clock was an astronomical model in which a celestial map on a flat disc was rotated by the water-machinery behind a fixed wire grid which represented the local co-ordinates of the place where the clock was to be used.

69 Fragment of a Late Roman Anaphoric Clock-dial, ?2nd century AD
Bronze, length of longer side 600 mm approx.
Photograph.
The fragment is engraved with several constellations, and along the circumference the zodiacal signs of Taurus, Aries, and Pisces. The curved parallel lines to the left of Taurus represent the tropic of Cancer which is drawn in stereographic projection. The large semi-circular hole was for the axis holding the dial in place, and also represents the North Pole. The indentations along the circumference are probably the remains of holes into which a sun emblem was 'plugged' each day.
Salzburger Museum Carolino Augusteum.

70 Gallo-Roman Fragment of a Calendrical Dial, 2nd - 4th century AD
Bronze, largest overall dimension, 220 mm x 177 mm approx.
Photograph.
The very thin sheet of hammered metal was clipped over a disc and a fragment of (?) iron still remains in the small section of the rolled edge that survives. The holes along the circumference, of which there would originally have been 182 or 183 were punched through and left unfilled on the reverse. Around the edge are the month names with their main divisions, Kalends, Nones, Ides, spaced according to the number of days they contain. Each hole represents two days. There was probably a large hole at the centre of the disc and there are pronounced circular scratch marks covering the whole device. For an interpretation of the disc as part of an anaphoric clock see François Maitzner in Tardy 17-21, for a different interpretation, Neugebauer.
Musée des Antiquités Nationales,
St. Germain-en-Laye.

Such clocks might be set up publicly

71 The Tower of the Winds, 1st century BC
Photograph.
Despite its name, the tower built by Adronikos of Kyrrhos was more than a wind-direction indicator. Although all trace of the machinery has disappeared, recent investigations of the interior of the tower have revealed that remaining drains and channels in the floor fit the pattern to be expected of a large outflow clepsydra. On the outside of the building, beneath the sculpted winds, are traces of eight sun-dials (Gibbs 5001). The tower was thus a public 24-hour time-teller and if the water-clock incorporated calendrical indications it may also have married the time of year with wind-directions and other meteorological phenomena as had long been a tradition in Greek science.
Bibliography : Noble & Price

But they also served to display private wealth and status.

What?', says he, 'don't you know at whose house it is today? Why the millionaire Trimalchio's! There's a clock in his dining-room with a trumpeter complete so that he can know at any moment how much older he's getting.
Satyricon II, XXVI

I cannot tell you the time for sure since it is easier to find agreement between philosophers than between clocks.
Seneca, *Apocalocyntosis,* 2.2.

74

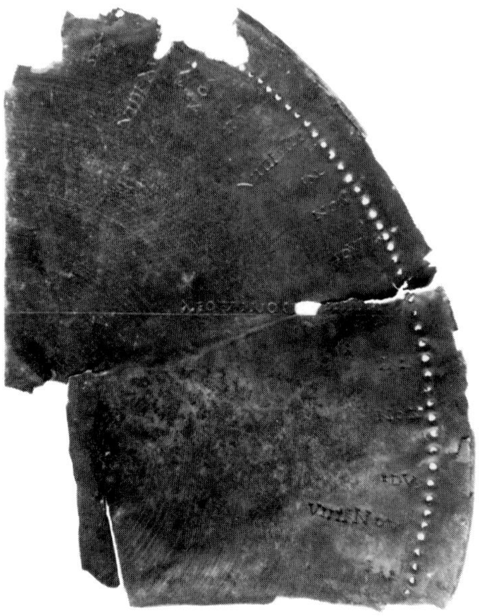

70

Islam

One moment in Annihilation's Waste
One moment of the Well of Life to taste -
The stars are setting and the Caravan
Starts for the Dawn of Nothing- Oh make haste'.
Omar Khayyam, *Rubaiyat* xxxiii
(tr. Fitzgerald).

72 Greek Technology Preserved in a Arabic Text, 3rd century BC - 8th or 9th century AD
A two-thirds scale reconstruction by André Sleeswyk of those parts which are most probably of Hellenistic origin of a water-clock ascribed in Arabic texts to Archimedes.
Overall height 1 m.
The mechanism shown here, in which time is indicated by the release of balls from the crow's beak and by a change in the colour of the person's eyes, as well as by the movement of a ring over two vertical hour scales, represents that part of the *Arshimîdas* text that may reasonably be ascribed to the Hellenistic period. The basic water machinery and the ball release mechanism may be ascribed to Archimedes himself or to his period. The head and the column with moving rings are probably slightly later, added by Philon or by craftsmen contemporary with him. A third section of the treatise, which is fully entitled *Kitâb Arshimîdas fi 'amal al-binkamt* (The Book of Archimedes on the construction of Water-clocks) was written by a Byzantine or Iranian scholar, but in either case incorporating Persian ideas, and a last part was probably added by an arabic craftsman. The eclectic nature of this text illustrates with immediacy the way in which the hellenistic technology of time-measurement was developed and changed in the christian and islamic Near East. The action of the device is as follows.
A float driving the time-indicating mechanisms is made to sink at a constant velocity. The reservoir in which it is contained empties itself in a lower reservoir at a constant rate of flow. That is engendered by a regulating float valve in an auxiliary reservoir interposed between the two main reservoirs. It is probably the first feed-back device in history.
The sinking float draws down a chain wound around a drum. On the same shaft are other drums with cords, and a large gear wheel. The latter drives a mechanism which every hour releases a metal ball that ultimately drops from the beak of a bird (perhaps a Phoenix) on a bell, producing a clanging sound.
The cords drive rings around two pillars with hour divisions on them, indicating the time, and a mechanism that causes the irises of the eyes of a (Medusa's) head to change colour every hour.

72 a

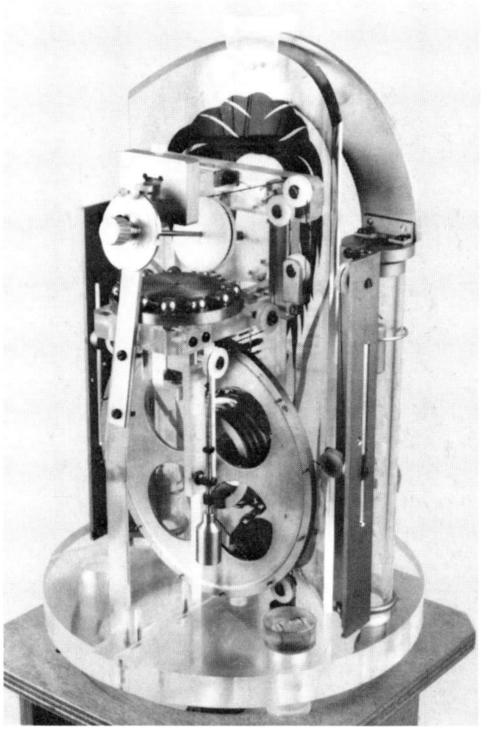
72 b

73 Two Views of the Time-indicating Mechanism of the 'Archimedes' Water-clock, 8th/9th century AD
Photographs.
1 A and B The crow's head from which the (24) balls fall with the circle of holes behind
C The head of the man
D The lantern pinion
E Wheel container
F Weight for the beak
G String
H Drum
J Toothed circle
K Chain
L Centre point of the lantern pinion
M Water-container
N Float

2 A Face whose eyes change colour
B Upper and lower millstones
C Drum container
D Drum container
E Weight for the beak
F Drum
G Toothed circle
H al-Shâhûrastûn, pinion
K Cymbal
L Ball recipient
Photographs : 1 the British Library ; 2 Bibliothèque Nationale.
Bibliography : Hill (2), 12-16 ; Hill (4), Ch 2.

They will question thee concerning the new moons. Say, 'They are appointed times for the people, and the Pilgrimage'.
Qu'rán II, 189 (Arberry's translation).

The number of the months, with God, is twelve in the Book of God, the day that He created the heavens and the earth ; four of them are sacred. That is the right religion.
Qu'rán IX.36 (Arberry's translation).

In AD 639, the Caliph 'Umar ordered that muslims should count the years from the day of Muhammad's migration (Hijra) from Mecca to Medina - 16 July AD 622. The years so reckoned in the new Hijra era (AH) were lunar years made up of 12 lunar months, i.e. approximatively 354 1/3 days.

The names of the muslim months are :

Muharram	30 days
Safar	29 days
Rabî'I (al-awal)	30 days
Rabî'II (al-akhair)	29 days
Jumadà I (al-awlà)	30 days
Jumadà II (al-ukhrà)	29 days
Rajab	30 days
Sha'ban	29 days
Ramadàn	30 days
Shawwàl	29 days
Dhu-l-Qa'da	30 days
Dhu-l-Hijja	29 or 30 days

79

LEO
LIBRA
SCOR
SAGI
CAPRI
AQVA

74 An Arabic-inscribed Lunar Volvelle Calendar, 18th or 19th century.
Brass, 83 mm.
The instrument is composed of three discs which may be rotated over each other. On the bottom, largest disc are the names of the signs of the zodiac, each divided into the usual 30° by groups of 5°. Over this moves a smaller disc engraved with the numbers 1 - 30 representing the days of the lunar month. The top disc has a circular opening showing the corresponding phase of the moon. Museum of the History of Science, Oxford.

Because the muslim year is 11.5 days shorter than the natural solar year, the months get out of step with the seasons, and pass through a complete cycle in 32.5 years. Intercalation to keep the calendar in place being forbidden, other solar or intercalated calendars such as the Julian, Syrian or Coptic, and the Persian Yazdigird, were used for civil affairs, the muslim lunar calendar being reserved for religious life.

75 An Ottoman Turkish Perpetual Calendar, Ramadàn AH 960 (AD August/ September 1553)
Manuscript of 12 folios written in red and black ink on paper, 240 mm x 160 mm.
The text is written in Arabic and Turkish. The calendar itself is enclosed within a gold frame, the oblique texts in the margins are commentaries upon it. On f.3v are circular diagrams of the Syriac months, zodiacal signs, hours and minutes and ff4r to 10r give prescriptions and precautions to be followed during each month. On f.11r is a concordance of the Persian, Coptic, Syriac, Roman and Greek calendars and on f.12r is a volvelle for dates, days of the week and the month in the muslim calendar. Such volvelles, although rare in Islamic texts (they were quite commonly used in Latin christian medieval calendrical manuscripts) were convenient ways of relating the muslim calendar with the solar Syriac calendar and the 14 month Turkish fiscal year. The present example may be compared with one of AH 1064 (AD 1653/4) now in Columbia University Library, New York, Ms Smith Or 24.
Musée de l'Institut du Monde Arabe, Paris, Inv. IA 88-o2.

76 An Ottoman Turkish Almanac Scroll (Rûznâme) 13th or early 14th century AH (19th or early 20th century AD).
Manuscript on vellum written in red, black and blue on a gold and water-coloured ground, 850 mm x 80 mm.

Scroll calendars of this kind, which were convenient for the pocket, were produced in the Ottoman world from at least the middle of the 18th century and probably

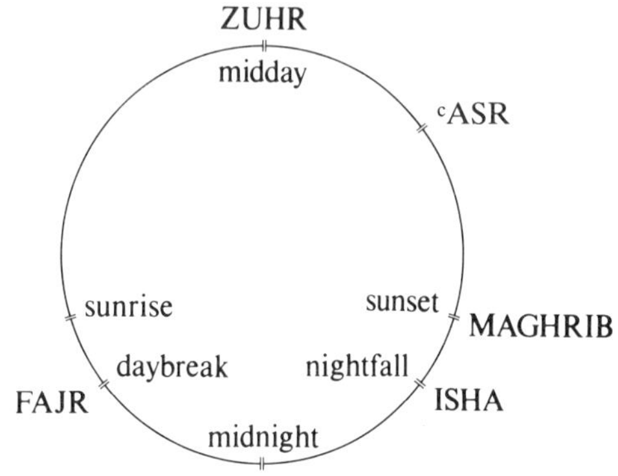

77

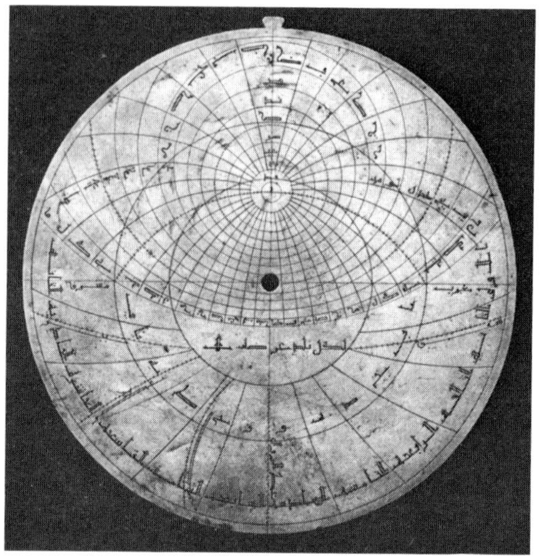
78

earlier. Often based on the calendar of Darendeli Mehmet Efende (Fl mid-17th century AD), they contain tables enabling the Muslim lunar calendar to be related to the days of the week and to the Rumi calendar which is the Syrian-Arab version of the solar Julian calendar. Further tables enable the times of the five daily prayers to be found throughout the year, the length of day and night, and the time of nightfall. Almanacs supplying a correspondence between the Muslim religious lunar year and the Julian solar year were essential for everyday life. Without simple tables as here, or a volvelle device as in item 75 above, lengthy calculations would be needed to determine seasons and dates.
Rogers Turner Books Ltd, London.

So glory be to God
both in your evening hour
and in your morning hour
His is the praise
in the heavens and earth.
alike at the setting sun
and in your noontide hour.
Qur'ân xx 17-18 (Arberry's translation)

The Muslim day, like the day of the week, begins at sunset. For the faithful it is structured by the times of prayer.

77 Prayer Time Structure of the Islamic Day
Diagram.
Since the Muslim day begins at sunset, the first prayer strictly is *Maghrib,* the period of which begins 4 minutes after sundown.

Popularly however the morning prayer is often called the first with a consequent displacement of all the others.
The second prayer period *Ishâ*, begins when dusk is over and the night is fully dark.
The third prayer period, *Fajr,* starts at the first gleam of light in the east.
The fourth prayer period, *Zuhr,* begins immediately after noon, as the sun begins to decline.
The fifth period *Asr* is midway between noon and nightfall. Bibliography : Lane 73-77.

As old learning was turned to the needs of the new religion, scales for the prayer times were incorporated into astrolabes.

78 The Prayer Times Marked on the Plate of an Astrolabe 18th century
Moroccan.
Brass, diameter 137 mm.
Plate engraved in western cufic on one side for latitude 31°, on the other for latitude 32°. The unequal hour lines drawn in the lower half of the plate are numbered in both *abjad* alphabetical numerals and ordinary arabic numerals. The times of the Maghrib, Fajr Zuhr and Asr prayers are marked on the plate by horizontally dashed lines.
Musée de l'Institut du Monde Arabe, Collection Destombes, Inv. AI86-10.
Bibliography : Mouliérac.

But although times could be found with the astrolabe, it has a greater importance as the first calculating instrument to use time as an input essential to its operation.

79 An Astrolabe from Islamic Spain which was Later used in Latin Europe AH 709 (AD 1309/10)
Signed 'Made by Ahmad b. Husayn b. Báso in the year 709'.
Brass, diameter 134 mm.
The rete for 23 named stars is fitted with four mudirs, small metal knobs with which it may be turned. The limb, which is graduated for each degree and numbered by groups of five, was cast together with the kursi and the back. Each side of the kursi is decorated with engraved foliate strapwork. There are five

plates of which three are engraved on each side with the projection for a specific latitude this being marked on the plate together with the names of the associated towns. Mecca, Oman, Aswan, Alexandria, Grenada, Malaga and Algericas. The fourth plate is engraved on one side with a latitude projection and on the other with the western form of universal horizon projection. The fifth plate, which is probably incomplete, is engraved only with lines for the unequal hours. On the back the astrolabe is engraved with a shadow square, scales for co-tangents and degrees, a zodiacal calendar (excentric type) with O Aries at 14 March, and the signature. The alidade is perhaps later.

Well and very carefully made, this astrolabe is the latest of the three known instruments constructed by Ahmad b. Husayn b. Baso, one of the best known makers of his epoch, a famous astronomer and muwaqqit of the Grand Mosque at Grenada under the Nasrides. His contemporary reputation was noted by the historian Ibn al-Khatib who underlined the fact that his works were esteemed for the beauty of the script, the well balanced execution, and for their precision. Ahmad b. Husayn b. Baso died in 709 so our instrument is probably his last work, a fact which may explain the incomplete state of the fifth plate. The historical interest of the piece however is increased by the fact that, perhaps not very long after the date of its construction, the names of the months and of the signs of the zodiac were duplicated in latin no doubt to facilitate use of the instrument by someone without knowledge of Arabic. The instrument thus offers a direct witness of the transmission of astronomical knowledge and techniques from Islamic Spain to the Latin West.
Provenance : Gautier Collection ; G. Prin Collection.
Private Collection.
Bibliography : Mayer 36, Brieux & Maddison AHMD HSYN BASO 3 ; C.C.A. 1203 ; Kurz 13.

80 What is an Astrolabe and How Does it Find Time?
Diagrams by David Penney.
The astrolabe shown here is a two-dimensional model in which the celestial sphere with a selection of stars is projected onto the terrestrial horizon-zenith co-ordinate system for a particular latitude. The parts of which an astrolabe is composed are shown in figure a. and the scales engraved on them in figure b.

To find the time by day or by night :

I Place beneath the rete the plate that corresponds with the latitude where the astrolabe is being used.
II Measure the altitude of the sun or star using the alidade and the degree scale marked on the back of the instrument.
III From the zodiacal calendar marked on the back of most western astrolabes, or from tables, find the sun's position in the zodiac for the date.
IV Turn the instrument over and turn the rete until 1° Gemini lies above the almucantar corresponding with the observed height of the sun. If a star has been observed place the pointer for that star on the rete over the corresponding almucantar.
V When using the sun, find the point diametrically opposite its position in the ecliptic using the rule and read off the unequal hour from the scale marked on the plate beneath the rete or the equal hour from the scale engraved on the limb. If the astrolabe has been set using a star find the sun's ecliptic position and read off in the same way.
In this operation from two given pieces of data - the sun's ecliptic position for a given date and its altitude (or that of a star) - time is found. But the operation can be changed. Knowing the time and the sun's ecliptic position, the date and altitude can be found and so can the positions of all the stars marked on the rete. In this case time is used as an input to the astrolabe which serves as a calculator. As such it was of the first importance to astrologers and astronomers enabling them to avoid long and tedious calculations and the high risk of error that accompanied them.

Various kinds of sun-dial were used in Islam. Some of them by their form recall their hellenistic origins.

81 Remains of the Oldest Surviving Islamic dial, c. AD 1000
Photograph.
The dial is an horizontal dial similar to the Greek pelecinum type (Cf. no 62c above). The signature in the upper right corner has been interpreted as that of Ahmad b. as-Saffar, a noted astronomer of Cordiba (d AH 426 AD 1034/5). Just over half of the dial (which was originally about 48 cm x 34.5 cm) survives. The two hyperbolic represent the summer and winter solstices, the nearly straight central line is that of the equinoxes. Across these are seven hour lines and, at the top, the arc of the zuhr prayer line. The gnomon, which was vertical to the dial surface, was set in the hole on the prolonged midday hour line which coincides with the meridian.
Photo Museo Arqueologico Provincial de Cordoba.
Bibliography : King 360-63.

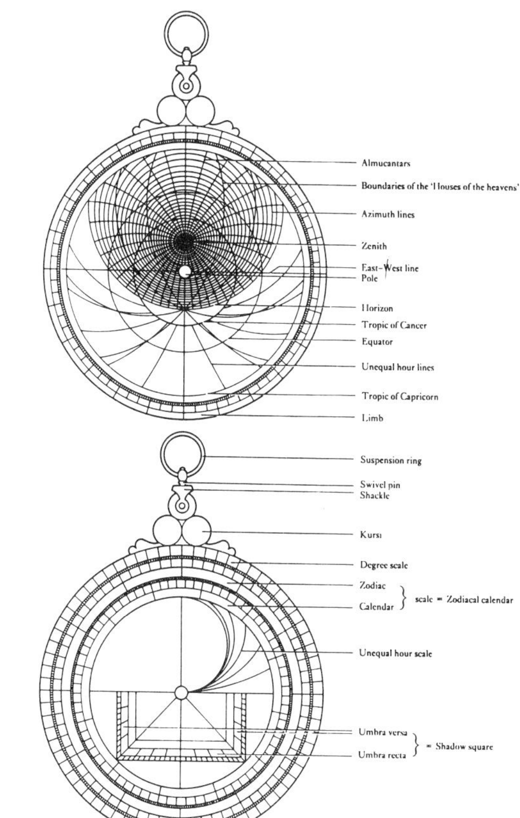

80 a

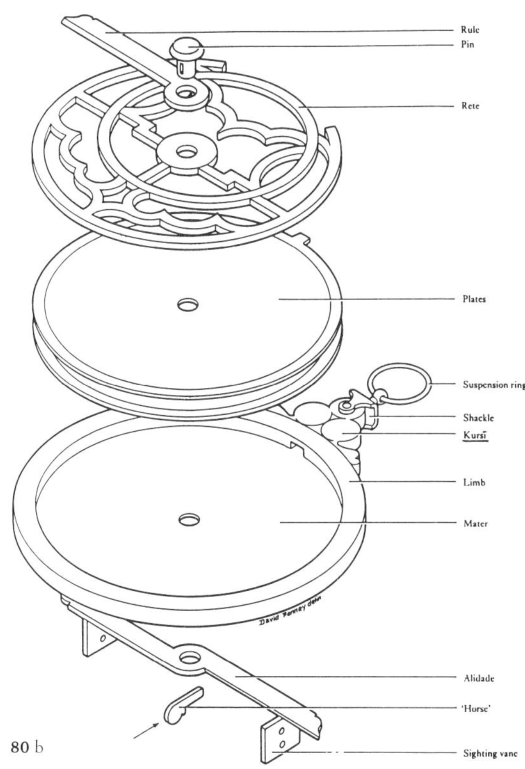

80 b

82

82 A Turkish Pillar Dial

?early to mid-12th (18th) century.
Wood, covered in paper, painted and lacquered in red, green, and yellow ; the inscriptions and numbers are written in black ink, and the lines of the scales in green, red, and black ; brass suspension ring, gnomon, and collar ; height 173 mm. Hollow cylinder with double domed top painted with spiral decoration. Immediately below the rim is a band inscribed with the names of the signs of the zodiac, two each for the six vertical columns (which are each sub-divided in three columns of 10 each) drawn down the sides of the cylinder. Immediately below this band is a collar which carries the collapsible gnomon. This collar may be slid round the cylinder to set the gnomon against the column corresponding to the position of the sun (declination) on the day of use. If the dial is then freely suspended, with the gnomon directed towards the sun, the time may be read off from the position of the shadow cast by the gnomon against the hour lines drawn as curves across the zodiac columns. Made for use in latitude 41°.
This type of dial was described by al-Marrkush. Pillar dials are self-orienting and simple. In Europe, they were made and used by shepherds in the Pyrenees until the first decades of this century (see Casanova, 292-3, & pl. XLVII).
Provenance : Collection Marcel Destombes.
Musée de l'Institut du Monde Arabe, Paris OA86-19.
Bibliography : Naffah.

83 Moroccan Quadrant and Time-calculator, AH 1197 (AD 1782/3)

Brass, diameter 220 mm.
The circular plate is cast in one piece with a double lobed *kursî* to which is attached a shackle, engraved with floral decoration. A small compass was originally set in the *kursî* (*Cf.* the illustration in Delphin). The obverse (i.e. the side from which the compass was read) is divided into four quadrants. In the upper quadrant is an inscription in cursive maghribi script.
'In the name of God, there is power and strength only in god, the Highest, the Magnificent. This blessed planisphere has been made with marvelous skill on the order of our master Abû-l-Hasan 'Alî son of our master the commander of the Believers Maulay Sîdî Muhammad b. 'Abd Allah b. Ism'âl (Muhammad I reigned AH 1171-1204 [AD 1757-89] may god make their high ranks be eternal, year 1197 [AH = AD 1782/3], for latitude 33° 40 [that of Fez]'. On either side of this inscription are two lists written in maghribi Kufic script each of eight stars with their co-ordinates (given in Abjad numerals). The lists in the right-hand quadrant is marked 'Southern stars, precession 16°', that in the left-hand quadrant carries a similar inscription for northern stars. Beneath each table is a short inscription written radially in cursive script. That in the right quadrant reads:
'It has been made on the order of our master 'Alî. The centuries do not show us a more perfect model.'
That in the left quadrant reads:
'May the days never cease to rise with you and may your kingdom never be divided'.
The lower quadrant of the instrument is entirely occupied by a sinecal quadrant marked with arcs of sines, 'versed sines', and the obliquity of the ecliptic (set at 24°). A hole at the apex of the diagram was used for attaching a plumb-line and bob with a sliding bead on the line.
On the reverse, on the outer edge of the upper two quadrants are two scales 0° - 90° used with the alidade for altitude measurements. Within this is a 'Table of settings of the sun for latitude 34°' (an average for Morocco). This consists of a circular table made up of twelve concentric rings, one for each sign of the zodiac, divided into thirty-one vertical sections by radial lines. The section in the upper centre of the diagram contains the names of the zodiac signs, and the remaining sections, each of which represents 1° of the ecliptic in that sign are numbered 1-3°. The numbers in the body of the table then give the time of sunset in terms of the number of minutes before 6 o'clock or midnight, as appropriate to the season. The numeration of the table is in ghubar numerals, that of the altitude scale is in abjad.
An unusual instrument which should be compared with a similar (although later c. 1850) instrument offered for sale in Paris in 1980 (Drouot, 9/10 October 1980 lot 184). No other instruments of this type are known. Presumably it would have been of use to the muwaqqit of a mosque, the official who was responsible for calculating the times of Muslim prayer which were astronomically determined. The degree scale on the reverse enabled altitude measurements of the sun and the stars listed on the obverse to be made. The sunset time provided by the circular table was useful in calculating the time of evening prayer, while the sinical quadrant on the obverse enabled problems of spherical trigonometry encountered when calculating the time from the sun's altitude and the latitude to be solved by graphical methods. Provenance : constructed at the command of 'Abû-l-Hasan 'Alî viceroy of Fez with authority over the mountain tribes of the Rif.
Acquired in Morocco *c.* 1880 by Captain Jules Erckmann, leader of the French military mission to Fez and nephew of the novelist Emile Erckmann (1822-99). Erckmann, who was highly esteemed by King Maulay Hasan III (reigned 1873-94) and was director of his astronomical service, wrote *Le Maroc Moderne*, Paris, 1885. The collection of Alain Brieux, Paris by whom it was lent to the Science Museum, London in 1976 for display in the exhibition 'Science and Technology in Islam'. The Time Museum, Rockford Illinois.
Museum of the History of Science, Oxford 88-13.
Bibliography : Delphin, Christies, 14 April 1988, lot 147.

84 ?Persian Horizontal Pin-gnomon Dial and Qibla-numa, 19th century

Lacquered wood, diameter 73 mm.
The upper surface is divided into four quarters by a meridional and an East-West line. Below the East-West line are a compass and a horizontal pin-gnomon sundial (gnomon missing) for Babylonian hours. Above the East-West line are marked the positions of various towns in relation to Mecca ; an index (now missing) rotating about the centre of the instrument was used in conjunction with these place-names as a Qibla-numâ. Underneath the instrument is a circular table of the inhirâf and jiha of 32 towns.
Provenance : Collection of Henri Michel.
Museum of the History of Science, Oxford, Billmeir Collection 57-84/44.

85 ?Persian Horizontal Pin-gnomon ial and Qibla-numa, 18th century AD

Brass, 129 mm x 99 mm.
The upper section of the plate is an horizontal pin-gnomon dial, engraved with hour-lines for Babylonian hours (i.e. hours counted from sunrise), and with circles to

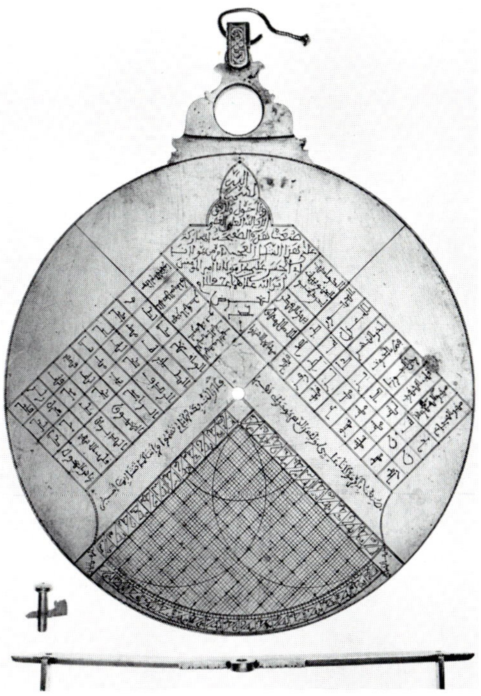

83 a

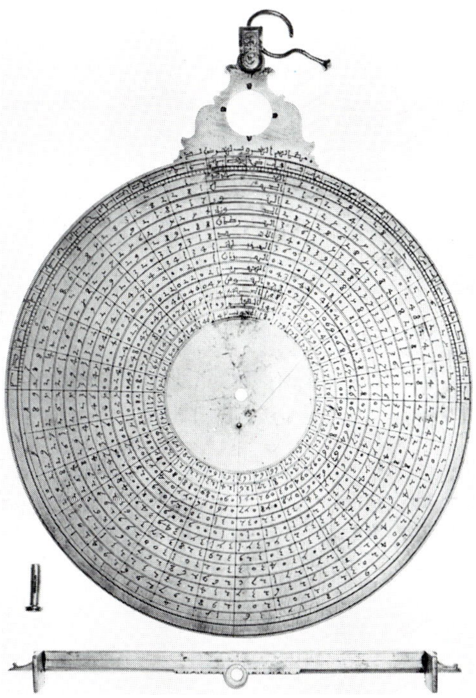

83 b

84

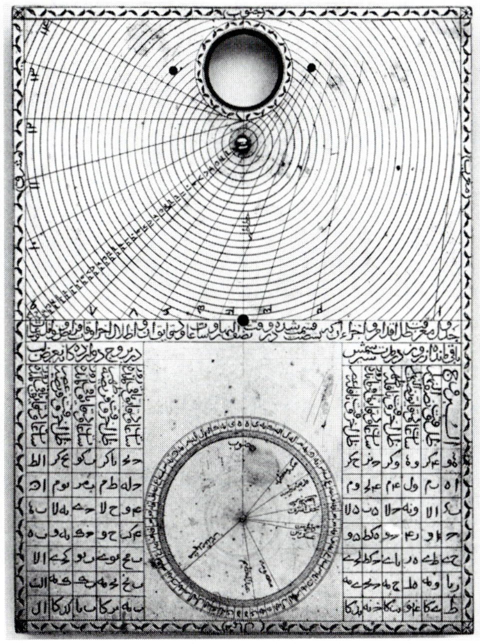

85

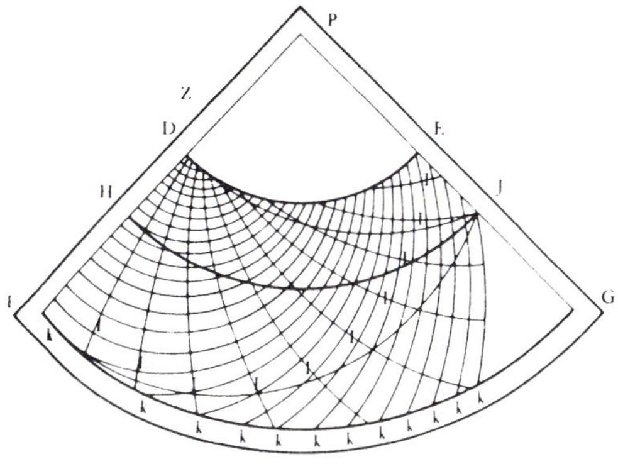

DE = Tropic of Cancer; P = North Pole; PE = Solstitial colure and meridian; PG = Equinoctial colure and line of intersection of equinoctial and prime vertical with the horizon; FG = Tropic of Capricorn; HJ = Equator or equinoctial; k = Almucantars; JK = Horizon; Z = Zenith; I = Azimuths; DJF = Ecliptic

86

show the declination of the sun. The hinged gnomon and a compass (set behind the hole in the plate) are missing.

In the lower section of the plate is a circle, divided into degrees, and showing, when the plate is correctly orientated, the direction of various places, including the azimuth of the Qibla. (The upper, the left, the right, and the bottom edges of the plate are marked 'South', 'East', 'West', and 'North' respectively.) The angle between the directions of Mecca and Medina, as shown in the circle, is 15 1/2° ; from this it has been calculated that the instrument was made for use at Shiraz. There was probably a rotating index attached to a projecting pin in the centre of the circle.

On the back is a table (?incomplete) giving the inhirf of 59 places.

This instrument may have been attached to a small stand, or may have formed part of an astronomical compendium. Museum of the History of Science, Oxford. Lewis Evans Collection 0.3a.

A particularly popular time-finding instrument in later Islam was the almucantar quadrant (rub al-muqantarat) derived from the astrolabe by the simple device of folding a standard astrolabe plate in half and this again in half. The rete is replaced by a plumb-line attached to the centre of the quadrant, carrying a bead which is moved to simulate the position of the sun or a star. The origins of this device have yet to be determined, although one Egyptian treatise describing it has been ascribed to the 12th century AD, and examples in brass have survived from the 14th century. The examples shown here however are of the lacquered wood type that largely replaced the astrolabe throughout the Ottoman Turkish world.

86 Scales on an Almucantar Quadrant

Diagram by David Penney

To find the time the user first measures the altitude of the sun above the horizon by means of the two sights set along one of the straight edges, allowing the plumb-line to hang freely. The point where the line cut the degree scale on the limb indicated the required altitude. Having determined the Sun's position in the ecliptic for the date of observation - from a set of tables or an almanac if no scale for this purpose was provided on the instrument - the user placed the plumb-line so that the bead or pearl sliding on it could be set against that point of the ecliptic equivalent to the sun's position. He then rotated the plumb-line from the apex, keeping the pearl always in the same position on the thread, until it lay upon the almucantar equivalent to the sun's observed altitude. This operation is equivalent to rotating the rete of an astrolabe. The position of the plumb-line on the limb then showed the time either directly in hours if an hour scale was marked on the limb, or in degrees which were to be converted into hours if no hour-scale was engraved.

87 Examples of Almucantar Quadrants from Late Islam

a. Lacquered wood, scales and texts written in red and black ink on a yellow ground, rubricated edges.
18th or 19th century.
215 mm x 190 mm x 10 mm

On the face is an almucantar quadrant with a 90° scale on the limb. On the back is a sexagesimal sine graph with the arcs of sines, cosines and the obliquity of the ecliptic. The chord and two further arcs departing from the right hand end of the degree scale allow the Sun's altitude to be found at moments for prayer. A double plumb-line is suspended from the apex and falls across each face of the instrument.
An almost identical instrument is preserved in the Time Museum, Rockford Illinois.
Tropenmuseum, Amsterdam 4321 - 17.

b. Signed above the 6 o'clock line of the horary quadrant 'Rasamahu al-faqîr Muhammad Abû Fadl' and on the back at the bottom of the table 'Rasamahu al-faqîr Muhammad Abû Fadl al-falaqî' (drawn [or marked out] by the learned in the law Muhammad Ab Fadl the astrologer).
AH 1220 (AD 1805/6).
180 mm x 180 mm x 5 mm

On the face is an almucantar quadrant drawn for latitude 30° (Cairo) with a double degree scale and a non-linear scale between them on the limb, and above a scale of 60° and the six numbered hour lines of an horary quadrant. On the reverse is a sexagesimal sine graph with the arc of the obliquity of the ecliptic, the arcs of sines and 'versed' sines, a chord and a prayer line.
N.G.Z.K. Schoonhoven, Inv. ZAZ 410.

c. Not signed, dated AH 1344 (AD 1925/6).
Lacquered wood with scales and texts drawn in red and black on a yellow ground and with lead plumb-bob, 89 mm x 114 mm.
On the face is an almucantar quadrant without hour lines. A degree scale is marked on the limb. On the reverse is a sine graph. The plumb-bob is stored in a cavity worked in the arc of the limb and protected by a screw-on cap.
Private Collection.

d. Brass.
Inscribed, on the back, 'Praise be to God and blessing upon our Lord Muhammad and his family - what God wills'. Made for the jurist and noble scribe Sayyid Muhammad, son of the holy and pious Sayyid Hajj 'Abd as-Salâwî [i.e. from Salé] in Rabat. Made by 'Abdallah Ahmad b. 'Alî (?mrsbl) al-Andalusî under the direction of the Mu'addil, the jurist Sayyid al-Mu'tî b. at-Tayyib, of Marînid descent, in the middle of the month Jumâdâ I, year 1219 (AH = 1804 AD)'.
Length of side approximatively 130 mm.
The front is engraved as an astrolabe-quadrant.
Museum of the History of Science, Oxford.

Water-clocks

Writing in AH 600 (AD 1203/4), Ridwan b. Muhammad b. Alî b. Rustam al-Khurâsanî Fakhr ad-Dîn b. as-Sa 'ati (d. between AH 620 (AD 1223/4) and AH 630 (AD1232/3)), stated not only that the Archimedes' clock (see number 72 above) was well known, but also indicated those parts of it which the Arabic tradition ascribed to Archimedes himself. These were the float chamber and float, the reservoir and its float, one cymbal, one bird's head, and one ball falling every hour. This was not however exactly the kind of clock familiar in Ridwan's day. He continues :
'... according to report, a man called Hormuz invented the twelve doors for the hours of the day and these doors rotated in both of them (i.e. by day and by night). And he invented for every hour two balls which fell, and he added the channels, and the heads of two birds. And he designed it ... a box, from which nothing except the doors was visible. The birds did not change from their positions, but only their beaks were protruding from the box, That (design) continued in the land of Fârs (Persia) for a long time, and was transmitted from there to the land of the Greeks, and its construction spread out in the land until it was transmitted to Damascus, where it was constructed up to the days of the Byzantines, and after that in the days of the Banû Umayya, according to what is mentioned in the histories. This clock attributed to Hormuz continued to be reproduced by one man after another on this pattern, and it was in the shape that we described above.'
Ridwan's account neatly sums up the way in which different influences united in the Islamic tradition of mechanical time-indication by day and by night, using aural and visual means. That such devices were wanted is perhaps less an indication of serious interest in time-measurement, than of the prestige and social status that they conferred on their owners.

88 A Monumental Water-clock by a Master Craftsman of the Late 12th Century

Full scale reconstruction of the Castle Water-clock of Ibn ar-Razâzz al-Jazarî built by Haward Horological Ltd following the translation and interpretation contained in Hill (1). Overall height approx. 4 m.
The mechanism of the clock is contained in a large box (the 'house') with all the time indication on one face. In al- Jazarî's own words (translation Hill 1, 18) there is a door which is closed by a wall of wood or bronze. Above the door, in a lateral straight line, are 12 doors, each of which has two leaves which are closed at the beginning of the day. Below these, and parallel to them, are 12 (more) doors, each with one leaf, which all have the same colour at the beginning of the day. Below the second set of doors is a frieze projecting from the face of the wall. At the side of the frieze is a crescent (moon) like a Dinar. This crescent moves along the ledge in front of the doors to the (other) end of the ledge. In either side of the wall below the ledge, is a niche like a mihrâb, and in each of these is a bird with outstretched wings, standing on its feet. Between the two niches are 12 roundels made of glass, which are so arranged that they form a semi-circle with ins convex side upwards. In front of each bird is a vase supported on a projecting bracket, and in each vase is hung a cymbal. Below the wall several figures are situated - two drummers, two trumpeters and a cymbalist. Above the wall is a semi-circle with its convexity towards the top. Around its circumference are six of the 12 Zodiacal signs (visible at a given time), and below this is a sphere carrying the Sun, a golden roundel and, below this, a sphere carrying the Moon, a glass roundel.
As to its significance : at the beginning of the day the crescent moves in its regular imperceptible way along the frieze until it has passed one door and is between the first and second doors, whereupon the two panels of the first of the upper doors open and a figure, made according to the choice of the

craftsman, comes out and stands as if he had suddenly emerged. Also, the first door which the crescent has passed turns over and changes colour, the two birds lean forward until they approach the two vases, and two balls are dropped from their beaks, each on to a cymbal, and the sound is heard from afar. The birds then resume their position. This happens at the end of every hour until the sixth, at which time the drummers drum, the trumpeters blow and the cymbalist plays his cymbals for a while. This occurs also at the ninth and twelfth hours.
The operation of the spheres is as follows : at the beginning of the day the centre of the sun will be in the appropriate degree of the Zodiac for that day, on the eastern horizon, about to rise. The opposite degree will be on the western horizon, about to set ; whenever a degree rises its opposite will set. The Sun climbs until noon, then descends until nightfall, when the Sun's station will set, and the six signs that have been visible will disappear, and the six that have been hidden will appear.
Suppose that it is the day of Cancer the Sun will reach its greatest altitude, while on the day of Capricorn it will reach its lowest. At night the moon is seen in the Zodiacal sign and in the degree corresponding to that night (and in its appropriate shape) : if it is appearing as a crescent it will change (as the month progresses) to a full moon and if it is a full moon it will wane.
The mechanism to operate all these movements was a mixture of elements taken from the mechanical tradition in which al-Jazar was formed and original ideas. The heart of the clepsydra was a reservoir tank connected by a tap and conical valve to a float-chamber with a constant head of water. The float chamber in turn was connected to a regulator which adjusted the water flow for each day and for day and night. The purpose of this was to compensate for the unequal lengths of the day and night period (each of which was to be divided into twelve hours) and their change throughout the year.
From a float in the reservoir, strings pass over pulleys set in the roof to two large driving wheels one of which controls the motion of the zodiac disc and the Sun and Moon rings, the other, the disc which uncovers the twelve glass roundels. At the same time, by means of a further string and pulley, the second wheel draws the 'cart' on which is mounted the lunar crescent which traverses the front of the clepsydra below the row of twelve doors. Attached to the cart is an iron rod which each hour tips the mechanism controlling the opening of the doors and the functioning of the falcons.
From the flow regulator, which is also the outlet for water from the reservoir, the water-stream falls onto a plate and thence through a pipe into one of three divisions of a tank. At the sixth hour a string, operated by the opening of the sixth door, opens a valve in the divisions so that water flows into the main trough and thence onto a water-wheel. As this turns iron pegs on its shaft operate the arms of the drummers and the cymbalist. From the water-wheel, the water passes through a pipe into an air-vessel equipped with a mechanical flute and a siphon. The end of the flute is set in the wall between the two trumpeters, thus producing a noise that seems to come from them. From the return of the siphon-tube the water runs out into a cistern set in the floor. From here it would be lifted back into the reservoir by the clock attendant when he resets the machine each day.
Provenance : Built for and displayed in 'Science and Technology in Islam', an exhibition held in the Science Museum, London 6 April - 31 August 1976 as part of the World of Islam Festival.
Tropeninstituut, Amsterdam.
Bibliography : Hill (1), Hill (4), 92-111.

89 A Copy of al-Jazarî's Treatise on Mechanical Devices, AH 969 (AD 1561/2)
Kitâb fî ma 'rifat al-hiyal al-handasiyya copied from a copy made by the scribe's master *Shaykh* Muhammad al-Mahwarî of the copy made in AH 891 (AD 1486) which is now in the Bodleian Library, Oxford (Ms Graves 27) being itself an excellent copy dated AH 742 (AD 1341/2) which was based on an autograph copy by al-Jazarî himself. Ms on paper written in ink in a *Naskhi* script. Chapter 1 of Category VI (description of a door) is omitted as being of no utility, and many of the diagrams are wanting.

Badî az-Zaman Abû al-'Izz Isma'il b. ar-Razzâz al-Jazarî tells us that when he wrote his book (completed AH 4 Jumadà ii 602 (AD 16 January 1206)), he had been in the service of the Artuqid kings of Diyr Bakr for some twenty five years. He had studied both the books and the products of earlier scholars and craftsmen and progressed 'by practicing it [applied mechanics], from the stage of book-learning to that of witnessing ... I assembled the divisions that they [previous scholars and technicians] had separated and put forth branches from roots where they had been correct, and devised specimens that worked splendidly.' Considering the researches he had made, al-Jazarî was anxious to pass on his knowledge but feared to do so until he was commissioned by the king. So he composed 'this book in which there are certain rents I have patched and basic matters the sub-divisions of which I have given and figures which I have designed.' (Translation Hill (1), 15).
Providing evidence, by the date when it was copied, for the continuing interest of al-Jazarî's work long after his death this manuscript also illustrates clearly how texts were transmitted from copy to copy across several centuries. That at least fourteen copies of the treatise have survived helps to confirm its popularity. In addition to water- and candle-clocks, the work deals with drinking vessels, pitchers and basins for hand-washing and phlebotomy ; fountains and mechanical flutes ; pumps and other water-lifting apparatus ; the casting of a metal door ; the making of a protractor ; of door bolts and of a combination-lock. This context, like the nature of the clocks themselves, suggests that it was the mechanical problems they posed, rather than a preoccupation with time, that was the motive for their construction.
University of Leiden Library, Ms Or. 656.
Bibliography : Voorhoeve 194-5 ; Hill (1), 4.
For the Arabic text see Hassan.

90 Clocks Designed by Ibn ar-Razzâz al-Jazarî, late 12th century AD
An audio-visual presentation.
Al-Jazarî illustrated his compendium of mechanical inventions with drawings made 'with the ruler and freehand'. Many of these illustrations which were often elaborate and well-coloured, were faithfully transmitted in subsequent copies of the manuscript (Weitzmann). Because of their artistic qualities the illustrations have long been better known than the text they accompany, and it is a scandal of western 'connoisseurship' that as a result two important manuscripts dated AH 715 (AD 1315/6) and AH 755 (AD 1354/5) were vandalised and the miniatures sold to collections throughout the world. What remains of the first of these manuscripts was sold at Sotheby's London 3 April 1976 being purchased by Spink & Son (Hill (3) passim). The remains of the second are in Sulemanniye Kütüphanesi, Istanbul. Ms Hagia Sophia 3606.

I The 'Castle' Water-clock
a. General view from the earliest known manuscript dated AH 602 (AD 1206/7).
Topqapu Serayi Müzesi Kütüphanesi, Istanbul. Ms Ahmet III 3472.
b. General view from the AH 755 manuscript.
Museum of Fine Arts, Boston. Goloubew Collection.
c. Mechanism to operate the falcons.
Bodleian Library Oxford, ms Graves 27 f.
d. The basic water mechanism of the 'Castle' clock.
Bodleian Library Oxford, ms Graves 27 f.

II The Water-clock of the Drummers
e. General view from the AH 602 manuscript.
Topqapu Serayi Müzesi Kütüphanesi, Istanbul. Ms Ahmet III 3472.
f. General view from the AH 715 manuscript.

64

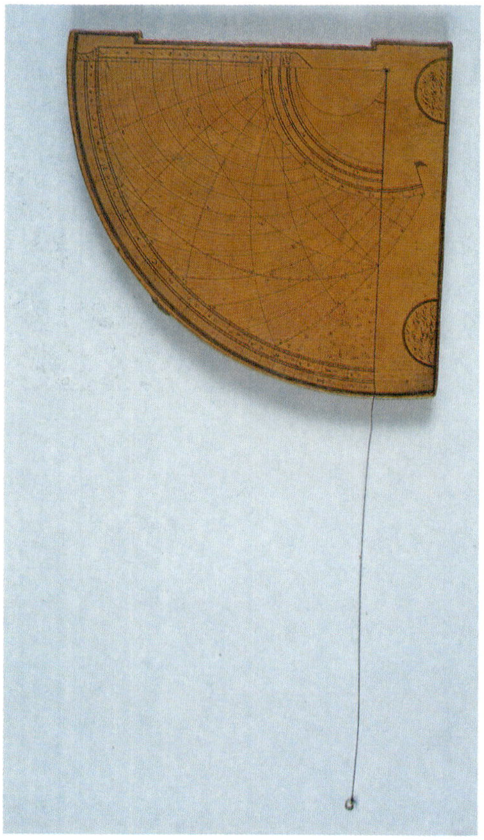

87

88

The Smithsonian Institution, Freer Gallery of Art, Washington D.C.

III The Water-clock of the Boat
g. General view from the AH 602 manuscript.
Topqapu Serayi Müzesi Kütüphanesi, Istanbul. Ms Ahmet III 3472.
h. General view.
Bodleian Library Oxford, ms Graves 27 f.

IV The Elephant Water-clock
i. General view from the AH 602 manuscript.
Topqapu Serayi Müzesi Kütüphanesi, Istanbul. Ms Ahmet III 3472.
j. General view from the AH 715 manuscript.

V The Beaker Water-clock
k. General view from the AH 755 manuscript.
Museum of Fine Arts, Boston.
l. Sectional view.
Bodleian Library Oxford, Ms Graves 27 f.

VI The Water-clock of the Peacocks
m. General view from the AH 602 manuscript.
Topkapu Serayi Müzesi Kütüphanesi, Istanbul. Ms Ahmet III 3472.

VII Candle Clocks
n. The Candle-clock of the swordsman.
Smithsonian Institution, Freer Gallery of Art, Washington D.C.
o. The Candle clock of the scribe.
Bodleian Library Oxford, ms Graves 27 f.
p. The Candle-clock of the monkey from the AH 715 ms.
q. Candle-clock of the doors from the AH 715 manuscript.
r. Water-clock of the boatman.
Bodleian Library Oxford, ms Graves 27 f.
s. Water-clock of the boatman from the AH 715 manuscript.
Bibliography : Coomaraswamy ; Stchoukine ; Hill (1).

91/92. Remains of two water-clocks from the 8th (14th) century Maghreb
Photographs.

91. A rotatable astrolabe dial surrounded by twenty four roundels and an inscribed outer ring from a water-clock in the Qarawiyyin mosque, Fez, Morocco installed by Abu 'abd Allah Muhammad al- 'Arabî in AH 763 (AD 1361/2), which itself reused parts of an earlier clock by Abû 'Abd Allah Muhammad as-Sinhâgî.
92. Gongs and ball channels of a monumental water-clock. Erected on the Bu'anâniya mosque Fez on AH 14 Jumadà I 758 (AD 6 May 1357) by Abû-l-Hasan 'Alî b. Ahmad at-Tilimsanî.
Photograph 92a shows the full set of 13 gongs from a postcard of c. 1920. Photograph 92b is a more recent close-up view of five of the gongs.

91

92

Between the Indian and the Pacific Oceans

98 Two Calendars from Bali, 19th or 20th century
a) wood, 500 mm x 220 &
b) painted cloth, 167 (168) mm x 135 (138) mm.
Traditional Balinese calendars, which are based on those of Java, are primarily concerned with marking the lucky and unlucky-days i.e. those governed by benevolent or malevolent demons. Where the *Uku* year is used, as in calendar a), it is divided into six periods of thirty-five days with a seven-day week division and a further division of units made up of from 1 - 10 days. Diagonal lines another and images of constellations or of controlling spirits are drawn in the square grid that makes up the calendar.
Tropenmuseum, Amsterdam, Ser 281-1 & Ser 2015-1.

99 A Thai Calendar, 20th century
Wood Rectangular slat, with diamond shaped suspension piece, divided into 30 rectangles.
Tropenmuseum, Amsterdam, 4682-8.

100 A Javanese Calendar-divination Board, 20th century
Dark palm wood, 100 mm approx. x 40 mm. One side is divided with a grid for the 35-day period marked with dots, and lines at various angles giving the signification of individual days. The reverse carries geometric forms.
Tropenmuseum, Amsterdam, Ser 596.

101 An Horizontal Sun-dial and Calendrical Instrument from Gresik
(= Tandes), Java, 1722
Photograph.
The horizontal dial, known as a *Bencet* measures about 60 cm x 60 cm and is formed of two arms set perpendicularly to each other in the form of a 'T'. Each member is graduated with shadow length values, the shorter scale being used to measure shadows at noon, the longer the shado length at *asr*.

Division of the instrument is based on a change of 1 unit from 4 south to 2 north between the time of the mid-summer shadow and that of mid-winter. The period taken for a complete change of one unit (= a sixth of the distance travelled by the sha-

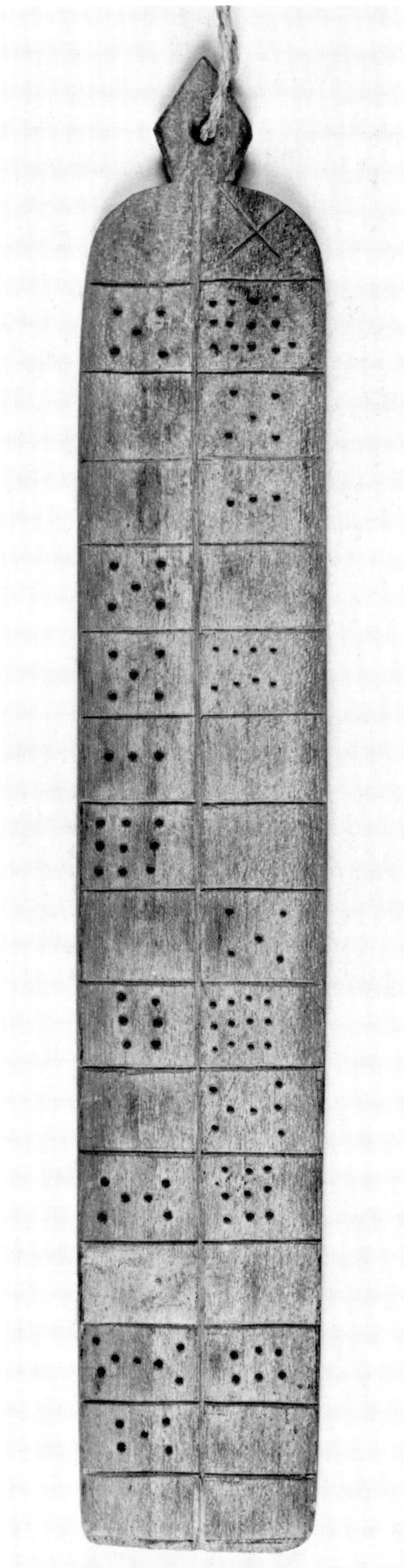
99

98 a

100

104

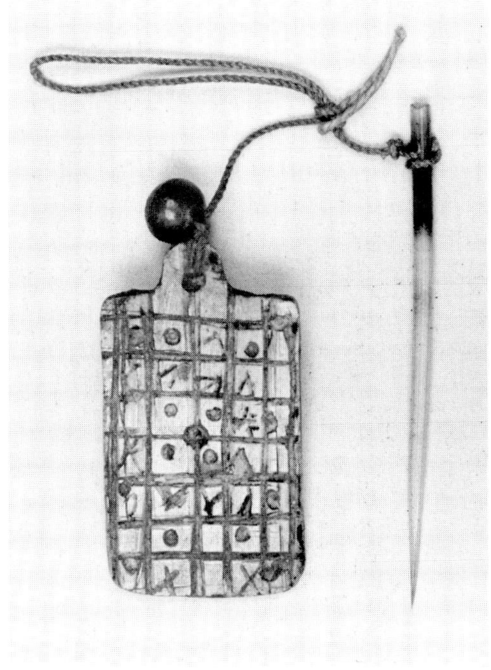

105

dow) is taken to representa *mangsas*, a year division which can vary from 23 to 43 days.

Bibliography : Negoro ; King, 239-40.

102 Two Cylindrical Batak Calendars, 20th century
a. bamboo, overall length 271 mm.
b. 285 mm.
Both in the form of open cylinders, a. with removable handle carved in the shape of a head.
Approximately half the surface of both is incised with a diagonal calendar showing lucky and unlucky days.
a. Private collection.
b. Tropenmuseum, Amsterdam, A4294.

103 A Batak Calendar in the Form of a book with 'Concertina' Folds, 19th century
Bark, 84 mm x 110 mm (closed).
30 'pages' written on each side.
Private collection.

104 A Twelve Rod Batak Calendar, 19th Century
Bamboo, each 180 mm long.
Primarily for divination. Six of the rods carry at the top a human face in profile.
Tropenmuseum, Amsterdam, Ser 1772 no 186.

105 A 'Month' Divination Calendar from the Batak Islands, 19th century
Bone, 55 mm x 30 mm.
A 35 division month calendar with porcupine quill stylus and small suspension piece ; index marks for the nature of the different days.
Tropenmuseum, Amsterdam, Ser 15-602.

China

Symbol of sovereignty and instrument of social organisation, calendrical science not only motivated Chinese astronomical research, but also linked the cyclical repetitions of the natural world and the Taoists, with the compartmentalised time of separately divided up seasons and dynasties, with a 'narrative continuity' history which spanned many centuries.
Between 370 BC and AD 1742, one hundred 'calendars' were produced in China.

106 Calendar Production 400 BC - AD 1900
Graph from Needham, Ling & Price, 174.
The white columns in thin outline give the number of new calendars per centruy when mere changes of name are excluded. The shaded, heavily outlined, columns show the result when calendars marking the beginning of new dynasties are excluded. Activity was greatest in the 6th to 8th centuries AD and again in the 9th to 13th. Activity in clock-making may be related to this as is shown by the names of horological innovators marked in the appropriate columns. Basic data for the calendar such as the length of the tropical year and the dates of the solstices, had been obtained in early times (?14th and 13th centuries BC) by gnomon shadow measurements.
A millennium later such measurements were still being made but now with sophisticated monumnetal apparatus.

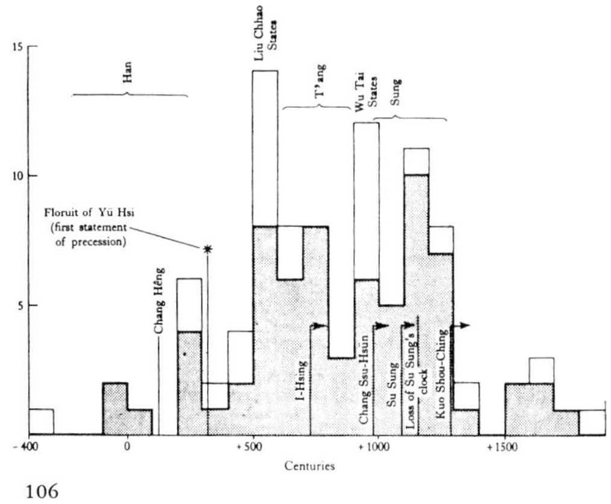

106

107 The Calendrical Tower of Dengfeng or Guo Shoujing's Tower for Measuring the Sun's Shadow, c. AD 1276.
Partly reconstructed during the Ming period (AD 1384-1644)
Photographs and diagrams.
The massive brick-built pyramidal tower is 8 m wide at the top, 16 m at the bottom and its observing platform is 9 m 86 high. On the north side is a further three-room structure 3 m 5 high with a central opening. One of the rooms contained a clepsydra and there was provision for measuring the meridian transits of stars. The large central opening allowed sunlight to fall on the top of a gnomon placed in a socket at the south end of the stone graduated shadow scale ('Sky Measuring Scale') which extends 36 m 60 to the north and is fitted with water channels connected at each end which serve as a level.
On observing instruments of this size, although the graduation can be effected with considerable accuracy, reading the instrument becomes difficult because the edge of the shadow becomes indistinct. To overcome this difficulty a 'shadow definer' was erected that could be moved along the horizontal scale. This ingenious device was described in the *Yuanshi* (History of the Yuan) as follows 'The shadow definer is made of a leaf of copper 2' wide and 4' long, In the middle of which is pierced a pin-hole. It has a square supporting framework, and is mounted on a pivot *(chi chu)* so that it can be turned at any angle, such as high to the north and low to south (i.e. at right angles to the incident shadow-edge). The instrument is moved back and forth until it reaches the middle of the (shadow of the) cross-bar, which is not too well-defined, and when the pin-hole is first seen to meet the light, one receives an image no bigger than a rice-grain in which the cross-beam can be noted indistinctly in the middle. On the old methods, using the simple summit of the gnomon, what was projected was the upper edge of the solar disc. But with this method one can obtain, by means of the cross-bar, the rays from the centre of the disc without any error.
(In +1279) on the 30th May, I observed the summer solstice shadow as 12.3695 ft. in length; and on the 11th Dec. of the same year found that the winter solstice shadow was 76.7400 ft. long'.
(translation from Needham III, 299 following Maspero).

Experience with calendrical gnomons may have stimulated the development of the sun-dial. Although references to dials are scarce in Chinese literature, they certainly made up part of the calendar specialists' equipment. In 104 BC calendar specialists '...determined the true east and west points, set up sun-dials and gnomons, and contrived water clocks. With such means they marked out the twenty eight hsiu (lunar stations) according to their position at various points in the four quarters, fixing the first and last days of each month, the equinoxes and solstices, the movements and relative positions of the heavenly bodies and the phases of the moon.'
Chchien Han Shu quoted from Needham III, 302.
Extremely few early Chinese sun-dials have survived.

108 A Chinese Sun-dial of Traditional Form, date uncertain
Bronze, 415 mm x 260 mm.
The dial face is marked with a semi-circular scale divided successively from the centre outwards into 12, 24 and 48 unequal parts. The manner of functioning of the instrument is unclear, but it is perhaps to be studied in relation with the earliest known chinese sun-dials from the Han period described by Needham (II, 302-9) and Brieux (2) no 12.
Deutsches Museum, Munich, Inv. 42910.

Later Chinese Sun-dials

Exactly when the double-faced equinoctial dial came into general use in China is not known, although it was known (because then invented or re-invented by Tsêng Nanchun) in the first half of the 12th century (Needham III, 309). Later dials show an interesting mixture of typical Chinese elements and western influences. They have been classified by Needham (III, 310-12) into three types.
Type A
Diptych dials with string-gnomons in which the shadow of the cord stretched between the two plates when they are opened gives a time reading on a scale drawn on either the vertical or the horizontal surface or both. A compass is incorporated in the horizontal leaf (base) to orient the instrument. Such dials are the result of western influence and were probably introduced by the Jesuits. Indeed one of the Chinese names for such instruments is 'the foreign sun-dial'.
Type B
Dials composed of a base plate at one end an inset compass, and at the other a scale of *Chieh*, the names given to divisions of the Chinese solar year equivalent to a fortnight (for a list thereof see Needham III, 405). Each of these periods corresponds with 15° of solar motion in longitude along the ecliptic and they could therefore be used as a solar declination scale. If it be assumed that each locality had a 'standard *Chieh*' given in the local calendar, such a list could be used as a latitude scale against which a central plate could be inclined by a hinged strut. Such dials seem to derive entirely from the indigenous Chinese tradition.
Type C
A rarer form of dial, often made of ivory, composed of a series of pin-gnomon plates showing both time and solar declination. Each plate is drawn for a different latitude.

109 Two Chinese Diptych Dials, 19th and 20th century
a. Wood and brass, with original lacquered case with floral inlaid mother-of pearl decoration, 90 mm x 62 mm.
Face 1a
Lunar dial and volvelle. When not in use the gnomon is screwed into the top edge of the plate. When used it is placed in the central hole, the lid is set at an angle appropriate to the latitude against the scale of *chieh* marked on the side, the volvelle is turned to correspond with the day of the lunar month and the hour is read off from the position of the

109 a

109 b

shadow of the pin on the scale on the volvelle.
Face Ib
Vertical dial having a common string-gnomon with
Face IIa
Horizontal dial with inset central compass.
Face IIb
Table.
The 24 *chieh* are inscribed along each side of the lower leaf.
Needham type A.
Museum of the History of Science, Oxford, 016.
b. Chinese Diptych Dial, first half 19th century of earlier
Wood and brass, 60 mm x 43 mm.
This example is exactly similar in layout to no 109a above.
Provenance: from the collection of Dr. John Lee (1783-1866) Q.C. F.R.S., antiquarian and man of science, at Hartwell Manor.
Needham type A.
Museum of the History of Science, Oxford, 050.

110 Chinese String-gnomon Dial, ?18th century
Gilt brass, 70 mm x 80 mm x 90 mm.
A rectangular brass plate is mounted on three screw feet and has an inset compass around which the hour scale is laid out. The string-gnomon is supported at its top end by a shaped open bracket hinged to the base-plate. A plumb-bob for levelling the instrument in conjunction with screw-feet originally suspended within the bracket is now missing.
An unusual form of Chinese dial which probably reflects European influence. For a similar instrument formerly in the Time Museum, Rockford, Illinois (Inv. 1723), see Christies SK 14 April 1988 lot 127.
Musée International d'Horlogerie,
La Chaux-de-Fonds, V-43.

111 Equinoctial Dial, late 19th century
Signed '[Made by] Fâng Xiûshuî (surnamed Zé), village of Xiû, district of Xîn'ân'.
Wood, marked in black and red ink, with brass fittings, 97 mm x 46 mm.
Rectangular base with the dial-plate hinged at the centre. The thicker end of the base has an inset compass surrounded by two concentric circles containing eight compass points and twenty-four directions. The thinner end of the base carries a scale of thirteen *chieh* for setting the equinoctial plate. The concentric scales on the equinoctial plate around the gnomon are (reading from the centre out):
1 Directions to orient the instrument.
2 The twelve terrestrial branches.
3 Division into the 12 chinese double hours.
4 Subdivision of the previous in eighth parts.
Needham type B.
Whipple Museum of the History of Science, Cambridge Wh 3188, Bryden 351.

112 Equinoctial Dial, late 19th century
Signed 'Jing Cheng-hui, Cai Jia'.
Ivory, 61 mm x 25 mm with wood case, brass fittings.
This dial is exactly similar in principle with the previous example. Instructions for use are marked on the underside of the base, and the case is inscribed with the name of the maker 'Master of the mountain with palm trees' and a short verse. Needham type B.
Whipple Museum of the History of Science, Cambridge Wh 3187, Bryden 353.

113 Two Horizontal Pin-gnomon Dials for Fixed Latitudes, ?19th century
Ivory, 203 mm x 105 mm.
Each side of each plate is inscribed with a set of radiating hour lines which cross the month arcs. Time was indicated by the position of the tip of the shadow cast by a short pin-gnomon which was set in a hole on the centre line of the dia. The dials are drawn for the latitudes of Bao Ding and Tia Yang (Shansi), and for Feng Tian and Sun Tia (Manchuria). The similarity of this diagram with some Greek and Islamic dials will be noted.
Provenance: Collection of Dr. R.S. Clay.
Whipple Museum of the History of Science, Cambridge Wg 3189a, b, Bryden 354.

Calendar verification stimulated much development in Chinese astronomy. Both activities were facilitated by the development of elaborate water-driven astronomical models which were also measured and indicated the time of day and night.
Such machines were developed over a period of more than a thousand years.
Particularly important for the innovations that it introduced to the control mechanism was a device which combined a weighing machine with the inflow/outflow water-clock.

114 A Tablet Sun-dial, 19th century
Ivory, 70 mm x 30 mm x 10 mm.
The Science Museum, London 1953-316.

115 The Great Steelyard Clepsydra of Kêng Hüsn and Yüwn K'ai, AD 610
A half scale working reconstruction by A.W. Sleeswyck.
The model presents the probable method of operation of the *T'ien Ho* (Celestial River) water-clock. The instrument uses a steelyard (an unequal arm, single beam balance with movable weight) to 'weigh' water entering into a container suspended on one side of the balance arm, and floating in a reservoir from which it is filled by a siphon.

115

When it is nearly full the clock's attendant moves the weight from its position near the fulcrum to one near the end of the beam. The change in the inclination of the balance that this produces causes the direction of flow in the siphon to be reversed and water now flows out of the container back into the reservoir. When the container is nearly empty the attendant again moves the weight towards the fulcrum, and the cycle is repeated. The reservoir is placed in a closed protective box, the chains attaching the container passing through small holes in the lid. The whole system is thus protected from dust and disturbance. The steelyard itself is suspended from an arch by an ingenious mount which both spreads the load carried by the hook and limits lateral rotation of the steelyard.
Time indication was on a vertical graduated rod suspended from the long arm of the steelyard. The rod was either drawn out of a box or let down into it, depending on the mode of operation. Time was read off from the middle of the rod. Seasonal adjustment was made by replacing the rod by one which was graduated differently.
Lent by A.W. Sleeswyk Esq.
Bibliography: For details, and the rationale, of the reconstruction Sleeswyk; for an earlier interpretation Needham III, 316; Needham, Wang Ling & Price, 94-2.

The 'Celestial River' clepsydra marks a turning point in water-clock technology for it is weight-driven and has a self-acting control system (i.e. an escapement). With the realisation of the astronomical tower of Su Song just over four and a half centuries later, a peak of achievement in this technology was achieved.

124 Table of the Sexagenary Cycle

125 The Zodiac Signs and the Hours of Day and Night

The first calendar reform based entirely on Japanese research was that of Shibukawa Harumi whose Jôkyô calendar was adopted in 1684. It was not replaced until 1843. The new calendar, the western-influenced but still traditional *Tenpo* calendar, was itself replaced by the Gregorian calendar in 1873.

Clocks

Although in the early 7th century a water-clock following Chinese models was constructed in Japan, and in AD 736 the gnomon was introduced, such instruments were strictly reserved for government use. Time-measurement by artificial means hardly existed in everyday life.
During the brief period of Jesuit influence in Japan (*c.* 1550 - *c.* 1630), a few western weight-driven clocks were brought into the country. In 1551 such a clock was introduced by St Francis Xavier and others were imported or received as diplomatic gifts. Such clocks were subsequently copied and adapted by Japanese craftsmen.

126 The Oldest Surviving Western Clock in Japan, 1581

In 1611 the Spanish Viceroy in Mexico presented a clock to the first Tokugawa Shogun, Ieyasu. This, a brass standing table clock approximately 203 mm high with a leather carrying case, is signed 'Hans de Evalo me fecit en Madrid 1581'. The clock is unusual in that the wheel arbors are set parallel with the dial, linkage being provided by a helical gear.
Hans de Evalo was probably of Flemish origin. Achieving success at Madrid, in 1583 he made a night clock for Philip II which has survived in the collections of the Escorial.
The clock presented to Ieyasu is preserved in the Toshogu temple shrine at Kunozan near Shizuoka.
Bibliography : Asahini ; Howard ; Yamaguchi 5.

127 Table Clock, 19th century

Brass, with wooden box, diameter 75 mm.
The clock records unequal hours and the silver hour markers can be adjusted as necessary on the dial. The markers are not the customary Japanese number system of 6,5,4,9,8,7, but the Chinese signs of the zodiac symbolically associated with each hour of day and night.
Museum of the History of Science, Oxford.
Barnett Collection, B 189.

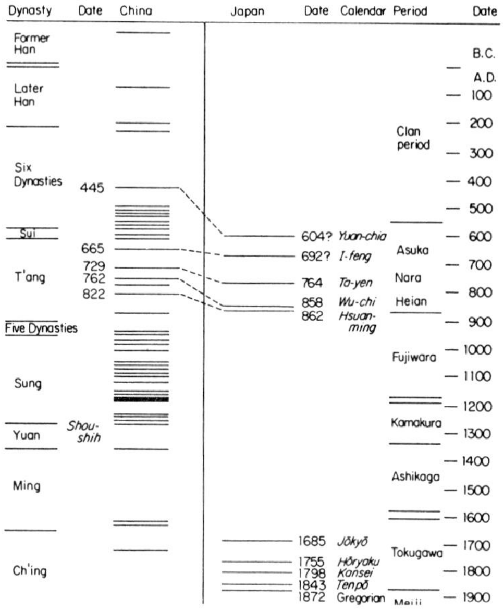

123

1.	甲子	*wood*-rat	21.	甲申	*wood*-ape	41.	甲辰	*wood*-dragon
2.	乙丑	wood-bull	22.	乙酉	wood-cock	42.	乙巳	wood-serpent
3.	丙寅	*fire*-tiger	23.	丙戌	*fire*-dog	43.	丙午	*fire*-horse
4.	丁卯	fire-hare	24.	丁亥	fire-boar	44.	丁未	fire-goat
5.	戊辰	*earth*-dragon	25.	戊子	*earth*-rat	45.	戊申	*earth*-ape
6.	己巳	earth-serpent	26.	己丑	earth-bull	46.	己酉	earth-cock
7.	庚午	*metal*-horse	27.	庚寅	*metal*-tiger	47.	庚戌	*metal*-dog
8.	辛未	metal-goat	28.	辛卯	metal-hare	48.	辛亥	metal-boar
9.	壬申	*water*-ape	29.	壬辰	*water*-dragon	49.	壬子	*water*-rat
10.	癸酉	water-cock	30.	癸巳	water-serpent	50.	癸丑	water-bull
11.	甲戌	*wood*-dog	31.	甲午	*wood*-horse	51.	甲寅	*wood*-tiger
12.	乙亥	wood-boar	32.	乙未	wood-goat	52.	乙卯	wood-hare
13.	丙子	*fire*-rat	33.	丙申	*fire*-ape	53.	丙辰	*fire*-dragon
14.	丁丑	fire-bull	34.	丁酉	fire-cock	54.	丁巳	fire-serpent
15.	戊寅	*earth*-tiger	35.	戊戌	*earth*-dog	55.	戊午	*earth*-horse
16.	己卯	earth-hare	36.	己亥	earth-boar	56.	己未	earth-goat
17.	庚辰	*metal*-dragon	37.	庚子	*metal*-rat	57.	庚申	*metal*-ape
18.	辛巳	metal-serpent	38.	辛丑	metal-bull	58.	辛酉	metal-cock
19.	壬午	*water*-horse	39.	壬寅	*water*-tiger	59.	壬戌	*water*-dog
20.	癸未	water-goat	40.	癸卯	water-hare	60.	癸亥	.water-boar

124

Month.				*Hour.*	
1.	子	ne	Rat (winter solstice)	Corresponding with	9 Midnight
2.	丑	ushi	Bull	do.	8
3.	寅	tora	Tiger	do.	7
4.	卯	u	Hare (vernal equinox)	Corresponding with	6 Dawn
5.	辰	tatsu	Dragon	do.	5
6.	巳	mi	Serpent	do.	4
7.	午	uma	Horse (summer solstice)	do.	9 Noon
8.	未	hitsuji	Goat	do.	8
9.	申	saru	Ape	do.	7
10.	酉	tori	Cock (autumnal equinox)	do.	6 Dusk
11.	戌	inu	Dog	do.	5
12.	亥	i	Boar	do.	4

125

128 A Portable Clock in the Form of an Inro, 19th century

Gilt brass dial with silver numerals, shitan wood case with gold lacquer decoration.
The striking movement has a verge escapement and the numerals in the chapter ring can be adjusted for the varying unequal hours (Cf. no 125).
Inro clocks, so-called because their cases are similar to an inro or medicine box, are usually carried hanging from the girdle of a kimono. A netsuke and an ojime were often attached to the cord which retains the lid.
Museum of the History of Science, Oxford, Inv. 59-107.

129 Lantern Clock, early 19th century

Brass and steel, 238 mm x 93 mm x 104 mm.
Plain case of standard construction (see Robertson 221) on low feet. Dial with the traditional Japanese 9-4/9-4 numbering using the zodiacal names of the hours. Inner dial with adjustable chapters, elaborate open-work hand, verge and foliot escapement, striking on bell.
'Lantern' clocks of this type were either mounted on floor stands (*cf.* 131) or placed in glazed cases hung on the wall (Robertson 239). The present example was probably such a *kakedokei.*
Rijksmuseum voor Volkenkunde, Leiden 2603-25.

130 Lantern' Clock, early 19th century

Brass and steel, 183 mm x 154 mm x 103 mm.
Marbled case with central double dial with traditional numeration, pierced hand, verge and foliot escapement, striking on a bell.
Perhaps used as a standing clock, *makura-dokei.*
Rijksmuseum voor Volkenkunde 2603-20.

131 'Lantern' Clock on a Floor Stand, 19th century or earlier

Brass and iron on black lacquered wood stand with red and gold edge decoration and brass mountings, height 1 m, 210.
Four leg pyramical case with glazed top encasing the movement Double chapter ring with traditional zodiacal numeration, centre alarm setting disc and pierced hand in the form of a chrysanthemum, date apertures below the dial.Three train movement with two verge and foliot escapements for the unequal hours of day and night set back to back and operating on a single crown wheel. The taller verge carries the day baiance. At the sixth hour a pin on the locking plate engages a star wheel or cam which rotates a rod with an horizontal arm set below the verge. On rotating this arm rises, pushing up the verge and so disengaging the pallets. As it does so, a second arm releases the night train verge so that it engages with the escape wheel. At the

128

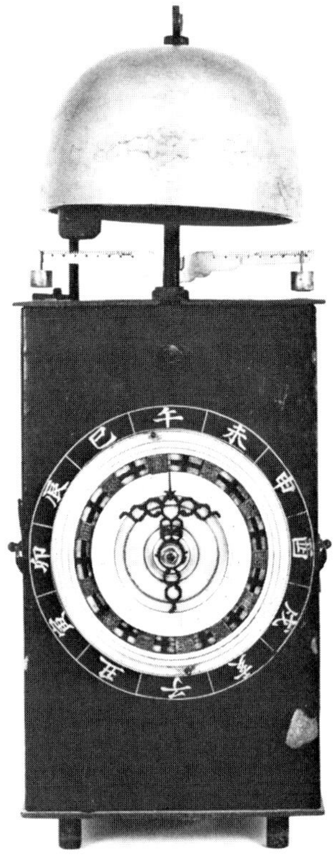

129

130

131 a

131 b

132

146

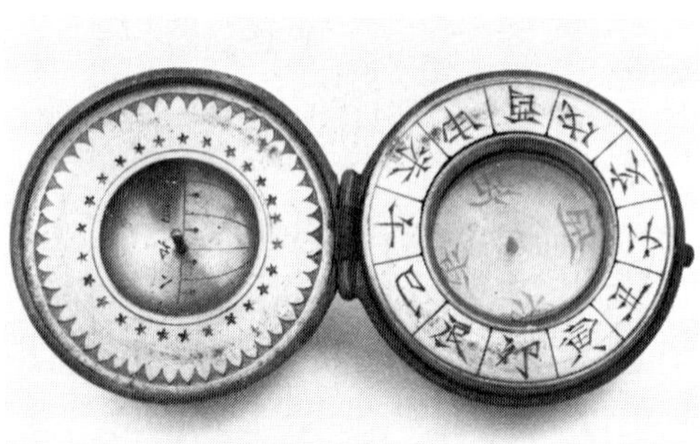

147 a

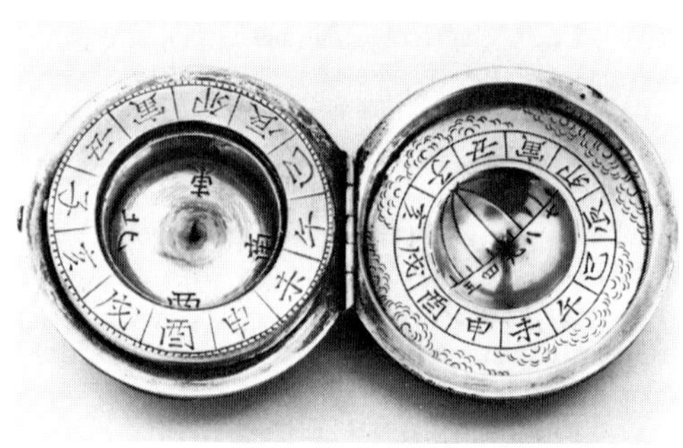

147 b

147 c

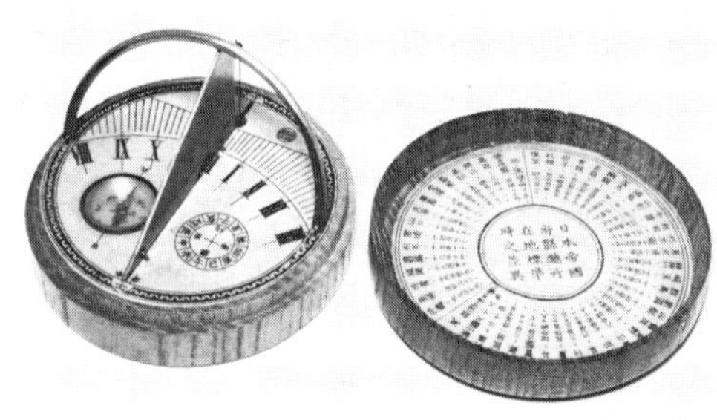

148

yuki Sukezaemon Tsuda, a blacksmith, who was probably one of the first Japanese craftsmen to copy a European clock. In the last quarter of the 16th century Jesuit clockmakers were teaching their craft to Japanese converts in the Kyushi and Kyoto districts. It was in such a centre that Masayuki probably learnt his skills which were then transmitted through nine generations of his family.
Successive members of it were accredited clockmakers to the Tokugawa Lords of Owari, Masanori occupying this post from 1830-1852 (see Yamaguchi 6 and 10-11).
Museum of the History of Science, Oxford, 65-112.

147 Three Examples of Japanese Combined Pocket Compasses and Scaphe Dials, c. 1900
a. Bronze and silvered copper, diameter 32 mm
Provenance : The Collection of T.G. Barnett, Birmingham, jeweller and antiquarian (1857-1935).
b. Silver, diameter 55 mm
The case is chased in relief with a symbolic scene on one side.
Provenance : Lewis Evans Collection.
c. Wood, 43 mm x 24 mm
Museum of the History of Science, Oxford 034, 033 & 031.

148 Japanese Pocket Horizontal Dial, 20th century
Wood, brass and paper.
The hour-lines, on the printed scale, are numbered with Roman numerals. In use, the folding gnomon is held vertical by the wire semi-circle. There is a plumb-line (the plumb-bob is concealed in the base) for levelling the instrument.
Museum of the History & Science, Oxford, 026.

149 Pocket Sun-dial and Compass, 19th century
Copper and silver, 25 mm x 34 mm x 37 mm (closed).
Rijksmuseum voor Volkenkunde, Leiden 2602-484.

150 Japanese Pocket Scaphe Dial and Compasse, c. 1900
Ivory, tortoise shell and wood, 37 mm x 24 mm.
The rectangular block containing the scaphe dial and the compass slides out from the protective, decorated box.
Provenance : Singapore.
Museum of the History of Science, Oxford, Lewis Evans Collection, 028.

In the Buddhist temples a quite different method of time-measurement was used.

151 Japanese Temple Incense Time-keeper (Koban Dokei), ?18th century
Photograph
Time-indication in Buddhist temples was performed with an incense time-keeper which, although of quite different form, worked on exactly the same principles as the Chinese examples (no 119, 120). The instrument illustrated here is 333 mm square and 460 mm high. Inscribed on each of the storage drawers is the name 'Kanyoro Isuka' and 'Kamincha Shinjuku' which is a district of Tokyo.
Photograph : The Time Museum, Rockford, Illinois.

After the revolution of 1866 when the Tokugawa Shoguns were dethroned, the policy of excluding all outside influences was reversed. Western ideas were now willingly embraced. On 1 January 1873 therefore the Gregorian calendar and western methods of time-keeping were officially introduced.

152 A Concordance of Non-Japanese Chronologies, 1867
Teibô bankoku futsûreki.
Ms on paper, height 83 mm.
The work provides corresponding dates between the time systems of Japan, Russia, the U.S.A. and European countries.
Rijksmuseum voor Volkenkunde, Leiden, 578 no 63.
Bibliography : Serrurier 474.

153 Seiyo, Jiki Teikoku-han (Methods of Time Telling with Western Clocks), Edo : Kanshutei, c. 1875
Japanese paper binding with label, 12 pp., 13.2 x 7.7 cm.
One of the earliest Japanese books devoted entirely to the western clock. Each page except the first shows a clock face in a frame surrounded by cherubs, with Japanese time notations printed in light red.
Roger Turner Books Ltd.

Korea

Time-measurement in Korea, as in Japan, finds its source in Chinese learning.
In the 6th century AD, the Chinese *Yuan-chiu* calendar was introduced, followed by the *Lin-te* system in 674 and c. 800 by the Hsuanming calendar. From the 11th century onwards numerous calendars were introduced until the mid 17th century, the new, Jesuit-reformed, Chinese system, the *Shih-hsien* calendar was promulgated. Revised and adapted, it lasted until officially replaced by the western solar calendar in 1894.
Records of sun-dials in Korea go back to the 1st century AD and examples have survived from the 6th/7th centuries. The earliest record of clepsysdras dates from the 8th century.

154 Remains of an Early Sundial, 6th - 7th centuries
Photograph.
The dial is made of granite and has a radius of about 33.4 cm. In form it recalls Chinese dials of the Han to Yuan periods and was probably equinoctially mounted with a central polar gnomon. As well as the hours corresponding directions are indicated and eight trigrams mark the eight cardinal points.
The dial fragments were excavated in 1930 at a castle in Kyongju.
Photograph : Kyongju Museum.
Bibliography : Jeon 43-4.

155 Japanese or Korean Scaphe Dial and Compass, 1810
Inscribed 'When the Emperor had ruled for seven years, in the last ten days of the second month of Winter, your servant Chiang Jun offered [this] to Your Majesty for criticism as made by himself'.
This may be interpreted as the year 1810, a Kêng-wu year, and the seventh of the Bunka reign period.
Ivory and brass, 73 mm x 36 mm.
Single block of ivory within which is carved an hemispherical scaphe dial with vertical hour lines and horizontal declination lines, the names of which are marked on the sides of the instrument. Inset compass with gold foiled ground, and a folding plummet holder. For a similar instrument described as Korean, see *Chine* no 39, p. 62.
Museum of the History of Science, Oxford, 027.

156 Korean Scaphe Sun-dial, 1880
Signed : 'Hsia Shang-Huan'. Dated : 6th year of the KuangHsü reign period.
Bronze, inlaid with silver lines and characters, diameter 287 mm.
This sun-dial is based on astronomical instruments made by the Yuan astronomer Kuo Shou-ching, as described in the astrological section of the *Yuan shih* (History of the Yuan Dynasty [1279-1368]). The term *angbu* (upward-looking bowl) describes the shape. Inside the bowl are marked 13 parallels, which indicate the 12 fortnightly periods between the Summer and Winter Solstices (23.5°N and 23.5°S respectively). A N-S meridian is drawn, and an inclined gnomon is fixed with its tip pointing to the North Pole. The time divisions are in 12 equal double hours, and 96 quarter hours. Around the rim are marked the cardinal points of the compass.
Scaphe sun-dials served as the first public clocks in Korea in the 15th century.
Bibliography : Jeon, 46-9.
Museum of the History of Science, Oxford, 78-41.

150

155

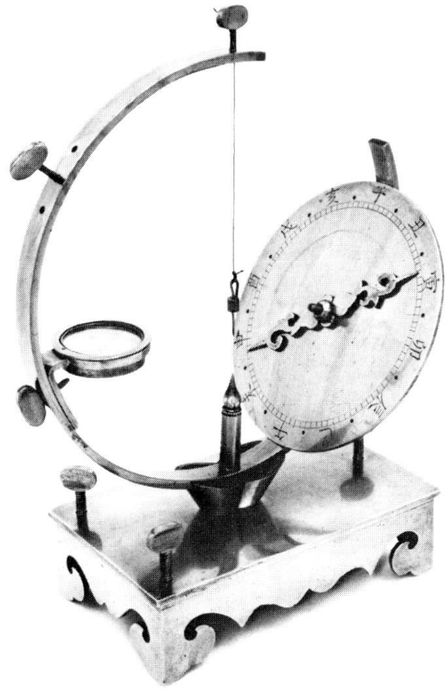
158

157 A Late ?Chinese Sun-dial in the Korean Tradition, ?18th century or later
Brass, diameter of the hour plate 90 mm.
A rectangular base with lobed decoration has three cinque-foil headed screw feet for levelling against a plumb-bob and vertical point index. From the base of the vertical point a meridian arc rises. This carries a compass and two tension-screws to which are attached threads which run to each end of a rule which may be adjusted over the hour plate which is set in the equinoctial plane. This is graduated in double hours sub-divided by four and again sub-divided into eight western quarter hours. The two threads together form a triangular gnomon which, because it is not solid, can operate even during the half year when the sun being below the equator, the hour plate itself is in shadow.

Although the latitude, 35°, of this dial suggests that it was made for use in China rather than Korea, the use of a double string-gnomon relates the instrument to a Korean tradition of such instruments which goes back to at least the 15th century. See Needham, Gwei-Djen, Combridge and Major, 44 & 5960, 63.
The Science Museum, London.

158 Another Example of a String-gnomon Equinoctial Dial, 18th century or later
Brass, 116 mm x 68 mm. Overall height 81 mm.
Museum of the History of Science, Oxford, Inv. 62-104.

As in China, considerable development took place in clepsydra technique. Striking mechanisms and visual hour indications were incorporated, and the basic water-mechanism adapted to drive astronomical models.
Later the European weight-drive, introduced from China, replaced water, and one example of this system has survived from the 17th century.

159 Remains of a Clepsydra with Automatic Time-signaling Apparatus, 1434
Photograph c. 1920 from Takabayashi, *Tokei Hattatsu* of the then existing remains of the clepsydra of the Ch'anggyong palace.
Built between 1432 and 1434 by Chang Yang-sil and Kim Pei, the clepsydra had four reservoir vessels, two inflow receivers, twelve indicator rods and an automatic rolling ball system for ringing a bell at each hour. The latter apparatus was removed during the 17th century, and the pipes and stone stands shown here were lost during the Korean war.
Photograph from Jeon, 57.

160 The Armillary Clock of Song IYong, 1664-69
Photograph.
Song I-Yong's clock consists of a weight driven pendulum movement with horizontal verge escapement, which drives an armillary sphere and an horizontal display of the double hours. There is also an hourstriking mechanism activated by rolling balls. The armillary sphere gives indications for both the sun and moon. Although built on a small scale (120 cm x 52.3 cm x 98 cm) the machine combines elements from the Chinese tradition of moving astronomical models with the western tradition of self-sounding bells and the weight drive.
Photograph : Koryo University Museum.
Bibliography : Needham, Gwei-Djen, Combridge & Major, Ch 4 ; Jeon 68-72, 162-66.

America

The best known calendars of pre-Columbian America, the Maya and Aztec, were inherited from Olmec and other earlier cultures.
Many of their major features are common to the calendars of other meso-American societies.
The Maya year *(tun)* has 360 *kin* with 5 *kin* *(uayeb)* added to make up 365.

The year was divided into 18 *Uinal* (equivalent to months) named as follows :

Pop	Yaxkin	Mac
Uo	Mol	Kankin
Zip	Chen	Muan
Zota	Yax	Pax
Tzec	Zac	Kabak
Xul	Ceh	Cumhu

Each of these 'months' was composed of 20 kin which are numbered from 0- 19, each kin having a name.

Imix	Chuen
Ik	Eb
Akbal	Ben
Manik	Caban
Lamat	Eznab
Muluc	Cauac
Oc	Ahau
Kan	Ix
Chicchan	Men
Cimi	Cib

The day names run without interruption through the months, years and *uayeb* in a cycle of four years. Each day name can therefore fall on only 4 possible month days.

156

Twenty *tun* make up a *kalun* of 7200 *kin* and twenty *katun,* a *baktun* (144,000 *kin*) or one cycle.

This reckoning is known as the long count. In the post-classic period it was abandoned in favour of the *tzolkin,* a cycle of 260 days probably introduced for ritual purposes.

The *tzolkin* is made up of two interlocking groups of numbers 1 to 13 and 20 names. Because of the shared value of 20 in both the long and the short count, it was possible to name *katuns* in the long count by the short count formula. Such synchrony reinforced the Meso-American view of time as cyclical, and thus its ritual importance.

161 **A Noted Aztec Calendar Stone**
Photographs.

Outside the sphere of Maya/Aztec influence, calendars were generally lunar, although the Incas recorded the movements of both Sun and Moon.

The month names of an Inca calendar underline its close relation with agriculture.

Small Growing Moon
Great Growing Moon
Glower Growing Moon
Twinears Moon
Harvest Moon
Breaking Soil Moon
Irrigation Moon
Sowing Moon
Moon of the Moon Feast
Moon of the Feast of the Province of Uma
Moon of Feast, Province of Ayanmarca
Moon of the Great Feast of the Sun

North American Indians had similar concerns. The lunar months of the Ojibway Indians were :

Cold Moon
Snow Moon
Warm Moon
Moon of Plants
Moon of Flowers
Hot Moon
The Buck Moon
The Sturgeon Moon
The Corn Moon
The Travelling Moon
The Beaver Moon
The Hunting Moon

Among the Hopi Indians of Arizona, the year was divided into two seasons, subdivisions being marked by recurrent events of day and night, the Moon's place, and the solstices. Ideas of past and future were mingled with those of 'being present' or 'not being present' and, as in most indigenous American tribes, beyond a limited reckoning of lunations, time was neither subdivided nor counted. Time in such societies was experienced, time was lived, its passage being marked only by particular repeating events.

On the west coast of Vancouver Island, the time-system of the Nootka Indians was specifically related to their hunting and fishing economy.

The appropriateness of certain periods of the year as the day for particular activities was understood, but such periods were not delimited. Nor was there any chronology. 'They could not talk about the Summer of 42, but could refer to the Winter of Our Discontent' (Folan, 94).

In the late historic period, the year was divided into two parts, defined by the northerly or southerly direction of the sun's shadow given by a gnomon. Shorter periods were determined by lunar periods of which 12, 13 or 14 made up a year.

All recorded moon period names relate to fishing or agriculture.

January	No (Food-Getting) for a Long Time (?) Moon
February	Bad Weather Moon (?)
February	False (Spawning) Moon
March	Herring Spawn Moon
April	Geese Moon
May	Stringing (Berries) Moon
June	Salmon Berries Moon
July	Wasp Moon
August	Spring Salmon (Run) Moon
September	Dog Salmon Moon
October	Rough Sea Moon
November	Elder Sibling (Moon)
December	Younger Sibling (Moon)

Reckoning the days and subdivisions of days are uncommon in North America. When such counts are made they are usually of nights. Pawnee Indians cut notches in sticks to reckon nights and even months, but reckoning in 'sleeps' as by the Kiowa, Dakota, Omahas, Hupos and Kaigan Indians was more common. Reckoning by 'suns' and days as by the Comanche was unusual.

Christian Time

For the things that are seen are temporal, but the things that are unseen are eternal
St. Paul

Christianity inherited from Judaism the idea that time is teleological and finite. Time is the field in which God's purpose is worked out. Within this time man should use his, infinitely short, life-time, to align himself with the purposes of God.
Regular prayer was a means towards this, and the times of prayer had been established in scripture. They lay at the basis of christian monasticism.
The early Christian monastery was a totally time-structured system. It is one of the earliest examples of the use of time to organise the life of a social group.

163 A Centre of Christian Time-discipline and Learning, the Abbey of Ripoll
Photograph.
The Benedictine monastery of Santa Maria de Ripoll founded at the foot of the Pyrenees and patronised by the Counts of Barcelona and Gerona, was a prosperous house which in the mid-10th century was a major centre of learning.
In its *scriptorium* not only were copies made of texts collected in Italy by the abbot Arnulf (948-70), but translations of mathematical and astronomical works were made from Arabic. The text of the water-alarm described below (no 170) derives from Ripoll to which house, as to all other monastic houses, it would have been of interest as an aid in following the strict time-structure of monastic life.
Bibliography : Riché 158-9 ; Millás 138-42.

164 The Time-structure of Monastic Life, 6th century and later
Graph after Biarne 113.
The graph is divided vertically into hours (modern equal hours at the left, medieval unequal hours at the right) and the variation of their length throughout the year is shown by the horizontal bands of varied width which run across the year represented by the space between the two hour scales. In addition the various activities making up the monastic day are indicated in their proper places. In monastic life, submission to the 'rule' of the order, of the superior, was an essential part of submission to God. Free time, leisure, was lost time, distraction from God. To follow the dictates of a time-structure was thus both to contribute to the good order of the monastery and to advance in perfection. Punctuality was respect for God to whom all time belonged. The unchanging regularity of the time-tabled day helped the monk to surmount his human condition, but it also created the modern sense of the need to use time.
Bibliography : Biarne.

For finding and keeping time in the monastery a variety of methods were used.

165 A Monastic Star Time-table of the Eleventh Century
Photograph.
The rough notes make up a table of the position of certain stars in relation to specified monastic buildings on certain dates. When a monk specially charged to watch the stars and the buildings from a designated position saw a particular star in a particular position on a particular date, he knew that it was time to raise the brothers and light the lamps.
Photograph : the Bodleian Library, Oxford, Ms 38 ff 19v - 23v.

166 An Early Sun-dial from Northern England, 8th or 9th century.
Photograph and reproduction.
The sun-dial is incorporated into the shaft of a free-standing cross, one of a group of sculpted stone crosses produced in northern and central England during the 7th to 10th centuries. The dial would originally have had a straight gnomon perpendicular to the face marked with twelve hour lines radiating from the gnomon-hole. Such dials, which appear to have been transmitted from the Near East to north-west Europe with monasticism, were not intended to show the time accurately in terms of a numbered system of hours, but rather to indicate what proportion of the daylight period had elapsed so that the monastic offices could be celebrated at the appropriate periods. It is for this reason that although the hour lines on the dial are not numbered, the 3rd, 6th and 9th hours, which corresponded with the offices of Terce, Sext and Nones are marked with a cross.
Reproduction : Museum of the History of Science, Oxford.
Bibliography : Brown ; Page ; Clapham Ch 3 ; Green 495-6 ; Eden & Lloyd 49-51.

167 The Church Dial as Public Time-shower, 1055-65
The dial is carved on a stone 2.36 m by 50.8 cm high, which is set into the wall over the south porch of Kirkdale Church, Yorkshire. Like all Anglo-Saxon dials it is a distinct piece of architectural sculpture, in part carved in relief, and here makes up part of the foundation inscription of the church.
The times for the offices from 6 am to 6 pm are marked by the crossed lines, an intermediate division being given by the shorter non-crossed lines. The inscriptions are 'Orm, son of Gamal, bought St Gregory's church when it was all ruined and tumbled down and he caused it to be built afresh from the foundation (in honour of) Christ and St Gregory in the days of King Edward and in the days of Earl Tosti.
This is the day's sun-marking at every hour. And Haward made me, and Brand, priest (?).'
Bibliography : Green 506-7 ; Okasha 87-8 ; Hunter Blair, 192 ; Taylor & Taylor 22.

168 A Portable Sun-dial from the Late Saxon Period, 10th or 11th century
Facsimile copy by Leslie Durbin, *c.* 1950 of the original in the Treasury of Canterbury Cathedral.
Inscribed on the sides '[Sa] lvs factoris' and 'pa[x] possessor[i]' (Salvation to the maker, peace to the owner).
Gold, and sterling silver, inlaid with niello and glass, 61 mm x 16 mm.
The dial consists of a cast rectangular tablet which tapers slightly towards the top surmounted by a decorated triangular cap which ends in a suspension loop in the form of an animal head with blue glass eyes. A plain gold ring passes through the animal's mouth for suspension. Each face of the dial is divided into three vertical columns, each column is containing a pair of month names. At the top of each column is a circular hole and two blue glass studs are placed at varied intervals in the column below. The arrangement of the month names is:
Face 1
Jan/Dec, Feb/Nov, Mar/Oct
Face 2
Mai/Aug, Iun/Iul, Apr/Sep.
When the dial is used, the gnomon, an animal headed pin which is stored in a hole in the base, inserted in the hole at the top of the column containing the month name corresponding with the time of observation. The dial is suspended and turned to face the sun. The lower point in each column represents noon (probably for the middle of the month), the upper point a position mid-way between day-break and noon, and noon and sun-set. The three positions thus correspond with the 3rd, 6th and 9th hours, and so with the offices of Terce, Sext and Nones.
A convenient, portable ecclesiastical dial which indicates those moments in the day which were uniquely important to its owner.
The original of this dial was found about two feet below the surface of the cloister garden of Canterbury Cathedral by a workman Joe Wansdell during alterations to the surface level in 1938. Interpretation of the object as essentially ecclesiastical is thus reinforced, although its date remains somewhat uncertain.
The Science Museum, London,1950-10.
Bibliography : Okasha 116 ; Catherine Haith in Backhouse, Turner, and Webster, 94.

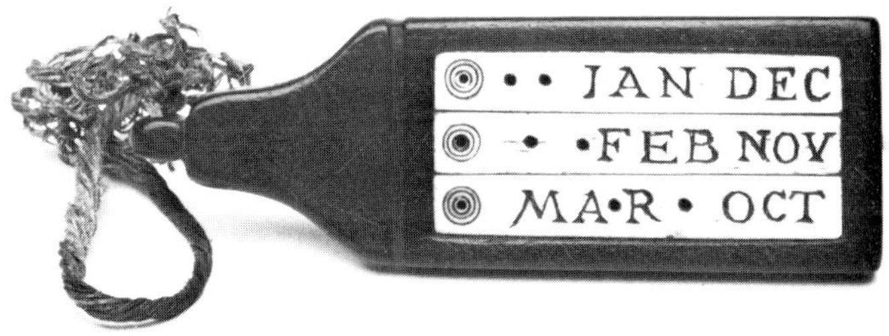

169 a

169 b

169 A ?Medieval Altitude Sun-dial of Unknown Provenance and Date
Bog oak and bone, 78 mm x 26 mm tapering to 25 mm x 6 mm.
Rectangular tablet with a slightly asymmetrical bell-shaped cap ending in a suspension hole. Each face is inset with a piece of bone engraved in three columns with pairs of month names:
Face 1
Jan/Dec, Feb/Nov, Mar/Oct
Face 2
Jun/Jul, Mai/Aug, Apr/Sep.
The method of operation is as that for the Canterbury dial (number 168 above), using the bone gnomon stored in the base.
Private Collection.

170 A Monastic Water-alarm of the 11th Century
Reconstruction by Edouard Farré-Olivé after the text of Ms Ripoll 225 ff 87v -93r. Only that part of the text describing the striking mechanism has survived. The water supply in this reconstruction is therefore the simplest possible to permit the apparatus to function. The description on which the reconstruction is based is contained in a codex containing various astronomical and mathematical treatises copied in the scriptorum of Ripoll during the 11th century (see above no 163). Unlike most of the other treatises in this collection which are translations or adaptations from Arabic works, the water-alarm seems to be of purely Latin inspiration. That the incomplete text describes a machine that was used in the monastery is not unlikely. That such devices had a relatively long life is shown by the recent discovery of a late 13th or 14th century drawing of a clepsydra with a very similar striking mechanism used to sound a carillon (see Farré-Olivé).
Edouard Farré-Olivé, Barcelona.
Bibliography : Maddison, Scott & Kent.

An elaborate body of calendrical knowledge - the computus - grew up to help solve the time problems involved in calculating the date of Easter.
The calendar used was the Julian calendar but years were counted in a new era - the Christian era, supposed to begin with the birth of Christ.
This era was first introduced in a table for calculating the date of Easter by the Roman Abbot Dionysius Exiguus in AD 525.

171 The Easter Table of Dionysius Exiguus

172 Bede's Codification of the Ecclesiastical Time-lore of the Early Middle Ages
Bedae presbyteri... Opuscula complura de temporum ratione diligenter castigata..., Cologne, 1537.
Printed book, in folio.
Two works on time, *De tempore* and *De temporum ratione* by the 8th century Anglo-Saxon monk and scholar Bede, provided authoritative treatment of time questions and were popular throughout northwestern Europe, especially in the Carolingian schools. Because of their prestige several apocryphal works became attached to them. They are shown here in the first edition of Johannes Noviomagus' (Johan Bronchorst 1494-1570) collected edition of Bede's computistical works, in which a number of undetected apocryphal writings are included.
Private Collection.

173 A Calendar for the Early 11th Century
Ms on parchment, 8ff written in a script with insular characteristics, 32-34 lines, initial KL Kalendas in green and blue, the calendrical tables coded in green, blue and ochre, 185 mm x 133 mm.
The first seven folios of this fragment of a larger work contain a Julian calendar with the days expressed in the Roman way in terms of Kalends, Nones and Ides. In addition the tables give dominical letters which enabled the day of the week (more especially sunaday) to be found, notes on the positions of Sun and Moon, notable saint's days and the length of the day and night in hours. The calendar was probably written in Liège and shows some north French influence. The tables of computistical information run from 1001 to 1063.
Provenance : Joseph Désiré Lupus, Brussels.
Koninklijke Bibliotheek, The Hague, Ms 130 E 9.
Bibliography : Catalogue no 397 ; 75.

174 A Classic Collection of Compotus Texts in a Printed Edition, 1519
Computus nouus & ecclesiasticus : totius fere Astronomie fundamentum pulcherricum continens (Venice, Petrus Liechtenstein, 1519).
The contents of the volume are
Prefatio
Cap. 1 De Aureo numero
Cap. 2 De ciclo Jnditionis
Cap. 3 De ciclo Solari
Cap. 4 De littera dominicali
Cap. 5 De Jnteruallo
Cap. 6 De concurrentibus
Cap. 7 De festis quinque mobilibus
Cap. 8 De forma celi et ordine spherarum
Cap. 9 De quatuor elementis
Cap. 10 De situ & motu firmamenti atque de equinoctijs & Solsticijs
Cap. 11 De aspectibus planetarum et de Eclipsibus Solis et Lune
Cap. 12 De motu Solis
Cap. 13 De motu Lune
Cap. 14 De coniunctione et oppositione Solis et Lune
Cap. 15 De initijs Mensium
Cap. 16 De Kalendis Nonis et Jdibus
Cap. 17 De initijs quatuor temprum
Cap. 18 Et vltimum De angarijs quator temporum

Canon (table for the length of day of the year, calculated for latitude 48 degrees)
The capita each start with the original text (in verse), followed by a Declaratio in prose.
Private Collection.

Water-clocks and alarms were used throughout the Middle Ages.

175 The Image of a ?Royal Water-clock, ?1223-26
Miniature from a French moralised bible (Ms Bodl 270b f 183v), probably written originally for Louis VIII. The miniature illustrates the miracle of the retrogression of the (shadow) of the sun on the 'dial of Ahaz'. Since horologium could mean any kind of time-piece in medieval Latin, it was not unreasonable for the artist to transport the miracle from the more usual (if mythical) sun-dial to a water-clock which may have been more of a novelty. Exactly how the clock functions is unclear from the illustration and a variety of interpretations have been put forward (see Drover 1 & 2 ; Sleeswyck 2; Combridge). Future interpretations will perhaps have to consider the possibility that the five bells are part of a carillon similar to that in the 13th or 14th century Catalan sketch of a water-driven carillon (see no 176), the striking mechanism being hidden in the tower above. It may also be noted that it is not self-evident that the mechanism shown is a 'monastic' water-clock as Drover described it. In the light of the miniature's provenance and of the subject matter, the artist may well have depicted a water mechanism existing in one of the French royal palaces.
Photograph : The Bodleian Library, Oxford.
Bibliography : Laborde ; Haussherr.

176 A Water-driven Carillon, late 13th or 14th century.
Photograph of a drawing in *Constituciones Cathalonicae de tempore de regis Petri I ad regem Jacobum II* (ACA Cancelleria, Caixes de legislacio no 2, II).
Diagram by David Penney.
The upper mechanism is a carillon which will play when the crack crank on the lower roller is released, and the driving weight above causes the striking mechanism to function. A possible mode of operation of the going train is shown in the explanatory drawing.
Photograph : Arxiv de la Corona d'Aragó, Barcelona.
Bibliography : Farré-Olivé.

But water could freeze - especially in northern climates - or evaporate. Efforts were made to find an alternative.

170

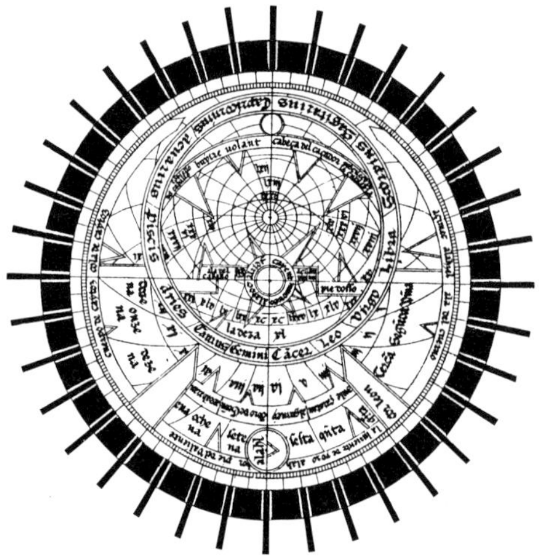

177

A reliable machine, self-functioning for 24 hours or longer, was of interest not only for sounding alarms, but as a motor to drive astronomical models at the speed of the firmament.

'...clockmakers are trying to make a wheel which will make one complete revolution for every one of the equinoctial circle, but they cannot quite perfect their work. But if they could, it would be a really accurate clock and worth more than astrolabe or other astronomical instrument for reckoning the hours....'
Commentary of Robertus Anglicus on the Sphere of Sacrobosco XI, 1271.

An obvious source of energy was that of a falling weight, but this needed to be controlled. One attempt was the mercury-braked clock of Isaac b. Sid.

177 Turning an Astrolabe Automatically, c. 1276
Reconstruction by A.A. Mills, of the mercury clock of Isaac b. Sid as described in the *Libros del Saber de Astronomia.*
The instrument consists of a wheel made up of twelve compartments. The wall of each compartment has a small opening into the next so that mercury, which fills half the compartments of the wall, may percolate from section to section as the wheel turns under the influence of a weight attached to the outer periphery of a drum surrounding the compartmented wheel. This wheel is mounted on an arbor which passes through a mounting board to a six-leaf pinion which meshes with the teeth of, and thus turns, an astrolabe dial which makes one rotation in 24 hours - 'just as the ninth heaven makes to produce day and night'. '...by the movement of this wheel let an astrolabe rete be moved over a plate...for any required place. The rete is to initiate the sky at all hours of the

day and night, indicating what is the ascendant, the positions of all the twelve hours, the altitude of the Sun and any bright star, and the hours passed.... And let this clock also have little bells, which can play by themselves at whatever hours are chosen by day or by night.'
A.A. Mills, Leicester
Bibliography : Mills.

The earliest successful device for controlling a clock, known as the escapement, was developed in the mid-13th century.
A form of it was used to control the clock and astronomical model of Richard of Wallingford.

178 The Planetary Clock of Richard of Wallingford, mid-14th century
Quarter scale reconstruction by Peter Haward.
Begun between 1327 and 1336, and completed after Richard's death during the abbacy of Thomas de la Mare 1349-96, Richard's creation was an elaborate astronomical model driven by clock-work. According to John Leland who probably saw the clock in c. 1540 and mentioned it briefly in his Itinerary 'one may look at the course of the Sun and the Moon or the fixed stars, or again one may regard the rise and fall of the tide, or the lines with their almost infinite variety of figures and indications.' This was to say that Richard's clock had an astrolabe dial (which would show the course of the Sun and the fixed stars, had a dragon hand for indicating lunar eclipses, incorporated gearing to show some planetary movements and had a tide dial. Besides these the clock struck the hour and may have incorporated a wheel of fortune. Power to drive all this was supplied by a falling weight, and this was controlled by a form of verge escapement (now known as the strobe from the Latin word *stroba*, pineapple, that Richard uses of it) which seems to have been displaced from use during the 14th and 15th century by the verge and foliot escapement (see number 179).
Although perhaps made over a century after the invention of some form of escapement device, the St Alban's astronomical model remains important historically as the earliest complicated weight-driven mechanism of which a sufficiently detailed description has survived to allow its reconstruction. Given the complex nature of the manuscripts and that they are neither clear nor in order, it is not surprising that interpretations of the clock differ. The present model follows the interpretations of J.D. North, but for different solutions see Watson.
P.N. Haward, Barking Tye.

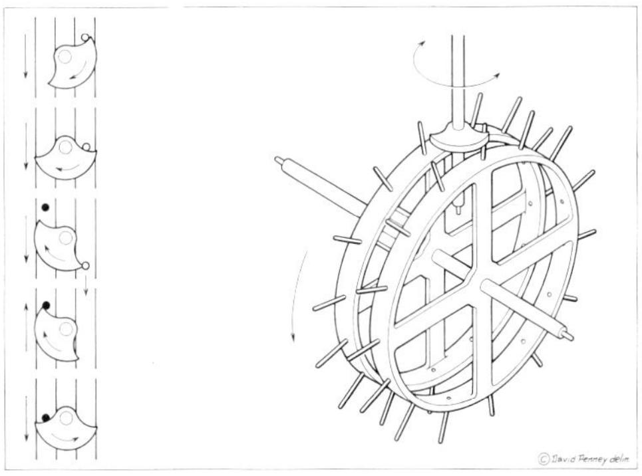

178
Another form of the escapement - the verge and foliot - became more popular.

179 The Action of the Verge and Foliot Escapement
Diagram by David Penney. It was to remain in use for the next three and a half centuries.
It was used to control the complex planetary model made by Giovanni dei Dondi.

180 An instrument ... for the celestial movements: the Astrarium of Giovanni dei Dondi, c. 1365-75
Reconstruction by Luigi Pippa, Milan *c.* 1964.
Dei Dondi's *astrarium* is a mechanical analogue computer for planetary and celestial movements. As his friend Philippe de Mezires wrote of it in 1385 it '... shows all the movements of the signs and of the planets with their circles and epicycles, and differences by multiplication, wheels without number, with all their divisions, and each planet of the said sphere is shown separately. By night one sees clearly in which sign and to what degree the planets and the stars appear in the sky' (cited from Bedini & Maddison 20). The seven faces of the instrument represent the movements, according to Ptolemaic theory, of the Sun, Moon, Mars, Venus, Mercury, Saturn and Jupiter. In addition the instrument carried a 24 hour dial, a dial showing the fixed eccesiastical feasts and another for the moveable ones dependent on Easter. In addition the nodes (points of intersection of the paths of the Sun and the Moon, at which time eclipses may occur) and the times of the rise and the setting of the Sun at Padua were indicated. The instrument was controlled by a verge escapement with balance wheel in the form of a crown and had a weight drive.
Dondi's *astrarium* was the wonder of the age. In 1381 it was purchased from dei Dondi by Duke Gian Galeazzo Visconti who installed it in his library in the Castello Visconteo, Pavia. Here it was protected by a heavy cover but nonetheless needed a certain amount of repair during the following century. By 1520 however the machine had fallen into such a state as to be irreparable, although a copy of it was made by Gianello Turriano. Thereafter all trace of it disappears from the historical record. Fortunately dei Dondi wrote a detailed description that reconstructions have been built. That shown here is one of four interpretations realised by different groups in Britain, Italy and France.
Private collection
Bibliography : Bedini & Maddison *passim* ; Poulle 511-50.

But it also controlled hundreds of other clocks set up in towers and churches through out Europe.

181 Some Clocks of Late Medieval Europe
Photograph.
The clocks illustrated are :
a. Bourges Cathedral
b. Duomo, Florence by Paolo Ucello
c. Padua
d. Prague
e. Rouen
f. Strasbourg Cathedral
g. Venice
h. Wells Cathedral

182 A Small Turret Clock Movement, early 16th century
Iron, 72 cm x 49 cm.
Four post 'bird-cage' frame composed of flat strips and with the corner standards riveted at 135°. The feet are extended in curved right angles as are the knob-ended finials and the verge suspension-arm. The two trains set end-to-end are pivoted in vertical pivot-bars attached to the cross-members of the narrow sides by wedged mortises and tenons. The going train is controlled by a verge and foliot escapement, the position of the verge and of the foliot weights being adjustable in slots in the foliot arms and the horizontal part of the verge suspension-arm. The striking train has an external locking plate, notched cam, and fly with curved vanes and ratchet. The fly locks against a pin on the locking plate arm which is split to pass either side of the pivot-bar. Winding is effected by two right angle lantern pinions into contrate wheels on each barrel. The weights of the clock are formed by two cannon-balls held in specially forged hemispherical cages.
Zaans Uurwerk Museum.
Bibliography : Van der Molen, 10.

183 Announcing the Time in the Middle Ages
Re-enactment.
The primary purpose of early clocks was to give an aural indication of time over as large an area as possible. In places where time-finding depended upon a sun-dial or, where the clock was disordered, the succeeding hours would have to be announced. The

135–136

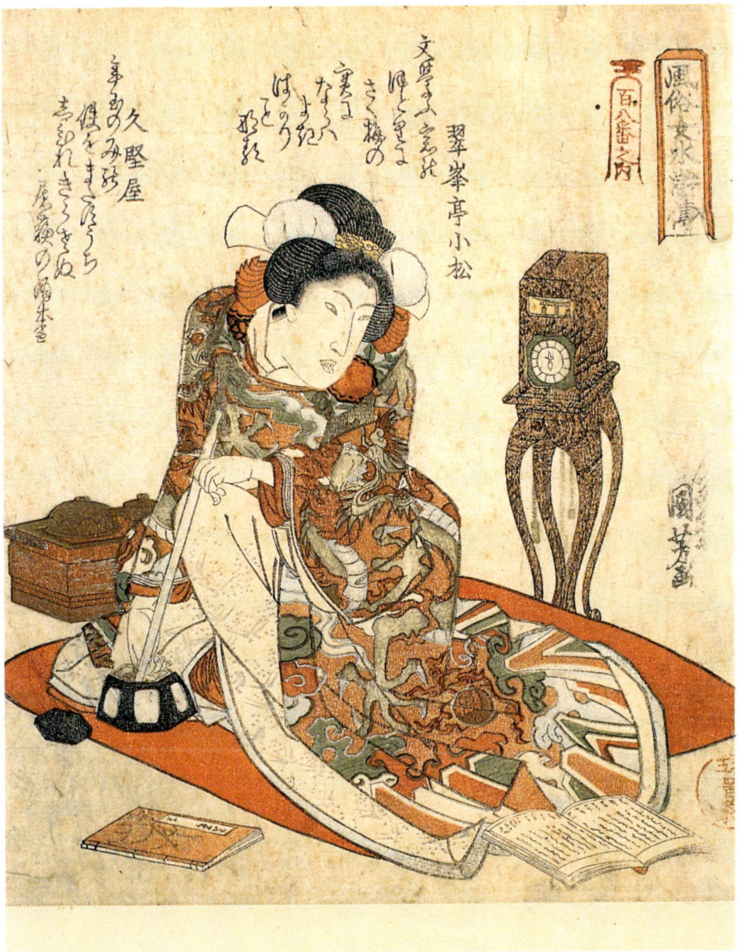

138

139

141

140

180

A m
now
pen
by a
fron
mor
the
two
a sh
degr
foldi
catir
arms
Mus
B 15

220
c. 15
Sign
for F
Gilt
170
The
partl
folia
colur
troph
mask
The s
Brig
depic
Prov
Gher
Lanja
carrie
inscri
Leyw
elabo
and a
engra
tids b
silver
The h
,conce
with a
back a
train
dial 1
square
barrel
and b
regula
attach
lever
the br
parate
plate
placed

Hans
1554
From
Albre
Privat
Biblio

182

221

221 A Clock Incorporating an Astrolabe, c. 1690

Signed 'Johann Leonhardt Bommel In Nurnberg'.

Gilt brass, steel, brass and silver, 472 mm x 342 mm.

There is a horizontal verge escapement. The main dial includes :

1. A chapter ring marked in hours, 1-12 and 1-12 ; the a.m. times are on the left, the p.m. times on the right.
2. An astrolabe *rete,* rotating within the chapter-ring, showing the ecliptic and 16 stars.
3. A solar index-arm which moves over the hourring and the *rete.* This index shows the hour on the chapter-ring and the position of the Sun on the ecliptic circle of the *rete.*
4. A lunar index-arm which moves above the solar index. This shows, on the chapter-ring, the time that would be given by a sun-dial used by moonlight. The other end of this index shows the age of the Moon, both by means of a pointer on a scale marked 1-29, and pictorially through a circular opening. It is also engraved with a lunar *aspectarium.*
5. A blue-steel index-arm (in the form of a dragon) indicating the nodes of the Moon. The coincidence of this index with those of the Sun and Moon indicates eclipses.

The clock is set, by means of the dial in the upper left corner, to strike either on a 12 or a 24-hour system. The last hour struck is marked on the dial on the lefthand side of the clock. The quarters struck are similarly indicated by the dial on the right-hand side.

The dial at the bottom left gives the dominical letter ; that on the right served to set an alarm mechanism.

On the back, the main dial shows hours and minutes. This is surrounded by a calendar giving the date and names of saints. The upper right dial gives the number of the year in the metonic cycle ; the lower left one the day of the week by its planetary symbol.

Although late in date this clock is typical of south German high quality, masterpiece production of the 16th and early 17th centuries.

Museum of the History of Science, Oxford 24-43.

Bibliograpy : Lunardi, 172-3 (for Bommel) ; Gunther, 538 ; CCA, 335.

222 A Small Drum Clock with Alarm.

Gilt brass and iron. Overall height approx. 65 cm.

The two cases are engraved with strapwork and foliate decoration, and the dial of the main clock is graduated I - XII with touch pins. Clipped to this box are three legs carrying a smaller alarm mechanism. This case may be slid round the lower element until a lever projecting vertically from the bottom of the upper part hangs above the

hour numeral when the alarm is to sound. When the single hand of the clock passes this time, it presses against the lever thus tripping the alarm mechanism into action.
Musée du Temps et de l'Histoire, Besançon.

223 Between a Clock and a Watch, mid 16th century
?German.
The outside of one of the pillars is carved with an 'I' and another 'Z'. Movement of iron and brass; dial for ordinary or Italian hours, with touch pins and with alarm setting dial in the centre; gilt brass tambour case, with hinged, pierced cover, decorated with animal and human figures, including boar-hunting scenes around the side. Verge escapement, with balance wheel; stackfreed; the movement has two trains, for going and for the alarm; no motion work.
Museum of the History of Science, Oxford, B1.

224 The Action of the Stackfreed
Diagram by David Penney.

225 An Early 17th Century Watch
Signed 'Iohn Tiese At London'.
Brass, gilt brass, steel and blued steel, 43 mm x 69 mm.
The gilt brass dial with touch pins is engraved in relief with fruit basket decoration. Single hand, turned baluster pillars, turned case with cover for the winding hole. Verge escapement with balance wheel, fusee with gut, ratchet and clock mainspring regulator, pinned cock with concealed pin; wound through the back.
Museum of the History of Science, Oxford B3.
Bibliography: *BCHE,* no 214; Maddison & Turner, W3.

226 The Working of the Verge Watch
Diagram by David Penney.

227 Time Trimming the Wings of Love, c. 1600
Ascribed to Otto Vaenius (Otto or Octavius van Veen, Leiden 1556-Brussels 1629).
Oil on canvass, 1.55 m x 1.06 m.
In a copse to one side of which a view of a house can be seen, a bearded figure of Time in a red tunic holds cupid across the throat on his knee and trims his wings with a sickle while trampling his quiver of arrows underfoot. To the left naked Venus starts from her couch but is restrained by a helmeted figure wearing a breastplate and wielding a truncheon (? Athena). To the right three cupids flee in horror from the scene. Inscribed on the side of Venus' couch is the phrase 'le cocur touiours ieune'.

Court painter successively to Alexander Farness and the Archdukes Albert and Isabelle, Vaenius was one of the most influential of the teachers of Rubens.
Musée du Temps et de l'Histoire, Besançon.
Bibliography: Benezit, VIII, 494-5.

228 A Compendium of Time and Astronomy for the Pocket, 1556
Signed 'Maigister [*sic*] Christophorus Schisler me fecit Auguste Vindelicorum Anno Domini 1556'.
Gilt brass or copper, 75 mm x 73.5 mm x 17.25 mm.
Octagonal box in three hinged parts engraved on each side.
1a Stereographically projected world map limited by the tropic of Capricorn fitted with a simplified rete and declination index.
1b Map of Great Britain and Ireland within a circle of compass points and fitted with an index.
2a Horizontal sun-dial for latitudes 49°, 52°, 55°, 61°.
2b List of the latitudes of 36 towns in Britain and northern Francc.
3a Universal rectilinear sundial (Rojas projection) and nocturnal with touch pins.
3b Luni-solar conversion volvelle also showing the age and phase of the moon.

An early instrument by the elder Schisler (*c.* 1531-1603) which was evidently intended for a British customer and incorporates an extremely map of the country.
Dudley Barnes Collection, Paris.
Bibliography: Brieux, 28.

229 A Sophisticated and Large Example of the Simple Ring Dial, 1572
Signed 'P. van Hooghendorp 1572'.
Brass, diameter 80 mm, width 25 mm.
The outside of the band, on which a suspension ring is mounted, is engraved with an hour and zodiac graph and is pierced with three holes, the spot of light from one of which (according to the season) indicates time on the diagonal hour scales market diametrically opposite the holes on the inside.
Nederlands Goud-, Zilver- en Klokkenmuseum, Schoonhoven, ZAZ 415.
Bibliography: Rooseboom, 75.

230 Time and Astronomy for the Pocket, 1579
Signed '+ Humfrey + Cole + 1579 +'.
Gilt brass, diameter of box 56 mm.
The circular box has a hinged lid an contains two hinged plates. Engraved on the outer top cover is a perpetual calendar market 'This Table beginith at Anno 1579 an so for ever' (in upper case roman) and contains circular scales of dominical letters, leap years, prime (= golden number) and the epact, the first two covering one complete cycle of 28 years (to 1606) and the second a

223

225

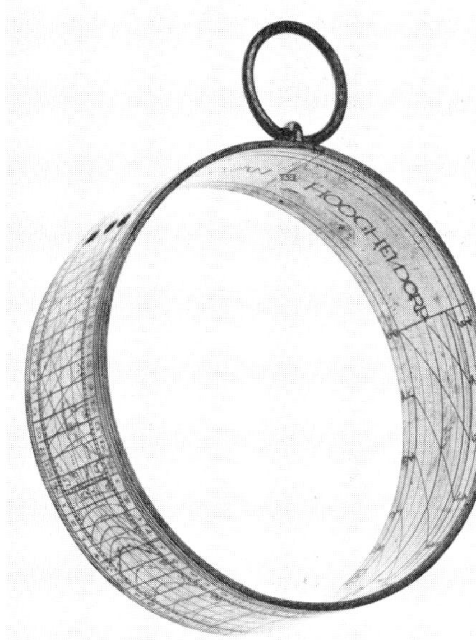
229

231

232

Metonic cycle of 19 years. The outer three circles containing month letter initial, date and abbreviated forms of saints' days and festivals. Within the box when opened is a thirty-two point hand-coloured compass rose with replacement needle. The outer face of the lower second box carries a degree scale in four quadrants reading to one degree, an unequal hour diagram, a shadow square and a alidade. The inside of this box is engraved with a circular table of the names and latitudes of 32 European towns. Hinged between the two halves of the case and using the same hinge is an arc supporting an hour ring for an equinoctial dial with folding gnomon and latitude arc. Hinged to the opposite side of the bottom half of the case is a plate carrying one side a nocturnal and on the other a table of the Sun's entrance into each zodiacal sign. Around the edge of the box is the motto 'Tempora mvtantvr & nos mvtamvr in illis sicvt vmbra dies nostri svper terram' (times change and we change with them as the shadow our day on earth). *Cf.* Eden & Lloyd 1322, pp. 422-3, ascribing the first half of the phrase to Matthew Borbonius).

One of four similar compendia out of a group of seven known within the surviving corpus of 26 instruments made by Cole. Cole (1530-1591) is the best known of the small group of instrument-makers working in late 16th century London.

Bibliography : Christies 29 March 1990, lot 231.

231 A German String-Gnomon Inclining Dial, 1586.

Signed '15 *V*S 85'(*i.e.* Vlrich Schniep), and inscribed 'viatorivm' (= the contemporary name for this type of instrument).

Gilt and silvered brass, 13 mm x 48 mm.

Under the point of the plummet is a scale of latitudes from 42° to 52°. The dial is adjusted to the latitude by tilting the dial-plate until the point of the plummet hangs above the dot on the scale corresponding to the correct latitude.

The compass can be rotated and is surrounded by a scale of degrees ; it may thus be adjusted to the magnetic deviation.

Inside the cover is a volvelle for converting ordinary hours into Italian hours.

On top of the cover are engraved the latitudes of 19 places.

The bottom of the instrument is engraved with a picture, inscribed 'ivdtha', showing her with the head of Holophernes.

Museum of the History & Sciences, Oxford, G. 386.

232 A Dial in the Form of a Cross for Ecclesiastics and the Devout, 1586.

Gilt brass, length 61 mm ; width of arms 32 mm.

The dial consists of a box shaped in the form of a cross with the cover hinged at the base. Engraved on the outer face of the box is a corpus christi. Engraved on each of the sides is an hour scale and a zodiac scale. There is floral decoration on the ends of the arms. The inside of the box was presumably intended to hold relics. On the inner face of the cover is a scale of degrees for latitude (*Gradus latitudinis* 0-75) against which the body of the dial may be set by means of a strut which hinges out of the box.

At the top of the cover is a small box for a compass and the latitudes of six places are engraved on the arms.

On the outer surface of the cover are a nonlinear altitude scale (*Gradus altitudinis* 20-70) and a shadow scale (*Partes umbrae versa* 6-30 x 6). Above these on the horizontal parts of the cross are the latitudes of a further 13 places.

At the top of the dial is a suspension piece and ring. The instrument is dated on the base of the cross above the hinge.

Private Collection.

Bibliograpy : Chapiro, Meslin-Perrier & Turner no 69, pp. 88-9 for similar dials by Zeelst.

233 Two Ivory Diptych Dials from Nuremberg, late 16th/early 17th century

In Nuremberg from the 16th to the mid 18th century a characteristic form of ivory diptych dial was manufactured by a relatively small group of *Compassmachers* who succeeded each other from generation to generation. Examples are shown here by makers from two such dynasties.

a. By either Hans Ducher (Tucher) II or III, late 16th century.

Ivory with brass fittings, 85 mm x 87 mm. Signed HD with between the initials the second form of their crowned snake mark.

1a Wind rose with the four cardinal points named and 32 points numbered ; central index in the form of a pointing finger ; inspection hole for the compass beneath.

1b Sun's position in zodiac and length of day by a pin gnomon contained within a circular vertical dial (vi-sii-vi x 15 minutes) sharing a common string-gnomon with.

2a Rectangular horizontal dial within inset compass with magnetic deviation mark and below two pin-gnomon dials for German and Bohemian hours sharing a common scale 6-24.

2b Lunar volvelle for phase of the moon and convertor. Below maker's mark and initials and in a frame round the edge 'Compassvs prope ferrvm non recte assignare potest' (near iron the compass cannot point properly) a favourite inscription of the Tucher family.

Private collection.

234

233 b

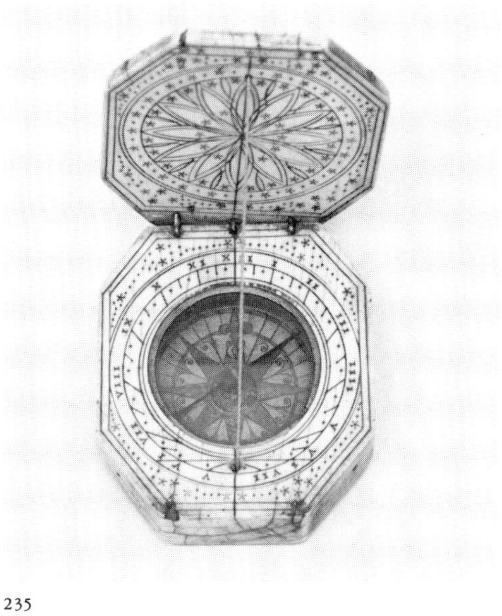

235

b. By Leonhart Miller, 1639.
Ivory with brass fittings, 105 mm x 64 mm.
Museum of the History of Science, Oxford, 57-84/85.
Bibliograpy : Gouk ; *Billmeir,* 85.

234 a A Flemish Polyhedral Dial, late 16th/17th century
Signed 'Elaborabat Antverpiae Michael Cognetvs'.
Gilt brass, diameter 146 mm.
The regular polyhedron has dials engraved on 25 of its 26 sides. The dials are drawn on a plain ground against set in a rich background of luxuriant foliate and floral entwining tracery inhabited by mythological, amorous and studious figures. The remaining (top) surface carries an inset compass box. The gnomons are replacements. The dial is made for latitude 51°31' (Antwerp) and carries the motto Omnes aequales sola virtute discrepantes (all men are equal and differ only in virtue), *cf.* Eden & Lloyd 891, pp. 356).

Michel Coignet (1544-1623) was a noted instrument-maker and schoolmaster who wrote several published books on commerce and navigation. At the siege of Antwerp he worked on the town fortifications and ended his career as mathematician to the Archdukes Albert and Isabella (*cf.* nos 227 & 234).
Private Collection.
Bibliography : Christies 10 July 1986, lot 302 ; Zinner, 281-2 (for Coignet).

234 b An Astronomical Ring Dial made for Albert and Isabella, joint Governors of the Netherlands, 1600
Signed 'Alberto Imperatoris filie Imp : Fratri Isabellae Philippi Regie filiae regis sorori aeternvm regere Ia D. Svcca 1600'.
Brass, diameter 157 mm.
A finely decorated universal instrument which is set for latitude by attaching the suspension piece to one of twelve set positions and then adjusting it more precisely before clamping it by a screw. In use such instruments, which were something of a Flemish speciality in the 16th and 17th centuries, were first set for latitude and for solar declination by placing the sights on the appropriate date. The instrument was then suspended and the meridian turned until shadow of the front sight exactly covered that of the back sign when the time could be read off from the equinoctial scale.

Of Jacobus de Succa little is known although he was probably related to Anthonis de Succa, *c.* 1557 1620, descendant of a noble Piemontese family who became portrait painter and genealogist to Albert and Isabella (see *B.N.B. ;* Thieme-Becker ; Benezit viii, 176). Three other instruments by de

220 a

220 b

221

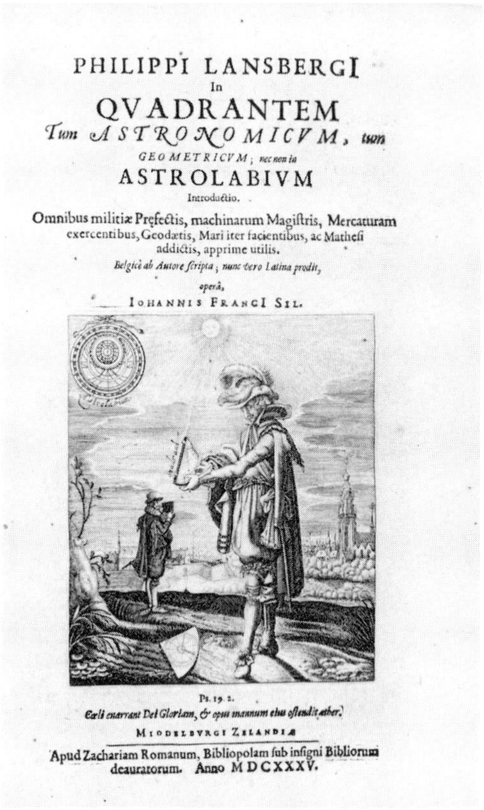
PHILIPPI LANSBERGI
In
QVADRANTEM
Tum ASTRONOMICVM, tum
GEOMETRICVM; nec non in
ASTROLABIVM
Introductio.
Omnibus militiæ Præfectis, machinarum Magistris, Mercaturam exercentibus, Geodætis, Mari iter facientibus, ac Mathesi addictis, apprime utilis.
Belgicè ab Autore scripta; nunc vero Latina prodit,
operâ,
IOHANNIS FRANCI SIL.
Ps. 19. 2.
Cæli enarrant Dei Gloriam, & opus manuum eius ostendit æther.
MIDDELBVRGI ZELANDIÆ
Apud Zachariam Romanum, Bibliopolam sub insigni Bibliorum deauratorum. Anno MDCXXXV.

237

Succa are known of which one is also dedicated to Albert and Isabella and another to Gaston Spinola, governor of Limbourg.
Museum of the History & Science, Oxford.
Bibliography: Maddison, 180, p. 38; Rasquin 138, 139 & 144.

235 An Octagonal String-gnomon Dial, 17th century
French. Ivory with brass fittings, 70 mm x 58 mm x 17 mm.
Octagonal block marked with a horizontal string-gnomon dial for latitude 45° around an inset compass with printed paper compass rose coloured blue and red. The inner face of the lid has geometric and star decoration.
Private Collection.
Bibliography: Rasquin, 82.

236 A Survey of Different Kinds of Quadrant, 1633
Philip Lansberg, *Verklaeringhe van het ghebruyck des Astronomichen ende Geometrischen Quadrants...*, Middelburg, 1633.
An important general account, first published in 1620, of all types of quadrants, horary, astronomical and geometrical (i.e. for surveying) by a noted local astronomer who lived from 1561 to 1632.
Private Collection.
Bibliography: Zinner, 425.

237 A Latin Translation of the Previous Work, 1635
Philip Lansberg, *Qvadrantem tum astronomicvm, tum geometricvm...*, Middelburg, 1635.
Latin translation of Lansberg's work by Johannes Franciscus Sil. The fine engraving on the title page shows a gentleman taking a back sight of the sun with Lansberg's own form of quadrant, a second quadrant on the ground displays its reverse side.
Private Collection.

238 A Quadrant According to Philip Lansberg's Design, 1645
Signed 'Ex praescripto P. Lansbergi Anthony Sneewins scvlp Anno Christi 1645'.
Brass, radius 200 mm.
On the face are two horary quadrants for equal and unequal hours, a zodiac scale used with them, a circular shadow square and a scale of degrees reading to 15 minutes of arc. Crossing the equal hour scale and concentric with the limb are arcs of circles corresponding with the tropics, the equator and the declinations of 26 stars. On the reverse are a zodiacal calendar, scales and radii enabling the stars marked on the obverse to be used for time determination, and scales of dominical letters, golden numbers and dates of full and new moon in the 19 year cycle.
Provenance: Utrecht Observatory.
Universiteitsmuseum, Utrecht 408.
Bibliography: Van Cittert, 26-32.

239 A Simpler Horary Quadrant, 1647
Signed 'Ioan. Melch. Low. Ritter'.
Printed paper on wood, 62 mm x 63 mm.
On the obverse is an horary quadrant for latitude 47° with a pictorial zodiac scale, a degree scale and a conversion diagram for converting from the double twelve to 24 hour time-reckoning surrounding, a sun emblem and surrounded by part of a compass point circle. On the reverse is a lunar dial and volvelle surrounded by a zodiacal calendar with scenes emblematic of the four seasons in the corners and with a table of epacts for 1646-64. The makers' initials and the date are inscribed in the centre below an aspectarium for the moon. With original sights, plumb bob and original red suede caselined fitted case covered with sheep blind-tooled on the lid with a central floral motif within a rolled border with corner pieces.
Private Collection.
Bibliography: Rasquin, 33.

240 An Instrument for Making Sundials, 1669
Signed Instrumentum Universale horologijs horizontalibvs verticalibvs inclinantibvs et declinantibvs delineandis i inserviens constructvm a Matthao Francisco Popell 1669' (a universal instrument serving for the construction of horizontal, vertical inclining and declining dials made by Matthew Francis Popell 1669).
Brass, approx. 300 mm.
A finely made instrument carrying an episcopal coat of arms, used for laying out the lines of sun-dials.
Popell worked at Passau whence he signed an instrument in 1696 and two other dials by him are known.
Rijksmuseum 'Nederlands Scheepvaartmuseum', Amsterdam S976(2).

241 French Polyhedral Sundial, c. 1700
Limestone with slate veneer; lines and scales picked out in black, red, and green.
40 mm long x 23 cm wide x approx. 30 cm high.
Carved stone incised with 14 dial faces and with subsidiary astronomical diagrams.
These are:

West face
1 On side of top oval, dial for Italian and Babylonian hours.
2 Eclipse diagram.
3 Diagram of Ptolemaic universe.
4 Direct west facing vertical dial for afternoon hours.

East face
Dials and diagrams corresponding with those on the west face.
North face
1 Slate veneer tablet engraved with the figure of an armillary sphere.
2 Vertical dials inset in the corners on either side of it.
3 Dial on inclined surface surrounding the top oval.
4 Central oval surface with 1 large square hole and 4 others.
The remains of construction lines perhaps for an equinoctial dial can be seen on this face.

South face
1 On oval, Babylonian and Italian hours.
2 Vertical dials for A.M. & P.M. hours completing those on the east and west faces.
3 Central book veneered with slate engraved with a polar dial and Latin inscriptions. Left page: a matines; right page: A vesperes and Latin exhortations to praise God.
4 On either side of the book are two staircases elegantly curved, the steps of which are numbered for hours.
5 Below the book is an horizontal dial the gnomon for which has been broken off. Below this slate veneer engraved with fleur-de-lys.

Although now much worn, lacking the top dial and some slate veneer, this remarkable dial nonetheless retains much of its original colour and all of its intrinsic interest. The dial derives from a religious establishment at Dijon close to the Ducal palace, and this conventual association is reflected in its design. The book on the South face is clearly intended to represent the Bible, while the staircases which flank it are surely intended to recall the miracle recounted in 2 Kings 20:8-11 and Isaiah 38:7-8 in which the Sun's shadow was brought by Isaiah's call and the Lord's command to return '10 steps by which it had gone down on the steps (or: dial) of Achaz'.
Private Collection.

242 A Portable Vertical Dial with Calendar and Tide Table, 1700
Signed 'W. Sneewins'.
Brass, diameter 59 mm.
Rijksmuseum 'Nederlands Scheepvaart-museum', Amsterdam, S3988.

243 A 'Butterfield-type' Dial by its Supposed Inventor, c. 1700
Signed 'Butterfield a Paris'.
Silver, length 60 mm.
Octagonal plate with underset compass and folding bird gnomon characteristic of these dials. The plate is engraved with three hour scales for 50°, 48°51', 46°, 44° and on the reverse the latitudes of 24 places in Europe arranged according to the dial scale with which they were to be read.
Rijksmuseum 'Nederlands Scheepvaart-museum', Amsterdam.

244 A Pocket Horizontal Dial for a Single Latitude, between 1662 and 1689
Signed 'H. Dufour Blois'.
Silver and gold, 48 mm.
The dial is set for 52° which would be appropriate for several places in the Netherlands. Henri Dufour, a clock-maker, was established in Blois by 1662 and died in 1699.
Rijksmuseum 'Nederlands Scheepvaart-museum', Amsterdam.

245 A Large Universal Equinoctial Dial, ?c., 1700
Signed 'Nicolaus Rugendas Augsp[urg] 48 Gr'.
Brass and silvered brass, diameter 165 mm, with original silk and suede covered wooden box linen with red velvet.
A very plain but unusually large example of this typically Augsburg form of dial. The hour scale reads from III - XII - IX and the dial may be set for latitude against a 90° scale on the folding latitude arc. Turned plum for levelling held in a folding stand plain four compass adjustable for declination within a range of 40° east or west of the meridian. For a dial of similar proportions see Bobinger, 295.
Three makers, successive generations of the same family named Nicolas worked in Augsburg. The present instrument is probably by the last of the three who was active 1665-1745.

246 A Variant Form of the Equinoctial Dial, c. 1700
Signed 'Johann Willebrand In Augspurg 48'.
Gilt brass, brass, silver with original fishskin covered case with red velvet lining, 118 x 118 mm.
Known as a crescent dial, this form is simply an equinoctial dial with the two halves of the hour ring set back to back and the two gnomons set above them adjustable in their mount for solar declination against a calendar scale. On the base of the instrument is a thirty two point windrose and on the reverse of the base is a perpetual calendar. Set in the lid of the case is a disc engraved with the latitudes of 67 places.
Private Collection.
Bibliography: Bobinger, 381.

239 a

239 b

230

230

233 a

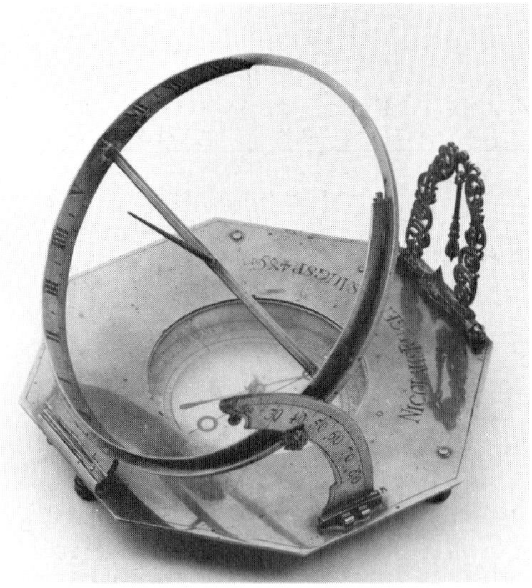

245

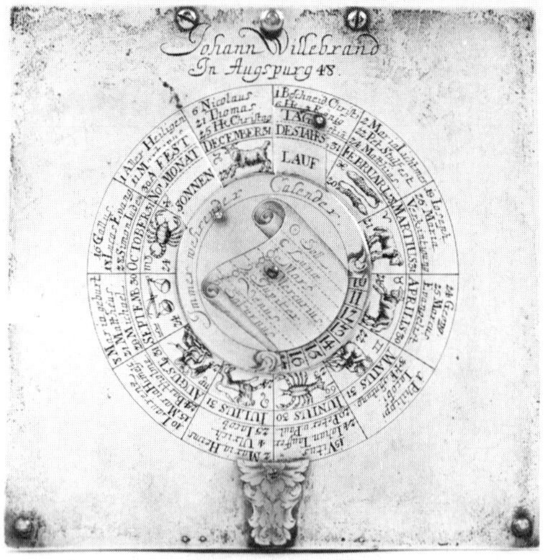

246

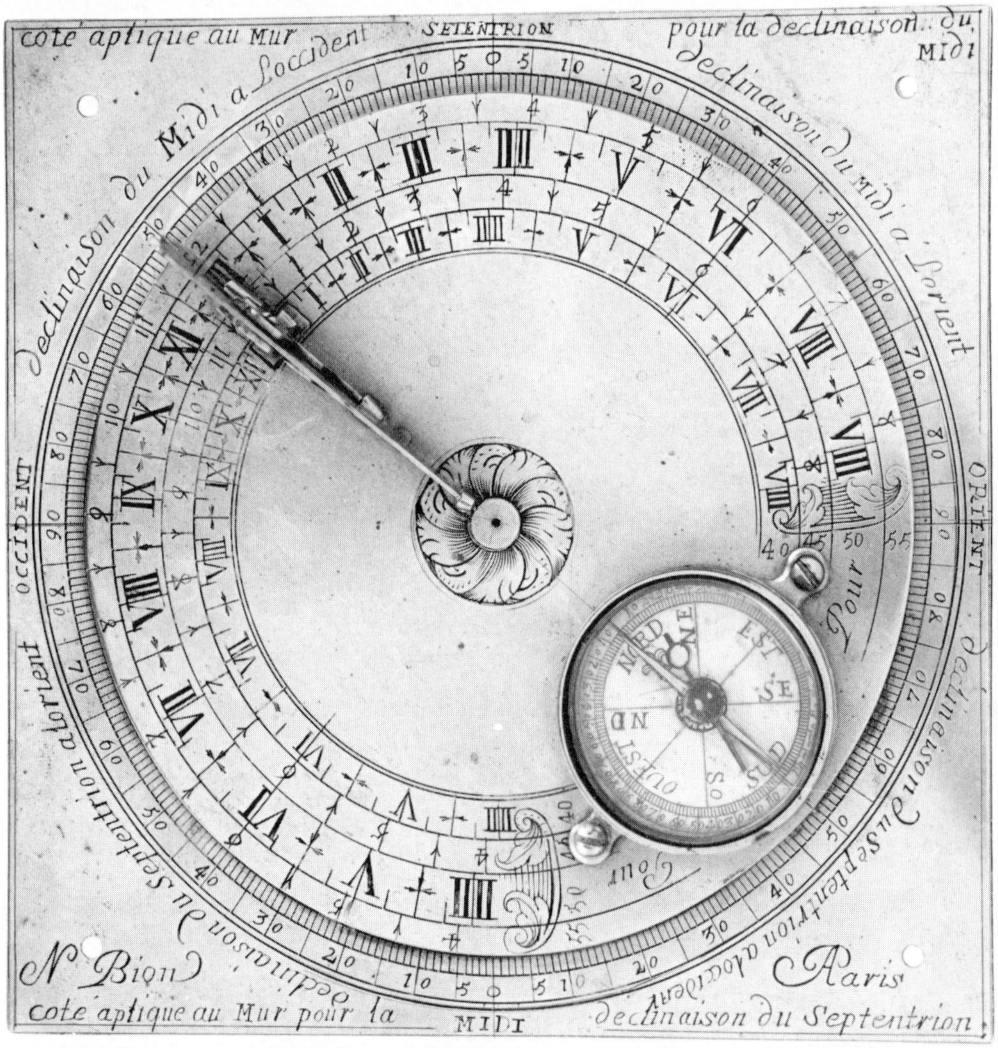

247

247 An Horizontal Sun-dial and Declinatory, c. 1700
Signed 'N Bion AParis'.
Brass, 128 mm x 129 mm.
Square baseplate with a circular scale of degrees in four quadrants. Free to rotate within this is an horizontal dial with hour scales for latitudes 55°, 50°, 45°, and 40° and a folding gnomon. Mounted on the same plate is a compass. This instrument was used to lay out vertical dials mechanically without having to calculate the declination of the receiving surface. An exactly similar instrument is described in the maker's own treatise on instruments (Bion 304-8, plate 25, fig. 3), and its use described in detail.
Private Collection.

248 The Self-orienting Analemmatic Dial, 18th century
Brass, 154 mm x 92 mm.
Two plates hinged together, one with an horizontal dial with collapsible gnomon, incorporating a plumb-bob which is set for latitude over a degree scale engraved on the bottom edge of the gnomon, the other adjustable over a calendar scale and used in conjunction with an elliptical hour scale. By combining a direction dial with an altitude dial, instruments of this type become self-orienting for the correct time is shown when the two dials agree.
Provenance : The Time Museum, Rockford, Illinois.
Private Collection.
Bibliography : Christies, 14 April 1988, lot 105.

249 A Mechanical Equinoctial Ring Dial, early 18th century (post 1714)
Signed : 'Made by J. Rowley Master of Mechanick to His Majesty'.
Brass and silvered brass, overall height 370 mm.
Influence from the clock is evident on this form of equinoctial dial which incorporates gearing. The dial having been levelled and correctly oriented. The alidate pivoted on the centre plate is set to the appropriate date on a calendar/declination scale. The whole pierced centre plate is then rotated until light from the sun passes exactly on through one vane onto the centre of the other. Time is then read off from the position of the hands on the clock face which are geared to the arbor providing the axis of the plate.
Private Collection.

250 An Inclined Plate Dial, c. 1700
Signed 'Chapotot A Paris'.
Brass, 71 mm x 60 mm.
Octagonal plate with inset compass. The plate is engraved on both faces with the names of places and their latitudes. Hinged to the plate is a second cut away around the compass carrying an hour scale and collap-

sible bird gnomon. Since the bird is fixed to the plate beneath, the hinged plate can be lifted up it until the beak of the bird indicates the appropriate latitude on a scale engraved on the gnomon. The instruments now functions as a horizontal dial for the place chosen. It is not an equinoctial dial.
Private collection.
Rasquin, 92.

251 An Horizontal Compass Sun-dial and Lunisolar time Convertor and Lunar Volvelle in the Form of a Barrel, c. 1700
Inscribed 'G C'.
Ivory with brass dial plate and moon dial.
Height 90 mm.
When not in use the instrument can be screwed together for protection and convenience of transport.
Rijksmuseum 'Nederlands Scheepvaart-museum' S2310.

252/253 Two Moulds for Casting Sun-dials, 1727
Signed 'Suyt Born Fecit'.
Stone, overall diameter 215 mm.

254 An Horizontal Compass Dial, first half 18th century
Signed on the compass rose 'I. Nemes Fecit'.
Brass with paper rose, diameter 55 mm.

255 An Italian Diptych Dial, 18th century
Wood and ivory with brass fittings, 118 mm x 81 mm x 29 mm.
1a. A 32 point wind rose diagram.
1b. Vertical pin-gnomon dial for 45 latitude for Italian hours and the sun's declination.
2a. Horizontal pin-gnomon dial for Italian and Babylonian hours and horizontal dial for ordinary hours around an inset compass.
2b. As 1a above.
The dial is in the form of a book with clasps and has the motto *Fili conserva a tempus* on face 1b.
Museum of the History of Science, Oxford.
Bibliography : Maddison 219.

256 Two Universal Equinoctial Ring Dials, mid-18th century
Both signed 'J. Sisson London'.
Silver, diameter 110 mm ; brass.
Two examples of one of the simplest, most effective and most popular of late 17th and 18th century dials. Although carrying the same signature this hides the succession of a father and son.
Private Collection.
Universiteitsmuseum Utrecht, A34.

248

255

250

252

253

234 a

ELABORABAT ANTVERPIE
MICHAEL COGNETVS
Horizont
Tropicus Cancri
SEPTENTRIO
Horizont
Æquator
Ab Occasus
Horizont
Ab Occasus
Æquator
Horizont
Æquator
Horizont
Æquator
OMNES ÆQVALES
SOLA VIRTVTE
DISCREPANTES

234 b

Made by I Rowley Master of Mechanick to the KING
The Sun Declination

249

256

268

274

271

277

270

272

274

257

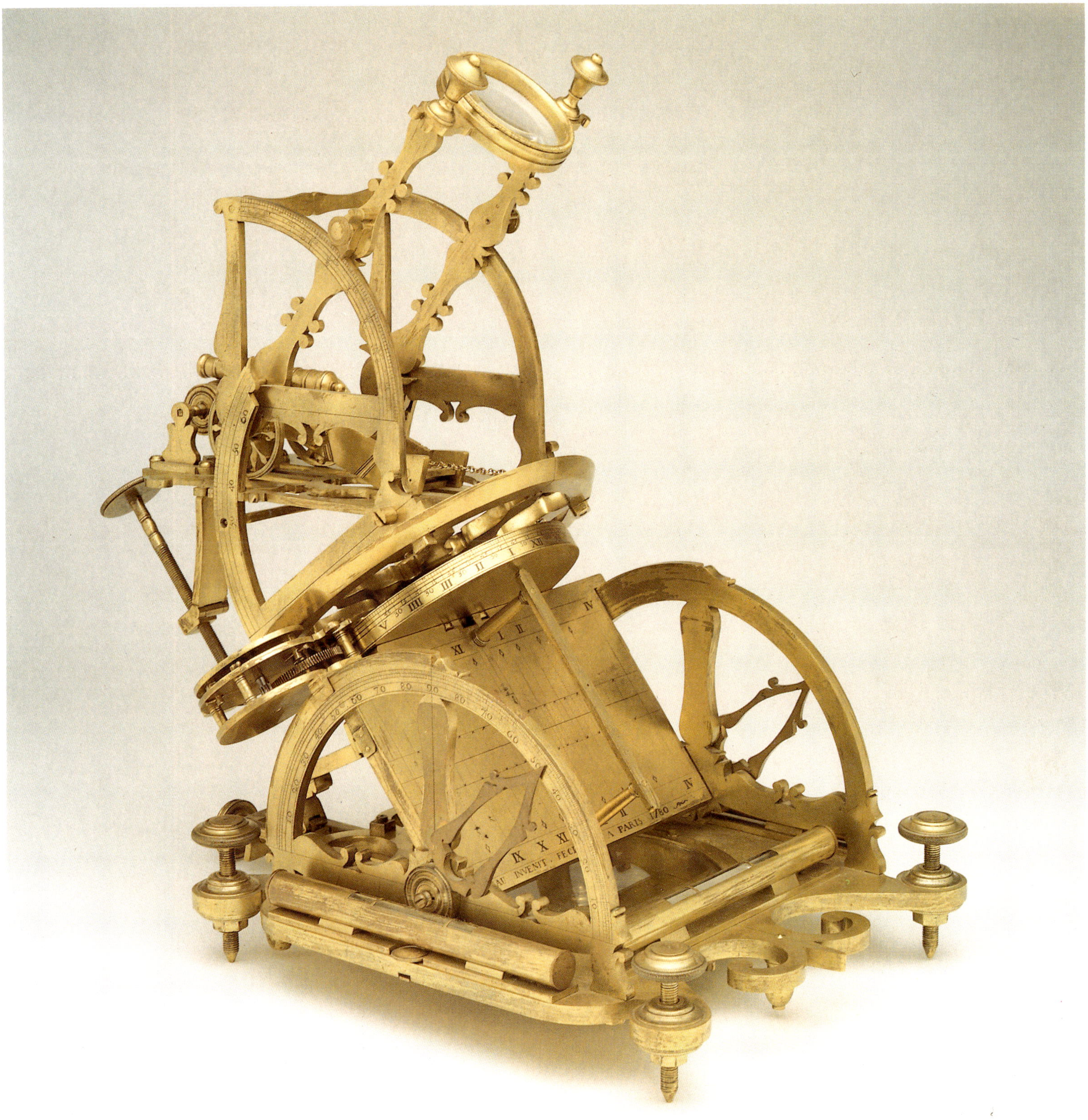

261

275 An Italian Night Clock, 18th century
Ebony, brass and steel.
Rectangular case with 'wandering hours' dial behind the hour aperture in the dial plate. Below is a painted scene of the figure of day dressed in white with the Sun in one hand and a sand-glass held aloft in the other announcing her coming to night who is seated reading beside a lamp and dressed in blue.
Private Collection.

276 A French Night Clock, c. 1700
Gilt brass, gilt bronze, iron & steel.
Height 370 mm, dial diameter 190 mm, depth 140 mm.
The clock is mounted on a turned bluster pillar with four shaped folding feet. The mobile dial has fretted numerals so that time can be read at night by the light of a candle placed on the case housing the movement behind. During the day they are read against an index fixed in what would be the 12 o'clock position to 15 minutes. The movement has a very large going barrel, for both trains, a verge escapement and simple wire pendulum with flat bob and silk suspension. Hour and half hour striking by a rack.
Musée du Temps et de l'Histoire, Besançon, D 47.2.2.

277 A Tabernacle Clock With a Later Movement, 17th century and later
Brass and steel.
Museum of the History of Science, Oxford.

278 Time and Astronomy Joined in a Planetary Clock, 1754
Copper plate engraving on paper of Jan van den Dam's copernican *Sphaera perfecta.*
Museon, The Hague E2-5.
Bibliography : King & Millburn, 214-6.

279 A Personal French Country Clock become a Memorial, 1772
Signed 'Fait pour l'usage de Pierre Joseph Vermot Des Roch Dieu veullie Avoir sone ame 1772' 'Emmanuel Mercie Horlogé à Morteau'.
Iron and steel with enamelled dial and brass frontons, 186 mm x 340 mm.
Movement of a 'Comptoise' or 'Morez' clock with four post frame and alarm. Short bob pendulum with inverted horizontal verge escapement. Flail and star wheel strike with the locking plate incorporated into the striking train barrel.
Musée du Temps et de l'Histoire, Besançon.

280 A Zaanse Clock, c. 1720
signed 'Cornelis van Rossen op de Koogh'.
A characteristic Zaanse clock with going, striking and alarm trains.
Zaans Uurwerkmuseum.

281 A Frisian Stoelklok, 1736
Signed 'Bauke Haanstra Me Fecit Ao 1736'.
Brass and steel with painted metal dial and wood mount.
The clock indicates the hour, minute, day, date, month and the moon's age and phases by a volvelle. Striking and alarm on two bells.
Zaans Uurwerkmuseum.

282 Inclined Plane Clock, late 18th century.
Signed on the case 'C.W. Klute te Nijmegen' and on the watch movement 'Raym[en]t London'.
Brass and steel, inclined plane of slate in a mahogany case.
Lenght 550 mm, height 200 mm ; clock case diameter 70 mm.
The cylindrical base contains the watch movement and an eccentrically placed lead counterweight such that the centre of gravity of the system lies away from the axis of the cylinder. The motive power of the system is the turning of the watch movement and the counter-weight with respect to the case. This occurs when the clock is placed on an inclined, being prevented from slipping down it by dents in the rim when the centre of gravity being off the vertical axis the clock will function, the whole system gently descending the plane in 24 hours. This form of clock was first developed in the mid 17th century and has remained a popular curiosity.
Universiteitsmuseum, Utrecht.
Bibliography : Bos, 63-4.

283 The Interior of a Clock-makers Shop, 3rd quarter 18th century
Oil on canvass, 2 m 120 x 950 mm.
To the left the painting shows the workshop section of a large corner shop, well-lit from two adjacent windows. At the extreme left a workman is using a bow lathe mounted on a bench on which lie other tools. To his right the master seated at his bench turns towards the right in conversation with a customer. To the right in the sales area of the shop an assistant discourses about a large mantle clock to a group of four Turkish clients before moving on to the numerous other wall and mantle clocks standing behind them.
Musée du Temps et de l'Histoire, Besançon.

284 An Urn-shaped Clock with Revolving Dial, late 18th century
Signed 'Lepaute Paris' bronzes by 'Osmond'.
Gilt bronze, brass and steel.
Height 520 mm.
A two handled urn stands on a fluted column with a swag draped across it. The lower part of the urn had relief modelled foliate decoration from which protrudes the head of a snake which supplies the fixed index for revolving band carrying the individual enamel chapters of the hour scale. The winding hole is concealed beneath the swag of drapery.
Private Collection.

285 A Mantle Regulator with Calendar, late 18th century early 19th century
Signed 'Lepine à Paris'.
Gilt metal, brass and steel.
Rectangular fully glazed case on levelling feet, enamelled dial with a 31 day calendar in red, the hours and seconds in black. Half seconds beating with centre seconds, pin-pallet escapement and temperature compensated pendulum.
Private Collection.

286 A Treatise on Horology by a Leading French Maker, 1786
Ferdinand Berthoud, *Essai sur l'Horlogerie dans le quel on traite de cet art relativement à l'usage civil, à l'astronomie & à la navigation, en établissant des principes confirmés par l'expérience,* 2nd edition, Paris 2 vols, 1786.
Berthoud's two imposing volumes destined as much, if not more for his customers, as for other makers, were first published in 1763. This second edition was issued unchanged. In volume I Berthoud provides a general survey of the different types of clocks and watches and how to make them. In volume II he examines specific problems which affected precision such as wheel-teeth, train calculation, pendulum behaviour and temperature compensation.
Nederlands Goud-, Zilver- en Klokkenmuseum, Schoonhoven I. 10 Boek.

287 A Clock for Use at Night, c. 1810
Signed 'Bofenschen à Paris' and on the dial ring 'L 2243'.
Painted iron sheet, gilt bronze, enamel, glass, steel and brass, 480 mm x 230 mm x 190 mm, diameter of watch 52 mm.
The painted and decorated vase has a removable top for placing a light source inside. The movement has a hollow silvered dial with reversed figures at the back, inside the vase. Thanks to a lens incorporated in the vase an image of the dial can be projected onto a wall thus giving the time and lighting a room. Spring-driven verge and fusee movement, specially made for this system. This instrument corresponds closely with the 'Nocturnal Dial or chronometer' patented in England in 1808 (Patent specification No 3185) by the Danish clock-maker Johann Andreas Schmidt who, after

being captured at the battle of copenhagen (1801) and transported to London, was by 1808 established at St Mary Axe, London.
Nederlands Goud-, Zilver- en Klokkenmuseum, Schoonhoven PEN24.

288 Another Night Clock, late 19th century
Inscribed 'V.A. P. Breveté S.G.D.G.'.
Silvered composition metal, white milk glass, brass and steel, 190 mm x 100 mm x 90 mm.
The silvered stand is mounted on a single column foot. The cylinder-glass stands on the top of the case and turns round in 12 hours. The fixed dial shows the time (to 15 minutes) on the outside of the glass cylinder into which a light may be placed. Semi deadbeat escapement with horizontal crown wheel with fixed pendulum. Spring drive for one week.
Nederlands Goud-, Zilver- en Klokkenmuseum, Schoonhoven, WNK27.

289 The Pendulum Clock Turned Upside Down, c. 1820/40
?English. Brass and steel.
Overall height 1m 13., diameter of dial 240 mm.
The circular movement has an hour scale I - XII and a centre seconds hand. This movement supplies the bob of a flat-strip brass pendulum adjusted by a gnurled nut graduated 1 - 12 which is set immediately above the movement which is fitted with a deadbeat escapement.
Private Collection.

290 A Turret Clock of c. 1830/40 using a mid-17th century Bell
Movement signed (on an applied brass plate) 'C. Chasté Amsterdam'; bell signed 'Melchior de Haze me fecit' (Melchior de Haze made me).
Iron and brass and a wood stand.
Overall height 2 m 20, movement 71.5 cm x 33.5 cm.
The two trains are contained in an early version of the 'flat-bed' frame. This is composed of two parallel-sided long members, joined at each end by a cylindrical cross-piece bolted to them. The centre of each cross-piece is formed into a square, and to it is bolted a rectangular end-piece for the windingpinion arbors. The dead-beat escapement has a pin-wheel. The pallet fork is mounted at an angle on an arbor set above the train carried between 'A'-shaped supports. One of these supports also supplies a bracket for the pendulum which has a flat steel-spring suspension and lateral adjustment. The dial is now mounted above the clock with right-angle leading off work acting through two sets of bevel gears. The clock strikes the hours by a single hammer hitting externally from above the bell. There is an external locking plate and

273

345

266

281

308

309 Types of Sandglasses, 18th century
a. A 30-minute duration Sand-glass in a wooden mount with two colour, inlaid, symmetrical straw marquetry decoration.
Height 129 mm.
b. A 20-minute (?) glass with a hexagonal brass stand incorporating in each base coloured depictions of the Virgin Mary and the resurrection.
Diameter 65 mm.
c. A 30-minute traveller's glass composed of one double ampoule (2 x 15 minutes) and a single ampoule contained in a double black leather case, the outer section of which can be rotated over the inner to close the inspection aperture, and with a cylindrical carrying case with cover attached to the carrying thongs.
Height of glass 145 mm, diameter 54 mm; of carrying case 160 mm, diameter 75 mm.
d. A 60-minute glass composed of one quadruple ampoule (4 x 15 minutes) and one single ampoule contained in a hexagonal iron stand with lobed frieze.
Height 260 mm.
Private Collection.

310 Two Miniature Sand Glasses, 19th century
Bone and glass.
Height 20 mm and 30 mm.
The glasses run respectively for 5 and 10 seconds. Each has three turned pillars. Perhaps made as furniture for a dolls-house.
Musée du Temps et de l'Histoire, Besançon 953.7-1785/6.

311 A Sand-glass for a Tea Taster, 19th century
Brass and Glass, 250 mm x 90 mm.
The glass is composed of one single and one quadruple ampoule. Each section of the latter fills in 1 minute thus enabling the time of infusion of different kinds of tea to be timed.
The Science Museum, London, 1953-300.

312 An Alarm Sandglass, 19th century
Iron and wire mount 150 mm x 70 mm x 60 mm.
The glass is lightly held at an angle in the frame such that when the sand has run, its weight in the lower ampoule releases the glass and causes the clapper attached to it to swing downward and strike the bell can be set for up to 4 minutes.
The Science Museum, London, 1953-309.

313 A Norwegian Clog Almanac, 1648
Inscribed 'Ionaz Kelson' & Ioas 1648'
Wood.
Length approx. 935 mm.

314 A ?Danish Clog Almanac, 1660
Inscribed 'I C 1660'
Boxwood.
Length 120 mm.
One of four such almanacs presented to the Ashmolean Museum Oxford by John Heysig in 1683, the year in which the museum opened.
Museum of the History of Science, Oxford.
Bibliography : Duncan 345, p 133.

315 A Swedish Calendar in the Form of a Walking-Stick, 1784
Wood.
The dates 1784 and 1793 appear in parts of the calendar. The symbolic illustrations are carved in greater detail than is usual on such instruments.
Museum of the History of Science, Oxford.

316 Swedish Calendar, 1794
Signed '...Stokholm den 3 September 1794. F. Ekistubbe'. Wood with the scales written in ink and varnished. 215 mm x 78 mm.
The calendar runs from 1792-1909. On the face is a slide marked with the days of the week which is slid against the day number in the month scale.
Museum of the History of Science, Oxford.

317 A Calendrical Icon, 1865
Silver and wood, 460 mm x 530 mm.
In the centre is the ascension of christ, around this twelve phases of the life of the Virgin, and around the twelve months of the year represented by saints. In the border are 128 portraits of the Virgin and in the corners the four evangelists.
Private Collection.
Bibliography : Rasquin, 173.

318 Pages from a French Popular Calendar, late 17th/early 18th century
Paper printed in red and black with wood block illustrations, 128 mm x 88 mm open.
Pocket popular almanacs of this kind, in which words and numbers are reduces to a minimum, the saints' days, festivals and zodiacal signs being shown by images, can be traced back without break to the Middle Ages. They illustrate the need felt, even by the illiterate, of a manual to mark the ritual times of the church (saints' days) throughout the year which would relate them to the natural (and astrological) year represented by the signs of the zodiac. These are shown in the bottom line of each third of the month.
Private Collection.

319 A Popular Printed Almanac, 1684
Printed book, sm 8°.
Johannes van Keulen *Eeuwig Duerende Almanach. Nielykx Oversien en van voorgaende misslagen Gesuyvert,* Amsterdam 1684.
Universiteitsmuseum, Utrecht UM 127.

313

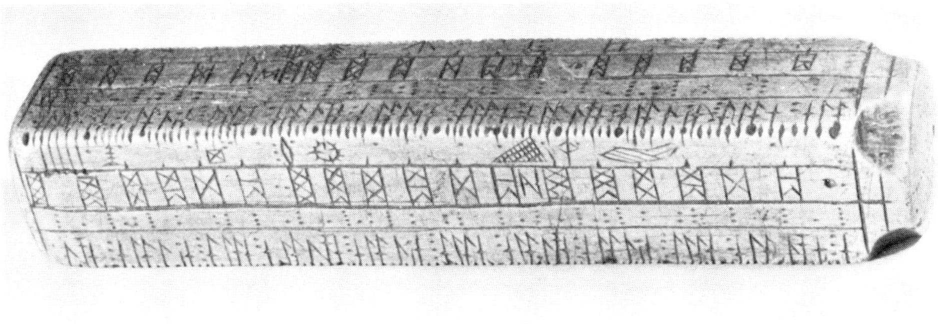

314

320 A Perpetual Calendar on the covers of an Aide Mémoire, late 17th/early 18th century
German, silver gilt brass, and ivory, 45 mm x 85 mm.
Rectangular notebook pivoting around a central point at one end with a clasp at the other, composed of the covers and six leaves. Each face is finely engraved with symmetrical foliate decoration and each carries a volvelle at the centre. One of these is a perpetual calendar for the date of the week with the governing planetary sign, the other supplies length of the month, feasts and festivals, sunrise and set, and the length of day and night.
Private Collection.
Bibliography: Rasquin, 171.

321 A Perpetual Calendar for the Pocket, 18th century
Brass.
Diameter 58 mm.
On the face is a fan shaped volvelle calendar for date, day of the week and zodiacal sign with at the centre a month index. On the reverse is a scene inside a smithy.
Rijksmuseum 'Nederlands Scheepvaart-museum', Amsterdam, S168(6).

322 Single Year Almanacs for the Pocket, 18th & 19th century
Copper and white metal alloy.
Various diameters between 35 mm & 55 mm.
a. For 1759 signed 'John Powell, Birmingham'.
b. For 1767 signed 'Visser fecit'.
c. For 1777 not signed, French.
d. For 1792 signed 'James Davies Birmingham'
e. For 1844 signed 'Naudin'
Calendars of this type seem to have originated in England in a perpetual form before spreading to France and the rest of Europe in the late 17th century. Exactly when the single year version was developed is however not known. Dutch and French examples are considerably scarcer that English ones.
Private Collection.

323 A Calendar for the Study, 1789
Printed paper, 370 mm x 525 mm.
Private Collection.

324 An Astronomical and Perpetual Calendar, 1861
Mlle GinotDesrois, *Calendrier astronomique et perptuel indiquant le quantième des Mois, les Jours de la semaine, les phases de la Lune, la place du Soleil et de la lune dans l'Ecliptique pour un jour donn ; le lever, le passages au Méridien, le coucher de ces deuz Astres et les Etoiles de la 1ère à la 4e grandeur, ainsi que les principles Eclipses de soleil visibles à Paris, depuis 1858 jusqu'en 1874.*
Planisphere and calendar lithographed by Choixeque and published, with an instruction booklet by Mallet-Bachelier Paris.
Musée du Temps et de l'Histoire, Besançon 53.7.1793.

325 Décret de la Convention National (No 1838). De 4e jour de Frimaire au second de la République Française... Sur l'Ere, le commencement & L'organisation de l'année & sur les noms des jours & des mois, Le Mans, 1793.
Printed, pamphlet, 4°.
The official promulgation and description of the decimal system of time measurement introduced in Revolutionary France. Sixteen articles describe the division of the year into 12 months having three deades each of 10 days, each day further subdivided into 10 hours of 100 minutes.
Private Collection.

326 A Watch Showing Decimal and Duo-decimal Time, c. 1795
Brass and steel with polychrome enamelled dial.
Diameter 60 mm.
In a simple landscape two columns support dials showing decimal time in 10 hours each of 100 minutes to the left and duodecimal time to the right. Between these two dials is a marbled pyramid carrying revolutionary emblems in the form of a trophy, and above it is a dial indicating the days of the Gregorian calendar (up to 31) and the revolutionary calendar (up to 30) in two concentric rings.
Universiteitsmuseum, Utrecht UM 376.

327 A Noon Cannon and Horizontal Sun-dial, early 19th century
Brass, 381 mm.
Following their appearance in late 18th century France (see number 261), sun-dials combined on the same plate with a noon gun became popular throughout Europe. The fixed horizontal dial of the present example is set for 52° 22' although the lens for igniting the fuse at noon may be set for a range of different latitudes.
Rijksmuseum 'Nederlands Scheepvaart-museum', Amsterdam.

328 Two Diptych Dials by Members of the Beringer Family, c. 1800
Signed 'Verfertigt von Phil: Beringer (a) David Beringer (b).
Hand coloured printed paper on wood.
An inexpensive attractive form of sun-dial, these instruments embodied a pair of verti-

317

339 a

339 b

315 a

315 b

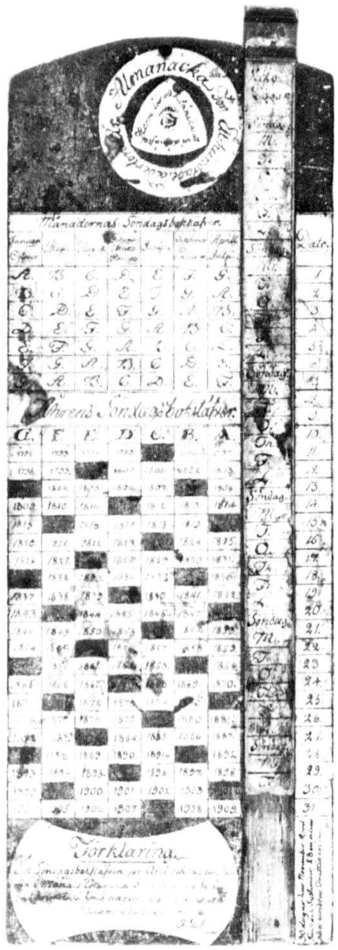

316

cal horizontal dials with common string-gnomon usable over a range of northern of latitudes ranging from about 35° to 55°. The lineal descendant of the more elaborate earlier ivory dials (see number 233) such instruments represent the last phase of the Nuremberg manufactory.
David Beringer (1756-1821) became a master *compassmacher* in 1777. Of Philip Beringer nothing is known.
Private Collection.

329 A Third late Nuremberg Diptych Dial, c. 1800
Signed 'J P Kleiniger'.
Printed paper hand-coloured on paper, 126 mm.

330 An Equinoctial Sun-dial from Augsburg, c. 1800
Signed 'And. Vogl[er]'.
Brass, 52 mm diameter.
A simple equinoctial dial for the pocket typical of Augsburg production in the late 18th and early 19th centuries.
Rijksmuseum 'Nederlands Scheepvaart-museum', Amsterdam, S4(3).

331 An Ingenious Compass Sun-dial, 1824
Signed 'Porter's Magnetic Sun Dial on the Floating Compass Card', 'Saml Porter Fecit Feby 16 1824' and inscribed 'Entered at Stationer's Hall'.
Cherrywood and printed paper.
Diameter 50 mm.
A pivotted ('floating') compasscard has a horizontal dial scale (VIII - XII - VIII) contained within the circle of a snake biting its own tail, and with a Father Time figure, printed at the centre.
Private Collection.
Bibliography : Rasquin 134.

332 A Sun-dial for a very Low Latitude, mid-19th century
Inscribed 'Ceylon Lat 8° N'
Brass, 229 mm.
National Maritime Museum, Greenwich, D364/A72-9.

333 A Sun-dial for Southern Latitudes, mid-19th century
Brass, 129 mm.
The hour scale V - XII - VII is a mirror image of what it would be on a Sun-dial for 33° north.
Rijksmuseum 'Nederlands Scheepvaart Museum', Amsterdam, S188 (21).

334 A Geographical Sun-dial by a Wandering Maker, c. 1855
Signed 'Rd Melville Fecit from London' and inscribed 'Horas non numerum nisi serenas' (I only count the bright hours) and 'Sol non occidit super iracundiam vestram' (let not the Sun go down on your wrath : *Eph.* ch 4).
Slate, 135 mm x 140 mm.
The rectangular plate is engraved with a circular equation of time scale and within this an hour-scale IV - XII - VIII reading to five minutes. Instructions for using the equation scale are given and between the two scales are the names of seventy places throughout the world written radially in positions correponding to their time difference from the place in which the dial is to be used (here London 51° 51'). Each corner of the plate contains a subsidiary dial respectively for New York, Alexandria, Isle of Borneo and New Zealand.
Richard Melvin or Melville came probably form County Down, Ireland. Dials of this style are characteristic of his production and he seems to have travelled throughout Britain and Ireland to make them. Examples are known signed for places in Ireland, Scotland, and northern and southern England.
Royal Museum of Scotland T1985.85
Bibliography : Clarke, Morrison-Low and Simpson, 210-16.

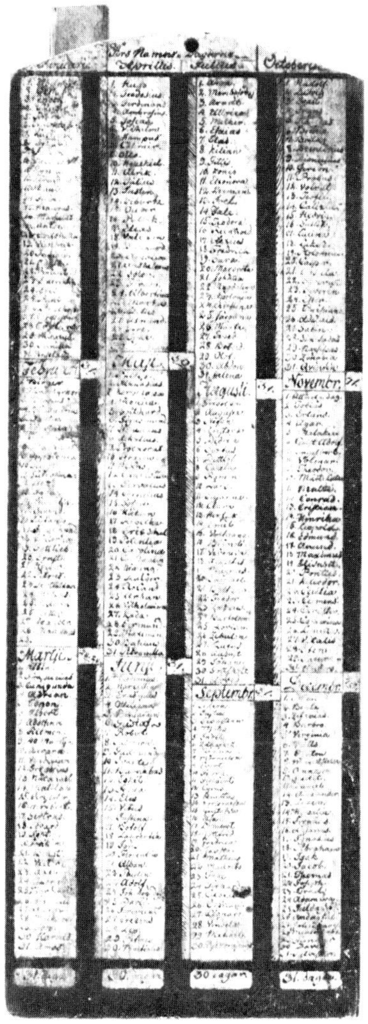

316

341

FEST UND
FEYERTÄG
Sonnen
Tag
Auf
gang
Unter
Nacht

Calendarium
Perpetuum

320

320

ALMANACH
Pour l'An 1777
Dimanches de l'Ann
Septuagesime 26 Jan.
1759
A CALENDAR
The Sunday Figures
John Powell
Birmingham
AMSTERDAMSE ALMANACK
1767
Visser Fecit

322

ALMANACH DE CABINET
POUR L'ANNÉE M. DCC. XC.

A POITIERS, chez FRANÇOIS BARBIER, Libraire-Imprimeur de Monseigneur l'Évêque & du Clergé.
AVEC PRIVILEGE DU ROI

323

326

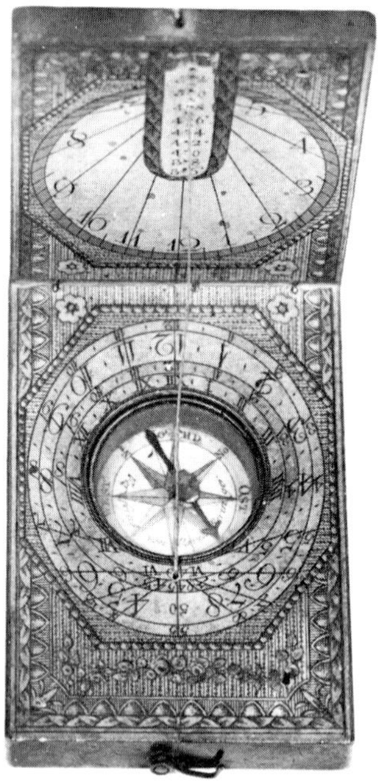

328

335

Steps towards precision

Go, wond'rous creature! mount where Science guides
Go, measure earth, weigh air, and state the tides;
Instruct the planets in what orbs to run,
Correct old Time, and regulate the Sun.
Pope, *Essay on Man,* Ep. II.

Time is of fundamental importance in positional astronomy and for the practice of navigation.
The interaction of demand for reliable time-keepers from astronomers and navigators, with the ambitions and curiosity of clock-makers, provided a potent motor for development.
Attempts to use high precision clocks in astronomy by Tycho Brahe were however unsuccessful.

335 An Early Depiction of Clocks in an Observatory, 1587
Inscribed 'Effigies Tychonis Brahe O.F. Aedificii et instrumentorum astronomicorum structoris A Domini 1587, Aetatis suae 40' (the picture of Tycho Brahe O.F. constructor of [this] building and of [its] astronomical instruments. In the year of our Lord 1587, of his age 40).
Copper-plate engraving printed on paper, with contemporary hand colouring, 270 mm x 418 mm.
Plate from the section devoted to Denmark and the Isle of Hven from Ioan Blaeu, *Atlas Maior,* vol. I, Amsterdam, 1662 (for which see Koeman), based on a woodengraving in Tycho Brahe, *Astronomiae instauratae mechanica...,* Wandsbek, 1598.
In the foreground Tycho's mural quadrant is being used by an assistant at the extreme right to make a transit observation. To the left a second assistant records the altitude, while to the right a third notes the time of transit which he calls out to the recorder. The clocks indicated seconds and there are two of them so that each may act as a check upon the other.
The scenes behind the quadrant, drawn merely to fill up the space, are a portrait of Tycho and his dog painted by Tobias Gemperlin. Above Tycho's head is a gilt-brass clockwork-driven celestial globe. Above this is a part of Tycho's library flanked by portraits of Frederick II and Sophie of Denmark. Behind Tycho is the terrace of the observatory showing his quadrant, one of his equinoctial armillary spheres, and his voluble sextant; his study with the 149 cm star globe and assistants working; and below this his chemical/alchemical laboratory. For a detailed description of these scenes see Raeder, Strömgren & Strömgren 28-31.
Private Collection.

Dramatic Advances in Precision Time-keeping were made by Jost Bürgi (1552-1632)

336 An Advanced Clock by Jost Bürgi, c. 1590-1600
Photographs
In July 1559, Bürgi was retained by the Landgraf Wilhelm IV as clockmaker at his court in Kassel, and here Bórgi remained until 1603. Among collections surviving at Kassel are two strikingly plain clocks which may have been experimental models or have been used for timing astronomical observations (*cf.* the appearance of Tycho's clocks in no 335). The second of these clocks is illustrated since it embodies Bürgi's two major innovations a) his crossbeat escapement, and b) his remontoire. The escapement beats seconds, the main wheel is extremely finely cut, and the remontoire operates every twelve hours.

337 The Action of the Crossbeat Escapement
Diagram by David Penney.

338 The Action of the Remontoire
Diagram by David Penney.

339 A Clock with Crossbeat Escapement and a Rolling ball Drive, 1660
Signed 'Niclaus Radellof Sleswig Anno Domini 1660'.Gilt brass and silver, brass and steel, ebony and tortoiseshell veneered case, agate balls.
Height 900 mm, 300 mm square.
Table clock in the form of a tower showing hours and minutes on dials which are modern replacements of the lost originals. The clock is driven by the descent, lasting twelve hours, of two agate balls down a spiral track. As they descend they press against vertical rods thus turning the whole cage within which the spiral track and balls are contained, and so transmitting power to the movement controlled by the cross beat escapement. When a ball ends its descent it is received by a pair of weak springs which break its fall through a trap into a container in the base where it trips a release for two hooked levers restraining the next ball in place at the top of the spiral. Hour and quarter striking takes place by two separate mechanisms on two bells, the brass cylinders that release the strike also acting as flies.
Three other clocks by Radelof of this design are known, all of them in Denmark (Von Bertele; Maurice & Mayr 60, p. 232). Radelof, a native of Saxony, was invited to Denmark by Olaus Worm, working there from 1647 until at least 1666.
Private Collection.
Bibliography: Lloyd, 97-8; Roberts, 36-40; Block.

The increased accuracy of Bürgi's clocks depended upon his exceptional craftsmanship, good design, and new mechanical ideas.
The pendulum, by contrast, intrinsically provided a natural time-standard.
Galileo recognised that the pendulum is isochronous, that is, the time taken by the pendulum to swing through its arc is independent of the size that arc.
In 1641 Galileo designed a mechanism for adapting the pendulum for use in time-measurement.

340 A Modern Reconstruction of Galileo's Pendulum Mechanism 400 mm x 250 mm x 250 mm.
In a letter to Prince Leopold dei Medici, Viviani described the action of the machine as it had been explained to him by Galileo. 'When the pendulum was at rest it prevented the fall of the weight; but, when raised and given free play, as it swing past the perpendicular, the longer of the two arms mounted on the pivot of the pendulum raised the decent which acts on and engages with the notched wheel, which latter being pulled by the weight revolved with its upper side towards the pendulum, so that one of its pins pressed on the shorter arm from above, giving it at the commencement of its return an impulse in such a way as to convey a definite motion to the pendulum, causing it to rise to the height from which it started; so that when it swung back naturally and passed the perpendicular position it once again raised the detcent, the notched wheel then at once through the force of the weight, resumed its movement, continuing to revolve and to impel the pendulum by the next following pin. And thus in regular action the to and fro of the pendulum was maintained until the weight had run down.' (translation from Robertson, 113-4).
Collection J.C. Sabrier.

Although both Galileo (1642) and his son Vincenzio (1649) died before realising a pendulum clock, a number of such were made in Italy before 1658 when Christiaan Huygens published the results of his own independent investigation.
Clocks following Huygens' designs were built by Salomon Coster in the Hague.

341 An Early Pendulum Clock, between 1657 and 1659
Signed 'Salomon Coster Haghe met privilege'.
Brass and steel, tortoiseshell veneered case, black velvet covered dial plate.
Case 216 mm x 275 mm, back plate 96 mm x 107 mm.
Rectangular case with applied chapter ring and medallion for the maker's signature below. Plated movement with going barrel, horizontal verge escapement, short pendu-

lum with silk suspension swinging between cycloidal cheeks, external locking-plate, separate alarm mechanism with horizontal verge control.
Zaans Uurwerkmuseum.

342 A Large Pair of Cycloidal Cheeks, late 17th century
Brass.
Width 85 mm, height 280 mm.
Used in the physics laboratory of Utrecht University during the 19th century, the pair of cycloidal cheeks appears to have come from the Domtoren.
Universiteitsmuseum, Utrecht UM 309.

343 The Theory of the Pendulum Clock, 1672
Christian Huygens, *Horologium oscillatorium sive de motu pendulorum ad horologia Aptato Demonstrationes Geometricae,* Paris, 1673.
Apart from describing his clock, the work explains the cycloidal form of the cheeks, republishes Huygens' equation of time table, proves the isochronism of the cycloid and discusses a variety of special cases including that of the conical pendulum.
The Royal Library of the Hague RL/447 B.28.

344 Further Investigations of the Conical Pendulum
Photograph of a sketch of a parabolic/conical pendulum and a reconstruction of such a device.
Photograph : University of Leiden Library Cod. 28, p 241
Reconstruction : Boerhaave Museum, Leiden A106.

345 An Early Clock with Long Pendulum, c. 1670-75
Walnut and black ebonised pearwood, tortoise shell, brass, steel, gilt copper.
Overall height 530 mm.
Inverted bell top case, almost square dial with engraved corner scenes representing the four seasons. The main dial shows minutes with each numbered, subsidiary seconds at 12 o'clock, subsidiary hours at 6 o'clock surrounding a calendar ring and a revolving moon phase indicator. Plated movement, three wheel train with endless rope drive, crown wheel with 14 tooth pinion, verge escapement, long pendulum with silk-suspension and cycloidal cheeks, arms of Olaf Roemer engraved on a shield anchoring the top pully to the case.
One of only three known clocks of this type. Roemer worked in Paris with Huygens from 1672-81.
Private Collection.

346 An Early Observatory Clock, c. 1670/75
Signed 'Anthonius Hoevenaar fecit Leyda'.
Gilt and silvered brass, brass and steel, ebonised case.
Height 1 m 910.
The chapter rings and corner and centre pieces are applied to the velvet covered dial plate in similar fashion to the contemporary Hague clocks. There are five dials for hours, minutes, seconds, date and phase of the moon, dat and day of the week. The seconds hand rotates backwards. Plated movement with short pendulum and horizontal verge escapement.
Anthonius Hoevenaar was primarily an instrumentmaker who was closely associated with the university and observatory in Leiden.
Boerhaave Museum Leiden & Zaans Uurwerkmuseum.
Bibliography : Van der Molen ; Morpurgo, 57, Rooseboom, 73 (for Hoevenaar).

Huygens' invention quickly became known in both France and England.
It was in England that a little more than a

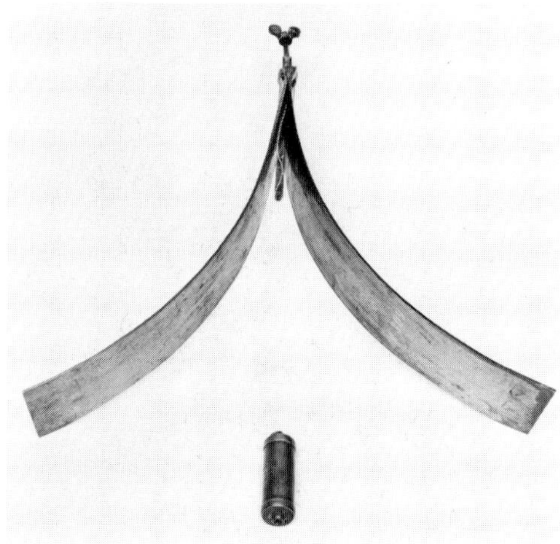
342

344

346

decade later further advance was made with the introduction of a new escapement design specifically for use with the pendulum - the anchor escapement.

347 **The Action of the Anchor Escapement**
Working demonstration model by Richard Good.

348 **The First Clock to be Fitted with an Anchor Escapement.**
Photograph
The clock of Wadham College, Oxford, built by Joseph Knibb in *c.* 1668 possibly with advice from Sir Christopher Wren.

Three decades later this was further improved by George Graham who removed the element of recoil from the escapement.

349 **Portrait of George Graham, 19th century**
Miniature portrait in oils on ivory after the mezzotint engraving by J. Hudson after a contemporary portrait by T. Hudson, 100 mm x 75 mm.
Rogers Turner Books Ltd, London.

350 **The Action of the Dead-beat Escapement**
Demonstration model by Richard Good.

Improvement in one part of the clock revealed defects in other parts. In particular changes in the length of the pendulum caused by temperature variation upset the regularity of precise clocks. In the course of the 18th century several solutions were found.

351 **Temperature compensated pendula of the 18th century**
Diagram by David Penney.

352 **A Small 'Gridiron' Temperature Compensated Pendulum, 18th century**
Brass and steel, height 495 mm.
The bi-metallic temperature compensated pendulum was first used in two clocks completed by John and James Harrison, *c.* 1726. By far the most successful of the various devices made, it remained in use for the next two centuries.
Museum of the History of Science, Oxford.

The combination of the dead-beat escapement, the temperature compensated pendulum and very high standards of workmanship had by 1730 produced a high 'precision regulator' which would remain fundamentally unchanged for over a century.

353 **The Development of Precision in Time Measurement**
Graph by F.A.B. Ward revised by David Penney & Anthony Turner.

354 **A Journeyman Clock, between 1793 and 1810**
Signed 'Onderdewyngaart Canzius Delft'.
Brass, silvered brass and steel with wood case.

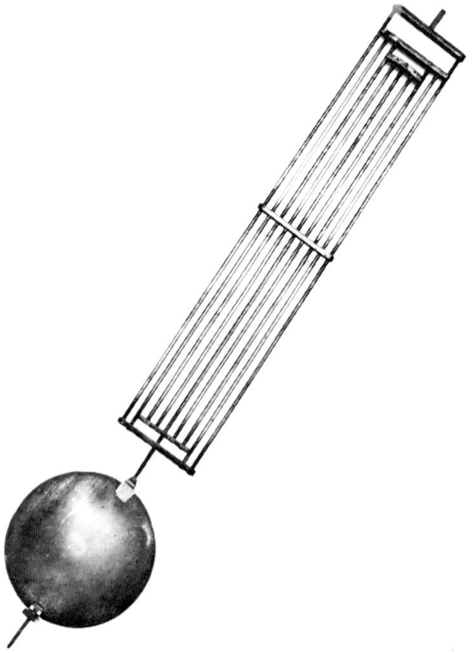
352

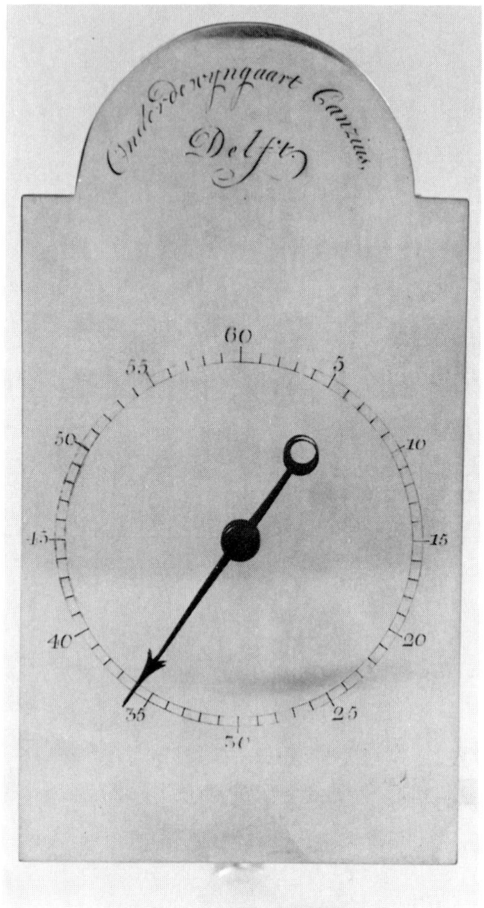

354

354

Dial size (arch included) 175 mm x 80 mm.
Plain dial with a single seconds dial and hand, pendulum with dead beat anchor escapement with direct drive to the hand; plated movement, knifeedge suspension.
Journeyman or 'slave' clocks which normally had a fairly loud tick were used by astronomers for counting seconds during observations.
Universiteitsmuseum, Utrecht UM 408.
Bibliography: Rooseboom, 4042 (for Canzius).

355 Regulator, mid 19th century
Signed 'A. Kaiser Hage No 34'.
Oak, mahogany, silvered brass, brass and steel.
Height 1 m 010, width 250 mm, depth 155 mm.
Plain case glazed on three sides. Dead-beat escapement, separated pendulum with spring suspension, temperature compensated, pulley suspension of weight with catgut line, maintaining power, duration 8 days.
Nederlands Goud-, Zilver- en Klokkenmuseum, Schoonhoven WA 9 94.
Bibliography: Erbrich, 30, pl. 50; Morpurgo, 69.

356 The Self-regulation of Time in the mid-19th century
Contents of the Utrecht Observatory transit room in 1853.
At a given time a known star will cross the meridian. When a star is on the meridian its right ascension is equal to sidereal time. If then the exact clock time of the transit is determined the right ascension being known, the error of the clock can be determined. Such self-regulation lay at the heart of the 19th century time services.
The main contents of the room are the fixed transit, a large portable transit, a regulator by 'F. Knobel Amsterdam', a regulator by 'Robert Molyneux, London', and a marine chronometer by 'A Lange & Sohne Glashutte b. Dresen, 48'.
Universiteitsmuseum, Utrecht.

357 A Complex Astronomical Clock, 1930-39
Signed 'Pouvillon Nogent sur Oise'.
Brass, silvered brass and steel with a blackened wood pedestal.
Overall height 1 m 474; 540 x 460 mm.
Clock of skeleton construction with dials on three of its four faces. The main dial at the front indicates the legal hour, minute and second. Above this true solar time is indicated on a mystery dial using Houdin's system (see no 294). On the right side of the clock are six concentric dials providing all the information necessary for computing the date of easter, viz the epact, dominical letter, indiction, golden number, day of the month of the coming 1st January, and solar cycle. These elements are mechanically summed to allow the Easter date to appear in an aperture in the centre dial. Above this is a 24 hour dial showing the time of sunrise and sunset by means of the position of two wings read against the dial graduations.
Above this are three further dials, added in 1948, indicating the month, season and sign of the zodiac. On the left side of the clock is a map on glass of the circumpolar constellations which rotates once per year. Above this is an indication of the age and phase of the moon, and above this three further dials, also added in 1948, for astrological aspects. The weight-driven movement with pendulum is laid out in three levels with the barrel

355

and bell in the lowest. The whole clock is surmounted by the planetarium with inclined path for the moon and provision for the precession of the equinoxes.
Paul Pouvillon (1878-1969) was a local clockmaker and repairer of high repute in the region of Nogent. Inspired by a close examination of A.L. Verité's astronomical clock at Beauvais Cathedral (1865-68) which was restored by one of Pouvillon's friends in 1929/30, Pouvillon conceived the plan of the clock which he built for his own recreation during the following decade. Exhibited first at Beauvais in 1939 and then in Paris where it gained him a silver medal, Pouvillon was named 'Meilleur ouvrier de France', and in 1942 in his turn worked on the Beauvais cathedral clock damaged by bomb blasts in the vicinity. In 1948 Pouvillon was named Chevalier of the Légion d'Honneur.
Following Pouvillon's death and that of his wife three days later, his belongings were dispersed. The planetary clock was eventually given, in pieces, to the charitable clothing trust the Compagnons d'Emmaus, whence it was saved and rebuilt.
Private Collection.
Bibliography : Miclet ; Roberts.

356 a

356 b

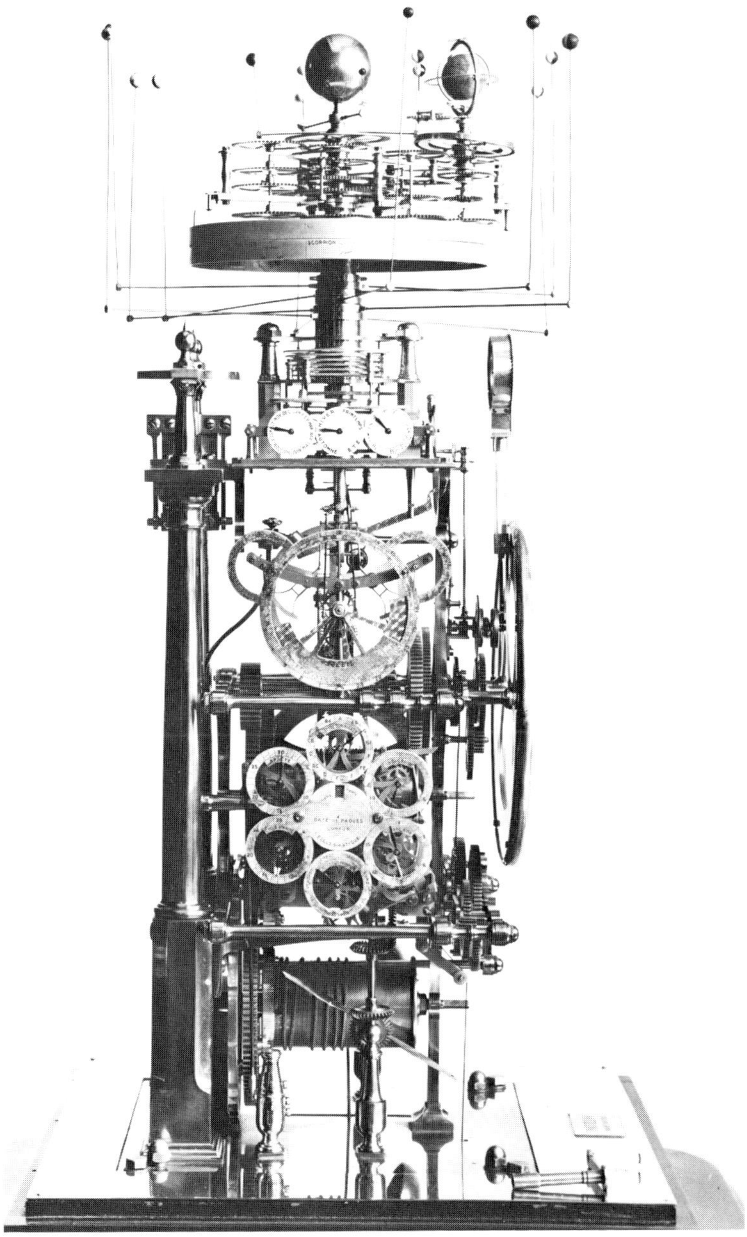
357

357

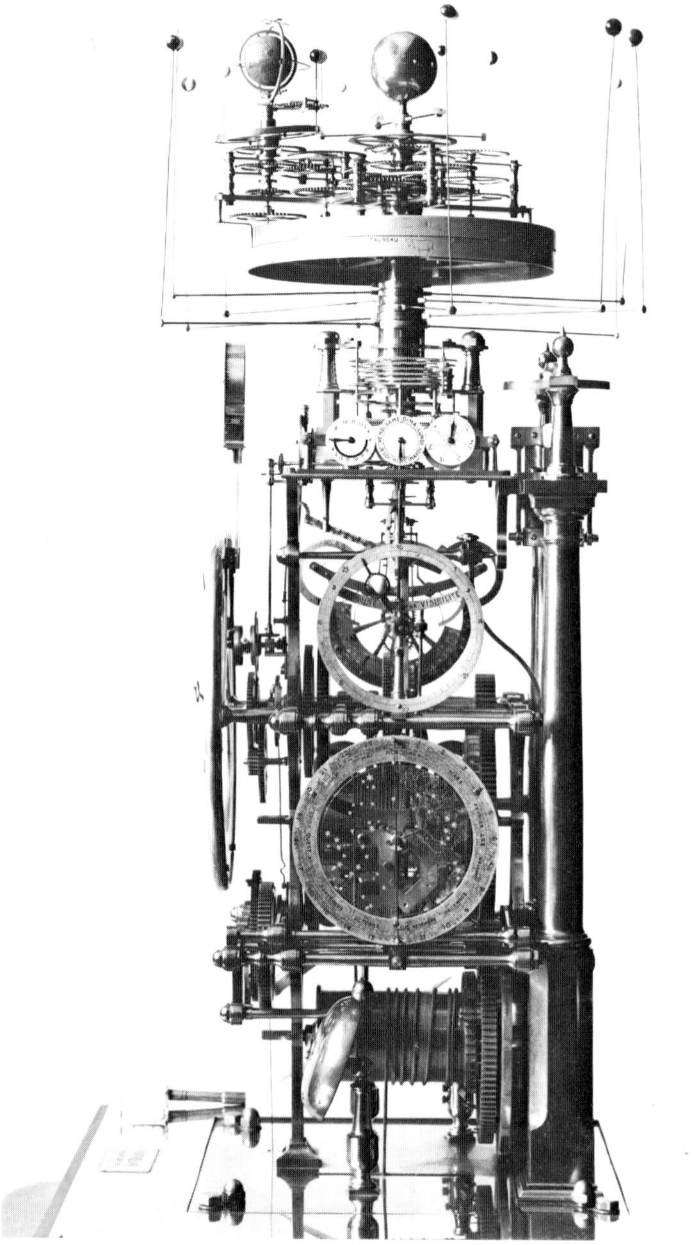
357

Precise Clocks and the Rotation of the Earth

Time, which measures everything in our idea, and is often deficiënt to our schemes, is to nature endless and as nothing.
James Hutton, *Theory of the Earth,* 1788.

A well-adjusted, precise clock, beats out a time which is invariable, each interval is exactly the same as the next. Because the Earth's axis is tilted, and its orbit around the sun is not circular, the length of a true solar day varies throughout the year.
Before the eighteenth century hardly any clock was accurate enough for this difference to be noticed.
With the introduction of the pendulum and anchor escapement however, the discrepancy became obvious. Now there were different kinds of time, natural solar time, which was variable, and clock time which was invariable.
Since clocks had still to be set by sun-dials and tables were drawn up to allow compensation for the difference.

358 An Equation Table for Use in Longcase Clocks, 1683
Photograph.
A Table of the Equation of Days... Printed for Tho. Tompion, Clock-maker,... 1683.
Equation tables of this kind were printed for, and supplied with their clocks by leading clock-makers from the later 17th century onwards. Very frequently such tables were pasted to the inside of the trunk door.
Private Collection.

359 An Instrument for Finding the Equation of Time for Any Date, 1816
Signed 'De la Combe Inv an 1816' and inscribed on the 'M. Lepaute horloger du Roi et de la Ville rue St Honoré No 24'.
Gilt brass, glass and steel with leather cover to the back of the case goldtooled with the ownership inscription, diameter 130 mm.
The dial is marked with a zodiacal calendar, a degree scale and an equation scale noon of which coincides with mean-time noon.
There are two hands adjusted by a winding button in the pendant. When the longer hand is set to a given date (and thus to the sun's place in the zodiac and degree) the smaller hand indicates on the equation scale how many minutes solar time is fast or slow in relation to mean-time.
Although invention of this device is claimed by the inscription 'De La Combe' placed on the dial, such devices existed earlier. One such was described by Hubert Sarton *Description abrègée de plusieurs pièces d'Horlogerie...* Liège, *c.* 1790 (see Baillie, 335) and another, though different, system was proposed by Du Fay in 1727 (Baillie, 171). Of De La Combe nothing else is know than a decimal watch signed by him (catalogue CNAM, 227), but an equation dial strikingly similar to his was made by the Breguet company, *c.* 1850/60 (Maddison & Turner W66).
Provenance : If the Lepaute whose name appears on the case belonged to the clockmaking dynasty as seems probable then he is either PierreBasile Lepaute (1750-1843) or, as is more likely, J.J. Lepaute (1768-post 1830).
Edouard Gelis (1876-1955).
Musée du Temps et de l'Histoire, Besançon 537.1791 & 52.1.17.

But many men did not wish to follow the artificial time of the clock or watch. Solar time, it was felt was the only true time.
Throughout the eighteenth century clocks were made which showed both mean and true solar time. Clocks were also made incorporating an equation dial. A simpler solution was to reset one's timepiece each day.

360 A 'Regulateur' with Equation Dial, c. 1720
Signed 'Baillon à Paris'.
Black lacquered wood case with gilt bronze mountings, brass and steel.
Overall height approx. 1 m 850.
The going train of the clock is weight driven, but the striking mechanism is spring driven.
The equation dial around the outside of the circular dial is manually adjusted.
Private Collection.

361 A Very Small Mantle Regulator with Equation, c. 1780
Signed 'Robin à Paris'.
Gilt bronze, brass glass and steel with enamelled dial.
Height 300 mm.
Rectangular case glazed on all sides with fluted borders and relief chased swags and laurel corner pieces to the dial. Convex dial with concentric scales for hours minutes and the equation. Shaped circular plated movement (also signed), going train unusually with fusee and train, deadbeat anchor escapement with adjustable depthing for the pallets and gridiron pendulum with knife-edge suspension adapted for rigid pendulum construction and compensating for use of the clock on sloping surfaces ; micrometer beat-setting. Going barrel for the striking train with external pierced locking-plate. Complex equation and calendar train.
Private Collection.
Bibliography : For three similar, but larger, clocks by Robin (Tardy II, 138-41).

The increase in accuracy of clocks and watches produced a corresponding improvement in sun-dials. Some sun-dials were engraved with an equation scale.
Greater accuracy could be obtained by

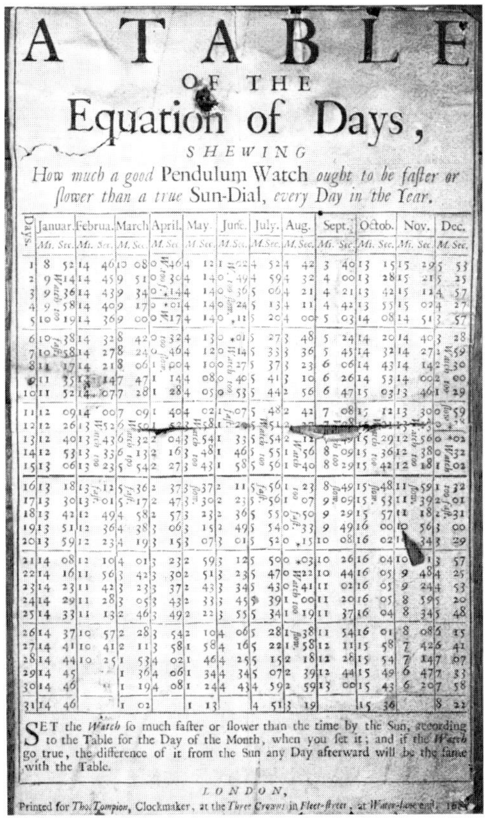
A TABLE OF THE Equation of Days, SHEWING How much a good Pendulum Watch ought to be faster or slower than a true Sun-Dial, every Day in the Year.

Days.	Januar. M. Sec.	Februa. M. Sec.	March M. Sec.	April. M. Sec.	May. M. Sec.	June. M. Sec.	July. M. Sec.	Aug. M. Sec.	Sept. M. Sec.	Octob. M. Sec.	Nov. M. Sec.	Dec. M. Sec.
1	8 52	14 46	10 08	0 46	4 12	1 22	4 52	4 42	3 40	13 15	15 29	5 53
2	9 14	14 45	9 51	0 30	4 14	0 49	4 59	4 32	4 00	13 28	15 21	5 25
3	9 36	14 43	9 34	0 14	4 14	0 36	5 06	4 21	4 21	13 42	15 11	4 57
4	9 58	14 40	9 17	0 01	4 14	0 24	5 13	4 11	4 42	13 55	15 02	4 27
5	10 19	14 36	9 00	0 17	4 14	0 11	5 20	4 00	5 03	14 08	14 51	3 57
6	10 38	14 32	8 42	0 32	4 13	0 01	5 27	3 48	5 24	14 20	14 40	3 28
7	10 58	14 27	8 24	0 46	4 12	0 14	5 33	3 36	5 45	14 32	14 27	2 59
8	11 17	14 21	8 06	1 00	4 10	0 27	5 37	3 23	6 06	14 43	14 14	2 30
9	11 35	14 14	7 47	1 14	4 08	0 40	5 41	3 10	6 26	14 53	14 00	2 00
10	11 52	14 07	7 28	1 28	4 05	0 53	5 44	2 56	6 47	15 03	13 46	1 29
11	12 09	14 00	7 09	1 40	4 02	1 07	5 48	2 42	7 08	15 12	13 30	0 59
12	12 26	13 52	6 50	[illegible]	3 58	[illegible]	5 51	[illegible]	[illegible]	[illegible]	[illegible]	[illegible]
13	12 40	13 43	6 32	2 04	3 54	1 33	5 54	2 12	[illegible]	15 29	12 56	0 02
14	12 53	13 33	6 13	2 16	3 48	1 46	5 55	1 56	8 09	15 36	12 38	0 32
15	13 06	13 23	5 54	2 27	3 43	1 58	5 56	1 40	8 29	15 42	12 18	1 02
16	13 18	13 12	5 36	2 37	3 37	2 11	5 56	1 23	8 49	15 48	11 59	1 32
17	13 30	13 01	5 17	2 47	3 30	2 23	5 56	1 07	9 09	15 53	11 39	2 01
18	13 42	12 49	4 58	2 57	3 23	2 36	5 55	0 50	9 29	15 57	11 18	2 31
19	13 51	12 36	4 38	3 06	3 15	2 49	5 54	0 33	9 49	16 00	10 56	3 00
20	13 59	12 23	4 19	3 15	3 07	3 01	5 52	0 15	10 08	16 02	[illegible] 34	3 29
21	14 08	12 10	4 01	3 23	2 59	3 12	5 50	0 03	10 26	16 04	10 [illegible]	3 57
22	14 16	11 56	3 42	3 30	2 51	3 23	5 47	0 22	10 44	16 05	9 48	4 25
23	14 23	11 42	3 23	3 37	2 43	3 34	5 43	0 41	11 02	16 05	9 24	4 53
24	14 29	11 28	3 05	3 43	2 33	3 45	5 39	1 00	11 20	16 05	8 59	5 20
25	14 33	11 13	2 46	3 49	2 22	3 55	5 34	1 19	11 37	16 04	8 34	5 48
26	14 37	10 57	2 28	3 54	2 10	4 06	5 28	1 38	11 54	16 01	8 08	6 15
27	14 41	10 41	2 11	3 58	1 58	4 16	5 22	1 58	12 11	15 58	7 42	6 41
28	14 44	10 25	1 53	4 02	1 46	4 25	5 15	2 18	12 28	15 54	7 14	7 07
29	14 45		1 36	4 06	1 34	4 34	5 07	2 39	12 44	15 49	6 47	7 33
30	14 46		1 19	4 08	1 24	4 43	4 59	2 59	13 00	15 43	6 20	7 58
31	14 46		1 02		1 13		4 51	3 19		15 36		8 22

SET the Watch so much faster or slower than the time by the Sun, according to the Table for the Day of the Month, when you set it; and if the Watch go true, the difference of it from the Sun any Day afterward will be the same with the Table.

LONDON,
Printed for Tho. Tompion, Clockmaker, at the Three Crowns in Fleet-street, at Water-lane end. 168[illegible]

358

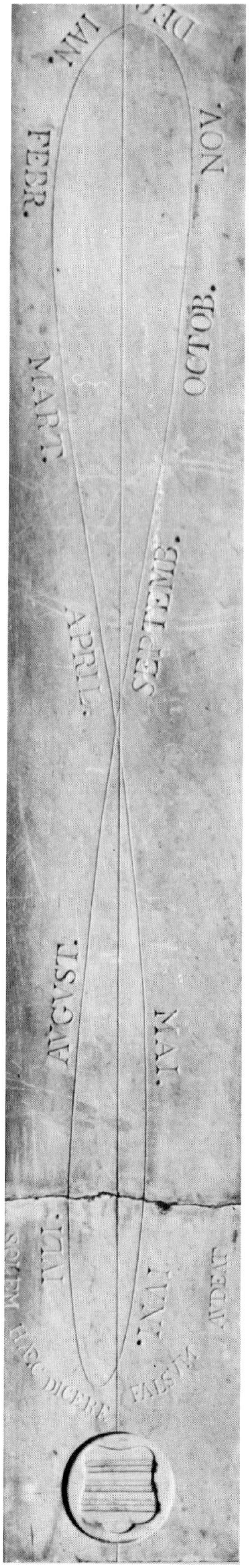

363

constructing large scale dials which marked only the moment of noon. Such meridians are characteristic of late 17th and 18th century Europe.

362 The Meridian of St Peter's Church in Bologna, 1726

E. Zannotti, *La Meridiana del Tempio di San Petronio Rinnovata l'Anno MDCCLXXVI aggiunge la ristampa del libro pubblicate l'Anno 1695 sopra la ristaurazione della Meridiana esoquita dai celebri Matematici Gio: Domenico Cassini e Domenico Guglielmini,* Bologna 1779.

Meridian lines combined with aperture gnomons had been constructed in churches and other large public buildings since the 16th century. That constructed by Cassini in San Petronio between 1653 and 1655 was one of the most famous and accurate in Europe. Revised by Cassini himself with Guglielmini in 1695, it was once more restored in 1726 and 1779, the occasion of the present publication.

Private Collection.

Bibliography : Riccardi.

363 A Stone Meridian, 18th century

Inscribed 'Solem Haec dicere falsvm avdeat'.

Marble.

The dial table of a meridian showing the noon line with the seasonal correction to be applied drawn as a figure of eight around it. A plate with an aperture-gnomon would have been mounted on arm projecting some distance at right angles from the dial.

Provenance : Utrecht University Observatory.

Universiteitsmuseum, Utrecht, UM 962.

364 Another Version of Julien Le Roy's Improved Dial and Meridian, c. 1735

Signed 'Inventé par Julien Le Roy de la Société des Arts fait par Jacques Le Maire de la Société des Arts au Génie A Paris'.

Brass, 240 mm x 170 mm.

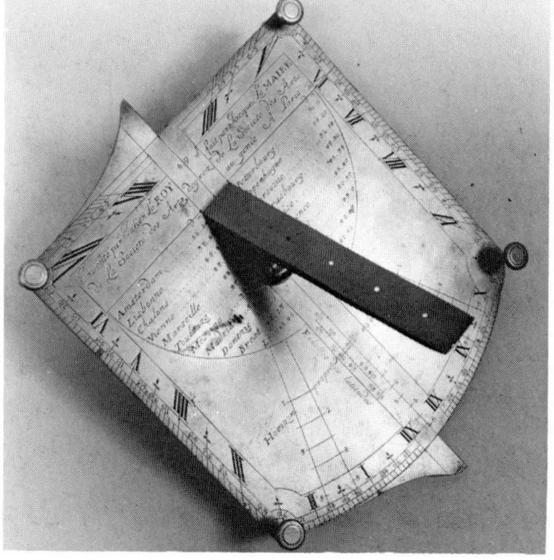

364

Meridians, even though public ones multiplied in 18th century cities, were not always available close to gentlemen's country houses. Le Roy's dial, both portable and easy to set up with accuracy, was intended to supply this want (see number 258).

Private Collection.

365 A Convenient Instrument for Finding Solar Noon, c. 1845

Three dipleidoscopes each signed 'E.I. Dent's patent meridian Instrument 82 Strand & 33 Cockspur St. London' and numbered 142 & 444 respectively.

Cast iron, brass and glass, a) length on side 150 mm, b) 85 mm x 60 mm square, c) 150 mm.

The dipleidoscope consists of a simple optical arrangement for finding the moment of solar noon. Two mirrors are placed at an angle of 60° forming an equilateral triangle with a piece of plain glass in front of them. If the instrument is correctly positioned and an observer take up a predetermined position such that his line of sight is parallel with one of the mirrors and intersects light from the Sun at the surface of the plain glass at 60° then by similar triangles and double reflection, a single image of the sun will be seen. If the observer remains in position, the rotation of the earth will cause a deplacement in the Sun's image such that the angle of entry changes, when a composite or double image will be seen by the observer. If the apparatus is arranged so that Signed 'S. Tracy Rotterdam'.
convenient way of checking the time of a clock or watch is provided, allowance being made for the equation and the difference between local and national time.

Edward J. Dent, a leading London clockmaker, patented the dipleidoscope on 20 June 1843 on behalf of J.H. Bloxam who had devised the optical arrangement and come to an agreement with Dent for manufacture. Instrument a) shown here is Dent's simple version. Instrument b), which lacks a telescope mounted on the bracket to the left of the mirror holder, is his 'mural' model, intended to be fixed in position.

The dipleidoscope was immediately successful, so much that Dent had difficulty in finding sufficient quantities of parallel glass. A booklet that he published in 1843 was reprinted in 1875, and a French translation was produced in 1845 by Lerebours & Secretan who distributed the instrument in France. The two instruments shown were purchased from Secretan by Utrecht Observatory in 1878 although they had been made earlier. Number b) costs 15 florins.

Provenance : Utrecht University Observatory (a & b).

Universiteitsmuseum, Utrecht (a & b); Rijksmuseum 'Nederlands Scheepvaartmuseum', Amsterdam, B191(8).

Foucault Pendulum, 1990

Hanging down from the vault of the Nieuwe Kerk is a 27-metre pendulum. It swings back and forth about once every 10 seconds, slow enough that you can measure the period against your wristwatch. As it swings, Earth rotates underneath it, quickly enough that you can see it knock over the markers on the floor, one after another.

366 The Foucault Pendulum, 1990
Lead sphere with brass mounts, diameter approximately 170 mm. Single-pivot suspension on steel wire.

A Foucault pendulum demonstrates the rotation of Earth. Named after Jean-Bernard Leon Foucault (1819-1868), who created his first pendulum in 1851, the instrument consists of a thin wire (0.6-1.1 mm) which carries a spherical weight of about twenty kilogrammes. The wire and the weight hang from a pivot which is as close to a point-suspension as could be managed. The theory of the instrument is that, once set in motion, inertial will tend to keep the oscillation of the pendulum in one plane. But to an observer the plane of the pendulum seems to change, because Earth is rotating beneath the pendulum. It therefore demonstrates the rotation of Earth. Thus, it demonstrates time. If the pendulum were set up at the pole, Earth would make a complete circle beneath it in 24 hours. If it were set up on the Equator, the rotation axis of Earth would be perpendicular to the pendulum suspension line. In that case there would be no apparent change of the pendulum plane at all. Since Amsterdam is neither at the pole, nor on the Equator, a complete cycle takes longer than 24 hours, because at intermediate latitudes the pendulum experiences an additional centrifugal force due to Earth's rotation. The principle however remains the same. The pegs which have been knocked over by the swinging pendulum indicate the movement of Earth beneath the pendulum since 10 o'clock this morning. According to Foucault's calculations, one complete cycle should take 30 1/2 hours. Because of air resistance, our pendulum slows to a stop in a few hours. See it reset at noon, 2 p.m. and 4 p.m.
Bibliography : Foucault.

Anything periodic can measure time. The Foucault Pendulum combines two periodic motions : the swinging of a pendulum and the rotation of Earth.
Watch the pendulum move : it measures time, in portions of about ten seconds. Inertia makes it tend to move in a plane which is fixed in space. Watch Earth move by the apparent motion of the Sun : it measures time, in portions of 24 hours. The planet is not fixed in space, and rotates

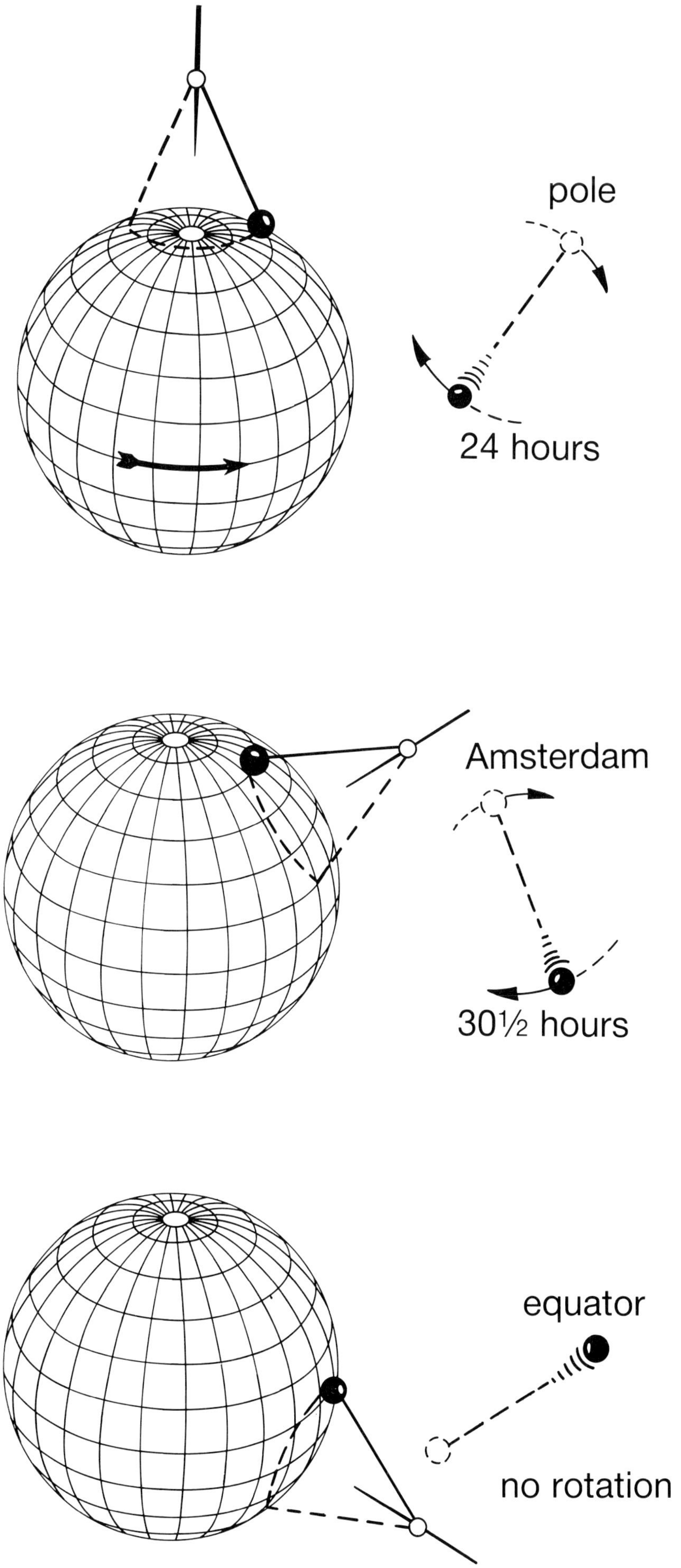

368

underneath the arc of the swinging pendulum.
These motions combined appear to us, fixed on Earth, as a rotation of the plane of the pendulum. Thus, the swinging ball sweeps past the markers set up on the floor, and knocks them down, one by one.

367 Foucault's Original Publication on His Pendulum, 1851
'Demonstration physique du mouvement de rotation de la terre au moyen du pendule', *Comptes Rendus de l'Academie des Sciences,* p.135. Foucault tried several times to perform this experiment, and finally succeeded with a pendulum which was just over twice as long as the one shown here.

368 The Foucault Pendulum at Different Places on the Earth
A scheme of what a pendulum on the North Pole would do; what it would do on the Equator; and what it does in the Nieuwe Kerk at 52 degrees Northern latitude. A pendulum at the Pole has its suspension point exactly on Earth's rotation axis. Therefore, the rotation of our planet cannot communicate itself to the pendulum, which behaves as if it were free in space. A spectator on the Polesees the pendulum rotate once in 24 hours. A pendulum on the equator has its suspension point exactly in Earth's equatorial plane. Therefore, the rotation of our planet produces a force perpendicular to the suspension line of the pendulum, which has no net effect. The plane of the pendulum would not appear to rotate at all. At a point in between pole and equator, there is a net force between the rotating Earth and the swinging pendulum, and the plane of the pendulum rotates with respect to the floor of the church. According to Foucault's calculations, it takes 30 1/2 hours for each rotation.

If the Nieuwe Kerk were located on the North Pole, it would rotate once every 24 hours around an axis passing vertically through the pendulum suspension. In that case, the pendulum would behave as if it were free in space. The church would rotate around it, and a spectator on the church floor would see the pendulum rotate once in 24 hours.
If the Nieuwe Kerk were on the Equator, the rotation axis of Earth would be perpendicular to the suspension point of the pendulum. In that case, the plane of the pendulum would not appear to rotate at all.

At the place where the Nieuwe Kerk actually is, it is somewhere in between these locations (at 52 degrees North latitude, actually). Thus, its behaviour is somewhere in between: the plane of the pendulum rotates with respect to the floor of the church, but a little more slowly than once every day. In fact, according to Foucault's calculations, it takes 30 1/2 hours.

Time and Navigation

Since early periods, time has played a part in navigation. By at least the 10th century a device had been developed for finding time at night.
Refined during the course of the Middle Ages (see a 15th century example incorporated in number 198), the nocturnal came to fully-fledged form in the late 16th/early 17th century.

369 A Navigator's Nocturnal, 17th century
Boxwood, height 229 mm.
The nocturnal enabled the time at night to be established from the changing position of the guards of the great or Little Bear in relation to the Pole Star. The present instrument can be used with either of the Bears and carries calendar and hour scales on its face necessary for setting the instrument.
Private Collection.

A knowledge of time is valuable to coasting vessels for calculating the tides.

370 A Month Calendar and Tide Indicator, c. 1720
Signed 'S. Tracy Rotterdam'.
Brass with blackstained wood case, 206 mm x 180 mm x 110 mm.
The instrument is mounted in a classically styled case with flattened sphere feet and two doric columns on either side. The lower three-quarters of the dial are occupied by a moon and tide volvelle with three scales for the age of the moon, time of the rising tide and time of high water. In the centre is a table showing the days of the week and the planet governing each. Above this are four apertures giving the name of the month, its length and the date upon which the Sun enters a new zodiacal sign. Above these are a further four apertures from which the time of rise and set of the Sun can be read in hours and minutes at 10 day intervals.
Historisch Museum, Rotterdam,
Inv. 50.703.

A means of keeping time is also essential for reckoning the distance a ship has travelled.

371 A Ship's Log-glass, 19th century
Inscribed 'Marine Imperiale No 425.
Wood frame, brass and glass, 69 mm x 69 mm x 137 mm.
30 seconds duration glass used for timing the distance run by a ship against knots in the log line.
Rijksmuseum 'Nederlands Scheepvaartmuseum', Amsterdam, RB 889 (24).

369

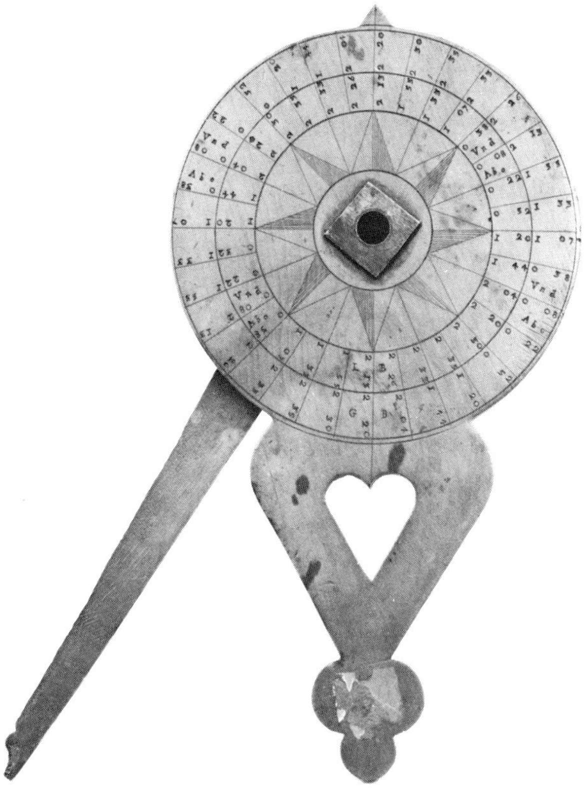

369

370

372

372 A Clock for Use in a Lighthouse, 1828
Signed 'James Clark Edingburgh 1828'.
Brass and silvered brass in a glazed case, 313 mm x 240 mm.
The skeleton movement is contained in a pair of open plates each with three verticals and has three dials. An hour dial (I-XII reading to 12 minutes) is mounted on the longer face with dials for the day of the week and day of the month on each shorter side.
The main-spring barrel is placed below the movement which has a fusee and chain, dead-beat anchor escapement embracing 5 teeth, a brass-clad flat wood rod pendulum and pin wheel transmission to the day of the month dial.
This clock was used in the most northerly lighthouse of Great Britain, that at Cape Wrath. The calendar information was included to overcome the tendency of lighthouse keepers in such remote stations to lose track of the year.
The Commissioners of the Northern Lighthouses and the Royal Museum of Scotland, T 1974-L10.

Longitude : The Problem
Longitude is position east or west of any given point on the earth's surface.
A first problem is that there is no obvious place from which to measure longitude.
Until the prime meridian of Greenwich was internationally accepted (in 1882) as having 0° longitude, different meridians were used on maps and charts by different nations.
A second problem is that longitude can only be measured by a comparison of the time at the place having zero longitude and the local time at the place of an observer.
Since the earth turns on its axis every 24 hours, successive meridians will pass through a given place at regular intervals. If meridians are imagined for each degree of the earth's circumference, then every 4 minutes a new meridian will pass through a given place, and so 15 in the course of an hour. Thus, if at point A on the Earth's surface it is midnight, at point B 45° to the West of it, it will be 9 p.m. in the evening, and at point C 45° to the east, it will be 2 a.m. in the morning.

Longitude
Diagram by David Penney.

From this two ways of determining longitude are possible. For if it is known that at a given time at point A a particular celestial event will take place, and the time when the same event takes place at point B can be measured, the difference of the two times may be converted into degrees and hence into the distance between the two places.
If a clock was set going at point A, kept going and not adjusted for the change of meridian when removed to point B, then a comparison of the time shown by this clock with the time determined locally from astronomical observation would equally enable the distance apart of the two places to be found.
In either case an accurate way of finding local time was also needed.

373 An Instrument for Altitude Measurements: the Mariner's Astrolabe, 3rd quarter of the 18th century
Signed 'Dollond London'.
Brass, 152 mm diameter, 11 mm thick, weight 368 gms with original fitted oak case.
The limb is divided into four quadrants of degrees and reads to 2 minutes of arc against a vernier at both ends of the alidade which has pin hole sights with cross-threads.
The mariner's astrolabe was developed out of astronomers' altitude measuring circles in the late 15 century. It remained in use until the early to mid 18th century. The example shown, which has clearly never been used, is remarkably late. Such instruments played a part in time determination when altitude observations were required for finding local time.
Nederlands Goud-, Zilver- en Klokkenmuseum, Schoonhoven
Bibliography : Stimson, 154-6.

In 1514 Johann Werner proposed the lunar distance method. The distance of the Moon from the Sun or a star was carefully measured at a known local time. The result was then compared with the time of the same event at the prime meridian, a value that could be found from tables. The difference could then be converted into degrees of difference of longitude.
Despite the theoretical simplicity of the method, in practice it required exact instruments and very careful observations, and lengthy calculations.
'In our century we see certain small and ingeniously constructed time-pieces being made which, because of their exiguous size, are a very small burden to the wayfarer'.
With these words, Gemma Frisius introduced the first description of the use of a timepiece for finding longitude in 1530. As he realised himself however, there were serious technical problems and little progress was made along these lines for nearly a century.
In 1660 Huygens' workmen completed the first machine specifically designed for use at sea.

374 Huygens' Sea-clock, 1660
Reconstruction by J. IJzerdraat, 1935.
Huygens' machine uses a horizontal verge and double string pendulum the suspension wrapping around cycloidal cheeks. The spring drive was exercised through a small weight-run remontoire.

Rijksmuseum 'Nederlands Scheepvaart-museum', Amsterdam, S1399.
Bibliography : Gould, 27-30.

Despite some initial success, Huygen's device, of which two examples were made and tested, was ill-adapted to sea use as were other mechanisms proposed.
In 1714 the British government proposed the enormous reward of £ 20,000 for a successful solution to the longitude problem.

375 **The Longitude Act, 1714**
12 Anne 12 November. 'An Act for providing a Publick Reward for such Person or Persons as shall Discover the Longitude at Sea'.
Original printed version of the famous act which gave an enormous stimule to the search for a solution. The gravity of the problem is indicated by the enormous size of the proposed reward of 20,000 equivalent to perhaps one and quarter million pounds of today's money or to the price of 13 or 14 300 ton ships in contemporary terms.
Collection David Penney.

The act stimulated many clockmakers to design new machines. Among the most promising were those of Henry Sully.

376 **Henry Sully's Marine Time-piece, 1724**
Photograph
The time-piece which is housed in a burr walnut case is signed 'Henricus Sully invenit et fecit' and the silvered dial is conventional with a separate upper centre seconds dial. The escapement is an adaptation of a frictional rest escapement devised by the French clock-maker Debaufre. It has two semi-circular pallets one above the other and a single escape wheel. The balance wheel is set vertically, its horizontal staff being supported on anti-friction rollers of Sully's design. From the end of the balance a staff behind the back plate near cycloidal checks were attached acting on a silk cord attached to a horizontal pendulum at the bottom of the clock. Motion of the balance thus caused the pendulum to rise and fall isochronism being assured by the checks. Problems of temperature change on a balance spring were thus avoided and Sully believed, perturbation from the ships movement.
Photograph : The Clock-maker's Company, London.
Bibliography : Gould, 35-8 ; Clutton & Daniels, 97-8.

377 **Sully's Description of his Marine Timepiece, 1726**
Description abrégée d'une horloge d'une nouvelle invention pour la juste mesure du temps sur mer...,
Paris, 1726, 4u
Detailed description with criticisms from George Graham. Sully's replies, and details of tests carried out on the time-piece.
Private Collection.

Sully's work was interrupted by his premature death in 1728. By this time however John Harrison had begun a marine timekeeper. After six years of work he completed his first machine.

378 **John Harrison's First Marine Time-keeper, 1735**
An exact facsimile reconstruction
Brass and lignum, 1.000 mm x 770 mm x 620 mm.
Harrison built this machine between 1729 and 1735. It uses an escapement of his own invention, the 'grass-hopper'. The pendulum is provided by two large dumbbell type arms mounted on antifriction rollers and linked by crossed wires. The influence of motion on either of the balance arms is thus compensated by the behaviour of the other. Drive by two springs with maintaining power, and there is gridiron temperature compensation.
National Maritime Museum, Greenwich.
Bibliography : Gould, 43-45.

379 **A Portrait of John Harrison, 1768.**
Mezzotint by P.L. Tassaert after T. King
400 mm x 300 mm.
Harrison is shown seated at a table indicating one of his diagrams on a sheet of paper held in place by one of his marine watches. In the background to the left is H3 in its outer case, and to the right the compensated pendulum of an unidentified work.
Collection David Penney.

Encouraged by its success, Harrison set about constructing a series of improved and smaller machines.

380 **Harrison's 2nd, 3rd, 4th and 5th Marine Time-keepers**
Photographs
National Maritime Museum, Greenwich.

After sea trials, Harrison's 4th machine was accepted as meeting the requirements of the 1714 Act on 9 February 1765, although the full amount was not paid to Harrison until 1772.

373

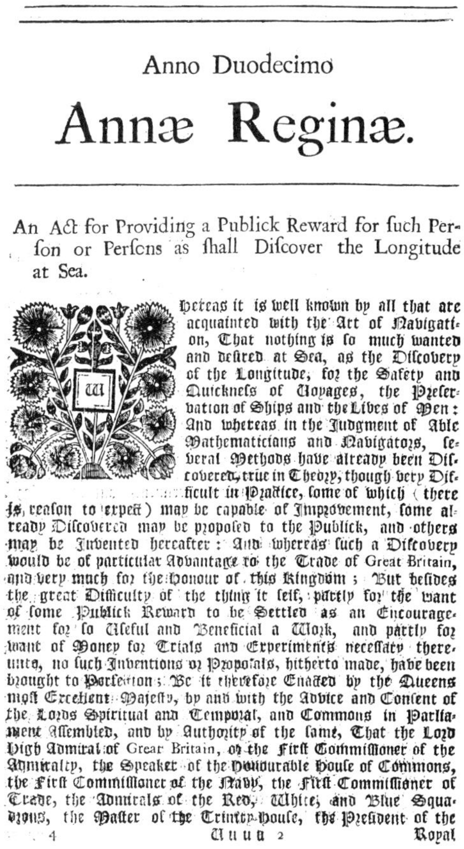

Anno Duodecimo

Annæ Reginæ.

An Act for Providing a Publick Reward for ſuch Perſon or Perſons as ſhall Diſcover the Longitude at Sea.

Whereas it is well known by all that are acquainted with the Art of Navigation, That nothing is ſo much wanted and deſired at Sea, as the Diſcovery of the Longitude, for the Safety and Quickneſs of Voyages, the Preſervation of Ships and the Lives of Men: And whereas in the Judgment of Able Mathematicians and Navigators, ſeveral Methods have already been Diſcovered, true in Theory, though very Difficult in Practice, ſome of which (there is reaſon to expect) may be capable of Improvement, ſome already Diſcovered may be propoſed to the Publick, and others may be Invented hereafter: And whereas ſuch a Diſcovery would be of particular Advantage to the Trade of Great Britain, and very much for the Honour of this Kingdom; But beſides the great Difficulty of the thing it ſelf, partly for the want of ſome Publick Reward to be Settled as an Encouragement for ſo Uſeful and Beneficial a Work, and partly for want of Money for Trials and Experiments neceſſary thereunto, no ſuch Inventions or Propoſals, hitherto made, have been brought to Perfection; Be it therefore Enacted by the Queens moſt Excellent Majeſty, by and with the Advice and Conſent of the Lords Spiritual and Temporal, and Commons in Parliament Aſſembled, and by Authority of the ſame, That the Lord High Admiral of Great Britain, or the Firſt Commiſſioner of the Admiralty, the Speaker of the Honourable Houſe of Commons, the Firſt Commiſſioner of the Navy, the Firſt Commiſſioner of Trade, the Admirals of the Red, White, and Blue Squadrons, the Maſter of the Trinity-Houſe, the Preſident of the

4 Uuuu 2 Royal

375

385a

387

381 **Publication of Harrison's Time-keeper, 1767**
The Principles of Mr Harrison's Time-keeper, with plates of the same. Published by order of the Commissioners of Longitude, London, 1767.
Spare sheets of the English edition bound with (as issued) the French translation by Esprit Pezenas together with additional matter by the same describing the trials of the timepiece and Maskelyne's report on it.
Rogers Turner Books Ltd.
Bibliography : Baillie, 272-4.

Although Harrison was the first to demonstrate conclusively that a solution could be found by horological means, other makers were working in parallel with him.
In France, Pierre Le Roy and his arch-rival Ferdinand Berthoud both produced successful time-keepers, Berthoud establishing a virtual monopoly of their manufacture.

382 **Marine Time-piece by Pierre Le Roy, 1766**
Photograph and explanatory diagram.
Photograph : Musée Nationale des Techniques ; diagram Jean Le Bot.

383 **Marine Time-piece No 8 by Ferdinand Berthoud**
Photograph and explanatory diagram.
Photograph : Musée Nationale des Techniques ; diagram Jean Le Bot.

All these products however were far two complex and expensive for general use. Simplification was essential before the world's fleets could benefit from the new invention. The work of simplication was carried out in England essentially by Thomas Mudge, John Arnold and Thomas Earnshaw.
By 1783 Earnshaw had produced an escapement for balance-wheel time-keepers which was to remain substantially unaltered for the next century and a half.

384 **The Action of the Spring-detent Escapement**
Demonstration model by Richard Good.

385 **The Classic Solution : Marine Chronometers of the 19th Century.**
a. Signed 'Arnold & Dent 84 Strand London. No 782', *c.* 1840.
Brass and steel in mahogany box
141 mm x 142 x 146 mm.
Printed instructions for use signed by Hohwü and dated 1852 are pasted to the inside of the lid.
b. Signed 'E & E. Emmanuel Portsmouth', *c.* 1870
Brass and steel in mahogany box, 190 mm.
Emmanuel were nautical suppliers in Portsmouth. This chronometer would have been bought in for resale.
c. Signed 'J.A. Seckel & Zonen Rotterdam No. 331', 1881
Brass and steel in mahogany box, 118 mm (diameter of cover).
d. Signed 'A Hohwü Amsterdam No 641', 1882
Mahogany, brass and steel, 170 mm square.
e. Signed 'A de Casseres Amsterdam No 682', 1887
Brass and steel in mahogany box
185 mm x 185 mm x 200 mm.
The similarity of the four different examples across half a century and two countries will be noted. All run for 56 hours.
a. Teyler's Museum, Haarlem NM11 ; b. Rijksmuseum 'Nederlands Scheepvaartmuseum, Amsterdam, A3950 ; c. and d. MEOB 000335 & 000455.

386 **A Deck Watch or Pocket Chronometer, 1871**
Signed 'T.S. & J.D. Negus New York No 53847'.
Brass and steel in gold case, diameter 51 mm.
Deck watches which usually employed the same spring detent escapement and temperature compensation as a chronometer were accurate watches used for transporting time about the ship or elsewhere, for the chronometers themselves were never disturbed.
The example shown is inscribed 'Presented by the President of the United States to H.O. Wichers, Captain of the Royal Navy of the Netherlands and Inspector of the Corps of Pilots. In acknowledgement of services in saving lives of Captain and Crew of the General Chamberlain wrecked near Texel on the Dutch Coast. Dec 18th 1871'. The President of the U.S. in 1871 was Ulysses S. Grant.
Rijksmuseum 'Nederlands Scheepvaartmuseum', Amsterdam.

387 **An Instrument for an Hydrographic Experiment : the Tropomètre , c. 1900**
Signed 'E. Delepine à St Nicolas d'Aliermont No 1723' and inscribed 'Marine service hydrographique'.
Brass and steel in mahogany box.
Despite the failure of decimal time in civil life, it continued to be used for some

scientific purposes throughout the 19th century. Agitation for an international standardised decimal division in astronomy, navigation and surveying increased in the closing decades of the century. An international conference on the subject having been arranged for late 1900, the French Ministry of Public Instruction set up some trials in order to have factual evidence ready for the discussions. From 1 March to 1 June therefore, the Commandant E. Guyou was at sea, navigating with decimal tables and instruments that had been developed thanks to a competition for manufacturers in 1898. The tropomètre replaced the traditional chronometer which, apart from its graduation, it exactly resembled. The basic unit is the *Grade* (= one hundredth part of a right angle) and the tropomètre shows décagrades on the dial marked 0 - 40, décigrades on the large dial 0 - 100 and milligrades on the subsidiary dial 0 - 100 divided by units of two. The interest of the system is that it gives a reading which is a direct measure of the angle through which the earth has turned in relation to the mean sun. Various conversion calculations involved when this value is measured in sexagesimal hours and minutes could thus be avoided. Guyou and his officers were unanimous on the greater convenience of the system.

Musée de St Nicolas d'Aliermont.

Bibliography : Guyou (1 & 2).

Precision Time in the Modern World and its Distribution

Les secondes maintenant sont fortement et solennellement accentuées, et chacune, en jaillissant de la pendule, dit: 'Je suis la Vie, l'insupportable, l'implacable Vie!'... Oui, le Temps régne ; il a repris sa brutale dictature. Et il me pousse, comme si j'étais un boeuf, avec son double aiguillon 'Et hue donc! bourrique! Sue donc, esclave! Vis donc, damme!'.
Baudelaire

National Time Services

388 Captain Robert Wauchope's Design of a Time Ball, from Edinburgh New Philosophical Journal, viii 1830, 290.
The use of the marine chronometer to find the longitude at sea required a precise knowledge of its rate. Navigators had regularly to work out the precise time, an elaborate process. Around 1830 Robert Wauchope launched a *Plan for ascertaining the rate of chronometers by instantaneous signal.* The device proposed was a time ball. By observing the ball dropping at pre-set times navigators could check their chronometers and thus determine rates on board. Wauchope's time ball was the beginning of modern time distribution.

389 Greenwich Observatory with Time Ball, and its meridian circle.
The first tests with a time ball were carried out in 1829 at Plymouth in England. A few years later, in 1833, a time ball was installed at the Astronomical Observatory at Greenwich.
Bibliography : Weider 325 ; Ambronn, 984.

390 Shepherd's Automatic System of Signalling from The Graphic, 8 August 1885.
England played a leading role in the development of time distribution. Around 1850 the Astronomer Royal, George B. Alry, implemented a plan using an electric clock made by Shepherd and a telegraphic link to drop time balls automatically in various harbours at the same time.

391 Model of a Time Ball, c. 1855
Oak and brass glazed case, brass mast with steel sphere and finial, overall height approx. 300 mm. Although not signed the apparatus was probably made by the Edinburgh clock-maker, Robert Bryson, as a window display for his Princes Street shop. The sphere would be dropped from the top of the mast at 1 o'clock precisely each day. The model would have been linked telegraphically with the observatory as was the City time ball on Calton Hill. Similar time ball models were made by other leading clock-makers such as Frodsham, London. Messrs. Hamilton & Inches, Edinburgh.

392 Leiden Observatory, and its Meridian Room in 1860
Annalen der Sternwarte in Leiden, Kaiser (3), plate 13.
The first time ball operated by Dutch naval officers was installed in 1839 in Batavia (now Djakarta) in Indonesia.
From 1846 onwards time balls were dropped from ships lying in Dutch naval harbours. A national time service was established in 1858. From then on the time determined at Leiden Observatory was distributed by telegraph to regulate clocks in the harbours.

393 Telegraph Connections Between Leiden Observatory and the Naval Harbours 1863
Beschrijving van de in Nederland gebruikelijke telegraaftoestellen, etc., Afbeeldingen, 1863, plate 43.Time signals were sent out from

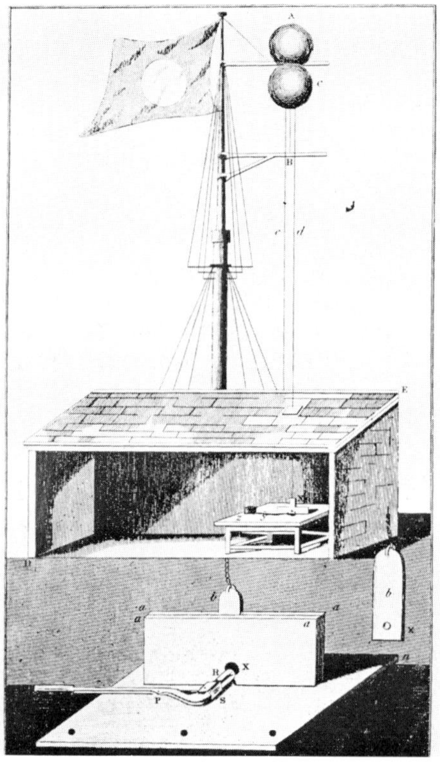

388

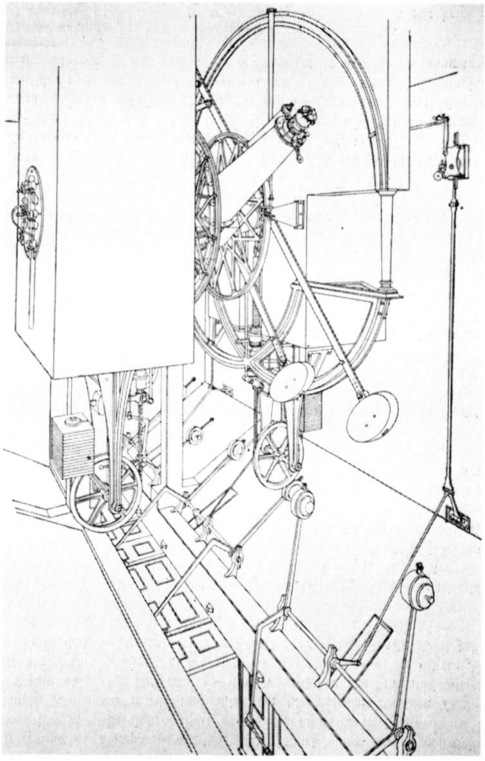

389

389

Leiden Observatory by the regular transmission of signals each Tuesday and Friday and if necessary on other days. Special connections were made, mainly by using the existing telegraph network, between Leiden Observatory and the Dutch harbour of Nieuwe Diep via Amsterdam and between Leiden Observatory and those of Hellevoetsluis and Vlissingen via Rotterdam.

394 Astronomical Clock with Mercury Compensated Pendulum, 1857
Signed : 'Andr Hohwü Amsterdam Nr. 15'.
Overall 1 m, 425 x 370 mm x 220 mm.
The Amsterdam clock-maker Andreas Hohwé earned his living mostly by constructing chronometers, but he also had a good reputation as a maker of precision astronomical clocks. The present clock was ordered in 1837 by the Dutch Navy. The initial purpose was to serve as a time keeper for time signals in Vlissingen When later that year F. Kaiser took over Dutch naval instruments this clock was moved to Leiden. There it served as the principal clock for the time service.
Transferred to Vlissingen in 1875, it is now at the KNMI in De Bilt.
Bibliography : Dekker, 37 ; Kaiser (2), 59-79.

395 Transmitting Key, 2nd half of the 19th century
80 mm x 130 mm x 130 mm.
In 1852 the Rijkstelegraaf started the use of this type of key for the transmission of signals. It remained in use until *c.* 1940. At Leiden Observatory the time signals were sent out by means of such simple instrument.
PTT Museum, Den Haag, 21071.
Bibliography : Kaiser (2), 21 ; Beschrijving, plate 7.

396 Alarm Clock (Time Signal Recorder) Used Around 1860 in Dutch Harbours for the Reception of Time Signals from Leiden Observatory
Made by Siemens and Halske.180 mm x 190 mm.
This electric alarm clock was placed in the harbour next to the pendulum clock and was triggered by time signals from Leiden Observatory. The rate of the pendulum clock was determined by the eye and ear method. The bell of the alarm clock is covered with red cloth in order to prevent overtones which could interfere with the verification of the pendulum clock.
Donated in 1948 by the Verificatie van 's Rijks Zeeinstrumenten (now a division of the MEOB) to the PTT.
PTT Museum, Den Haag, 16837.
Bibliography : Kaiser (2) ; Schell, 466.

397 Seconds Counter, c. 1820
Signed : 'F. Knebel, Amsterdam'.
320 mm x 200 mm x 90 mm.
In order to improve the accuracy of his time signals, F. Kaiser developed methods to verify clocks using the principle of the 'nonius of time'. One of his methods used a seconds counter that makes 51 ticks in 50 seconds. By comparing the ticks of such a counter with those of a pendulum clock, a precision is obtained of about 0.02 seconds. Thus, the moment of transmission of the time signals could be very precisely fixed.
Utrecht University Museum, Utrecht, UM 1833.
Bibliography : Kaiser (2), 29-31.

398 Chronometer, c. 1825 in a metal box
Signed : 'F. Knebel Fecit No. 56 Amsterdam'.
New Balance and Spring by Arnold and Dent in 1836.
Diameter of dial plate 80 mm.
The interest of this instrument lies in the adaption, made in 1860, of the seconds counter such that it makes 130 steps in a minute. The purpose is to obtain a higher precision in verifying clocks by applying again the principle of the 'nonius of time'. Kaiser converted some chronometers that were not accurate enough for use at sea into 'observation' watches with such a 130 seconds scale.
Rijksmuseum Nederlands Scheepvaartmuseum, Amsterdam.
A. 5826 (registered by the Dutch navy as TM Mar 38 and Wh Mar 74 respectively).
Bibliography : Kaiser (1), 21.

399 Model of the Signal Apparatus used in Dutch Harbours to Signal the Time at Pre-set Hours for Rating Ship's Chronometers, c. 1850, made by Van der Loo
This is one of two models made by the model-maker of the Dutch navy. It was designed by Kaiser, in 1845. By operating a lever system the circular blades are brought from a vertical to a horizontal position. The apparatus was first installed in 1850 overseas and a few years later also in Holland. Its true size was 100 Amsterdam feet.
Rijksmuseum Amsterdam, afdeling Nederlandse Geschiedenis.
Marinemodellen collectie MC 827.
Bibliography : Dekker, 35.

Electricity and Time Keeping

400 Illustration of the Applicability of Electricity in Horology by Alexander Bain, from Bain, 25, fig. 13
First attempts to apply electricity to horology were undertaken in the first half of the nineteenth century. Among the inventors were C.A. Steinheil in Germany (1839) and Alexander Bain in England

392

392

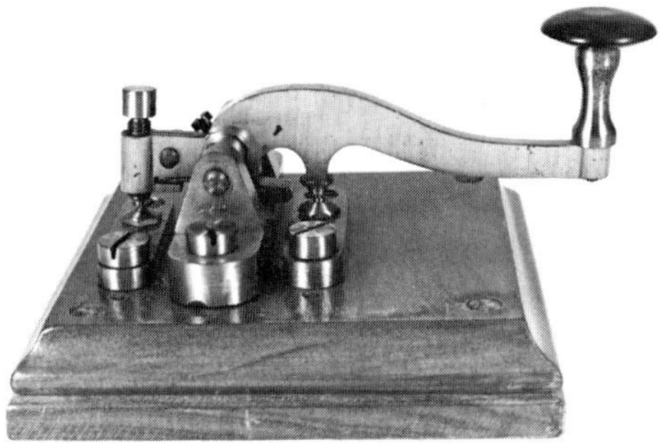

395

394

(1841-5). Bain studied electrically-driven pendulums, searched for a handy electrical power supply for clocks and worked out the use of electricity for the transmission of time over considerable distances.

401 Scheme of the Shortt Free Pendulum, from Hope-Jones, 169
The optimal use of electricity in horology was made in 1921 when Shortt, in collaboration with Hope-Jones, developed their 'free pendulum'. It actually consists of two pendulums which are electrically coupled - by an ingenious 'hit and miss' synchronizer - to make them swing synchronously. One of them, the master pendulum, has a limited task : to swing as unperturbedly as possible and to synchronize the other slave pendulum. The latter does all the work. It drives the gear work by which the number of pendulum swings is counted, it activates the gravity levers every 30 seconds to keep both pendulums in motion and it take scare of the transmission of pulses towards external systems. Shortt free pendulums have been in use in many observatories over the world and were for some decades the most precise clocks available : rates amount to about *c.* 0.0015 seconds per day only.

403 Regulator with Electrically-Driven 1/1 Second Grid Pendulum, 1866
Signed : 'F.C. de Jong Amsterdam'.
Overall 2 m 000 x 460 mm. The Amsterdam clock-maker F.C. de Jong presented this clock in 1866 at an industrial exhibition in Amsterdam in the Paleis voor de Volksvlijt (Palace for the Diligence of the People), being awarded a silver medal. Functioning is that every two seconds a wheel closes a contact, the electromagnets are activated and attract a pin. The impulse of the latter is transferred to the pendulum.
Nederlands Goud-, Zilver- en Klokkenmuseum, Schoonhoven, eau 108.
Bibliography : Electrische tijdaanwijzing, 8-11.

404 Clock with Electrically-Driven 1/2 Second Pendulum, c. 1925
Signed : 'Bulle-clock breveté S.D.G.D. patented'.
350 mm x 200 mm.
The Bulle system was invented in 1920 by Marcel Moulin and M. Favre Bulle in Paris. Its original feature is the arrangement of the permanent magnet with the north pole in the centre and the south poles at either end. When the pendulum swings to the right nothing happens but upon returning and passing the centre position a contact is made. The coil at the end of the pendulum then acts as a magnet with the north pole to its right and a south pole at the left. Consequently, the pendulum is pushed away from the centre and attracted towards the left, in the direction of its momentary motion. The Bulle clock compares with the 1843 design by Bain, but the latter used a very poor contact mechanism. In the Bulle clock one side of the contact is isolated so that contact is made only in one direction of motion.
Nederlands Goud-, Zilver- en Klokkenmuseum, Schoonhoven, eau 115.
Bibliography : Electrische tijdaanwijzing, 29 ; Hope-Jones 712.

405 Scheme of a System with One Master Clock and a Number of Impulse Dial Clocks in Various Forms. From Philpott, 39
The use of telegraphy made it possible to transmit time signals over long distances. Impulse signals sent from a master clock regulate several counting devices to indicate time at various places. Such systems rapidly became popular with the railways, the post offices and factories as only one expensive precision regulator was needed to display the right time on many simple clocks.

406 Master Regulator with Electrically Driven 1/2 Second Pendulum, c. 1925
Used to drive 4 groups of 20 to 25 impulse dial clocks.
Signed: 'Ancienne maison Hipp' and 'PEYER, FAVARGER & Cie NEUCHATEL (Suisse)'.
570 mm x 340 mm x 200 mm.
An iron armature is fixed to the pendulum and an ironcored electromagnet below. By an ingenious contact mechanism, invented in 1860 by Hipp and named by Hope-Jones 'the butterfly escapement', contact is made only when the amplitudes of the pendulum has fallen below a certain value. This is realized by a small, freely swinging vane. The vane trails forwards and backwards over a tiny notched block attached to the pendulum as long as the amplitude is sufficiently great. When the pendulum slows down the vane cannot ride completely over the cam but only reaches the notch and falls into it (see *figure*). When this happens a contact is closed and the pendulum is then sped up.
PTT Museum, Den Haag, Inv. nr. 12916.
Bibliography: Hope-Jones, 62; Hoefer, 92.

407 Master Regulator with Electrically, Driven 1/1 Second Invar Pendulum, c. 1920, made by Brillié of Paris
1330 x 270 mm.
The electrical device driving the pendulum of this master clock consists of a magnet fixed to it which moves through a coil. At the moment that the pendulum reaches a vertical position an electrical circuit through the coil is closed. The resulting magnetic force tries to accelerate the pendulum but is counteracted by the inductive force on the magnet which is generated by its motion through the selenoid. By carefully choosing the current in the electrical circuit the pendulum is kept swinging at a fixed amplitude. If the latter were to decrease and the motion of the magnet to slow down the inductive force decreases. As a result the pendulum speeds up. The amplitude increases until the speed of the magnet results in an inductive force that is just a little less that the magnetic force of the coil. Brillié's system is very close to the one designed in 1912 by Charles Fery, Paris.
Museum voor Onderwijs, Den Haag, 59517.
Bibliography: Hoogeveen, 191198; Hope-Jones, 22.

408 Regulator with Electrically Driven 1/1 Second Pendulum Compensated by Mercury in Low Pressure Glass Tube, 1877.
Signed: 'M. Hipp Neuchatel Suisse. Brevé S.D.G.D.'.
Of all electrical clocks invented in the nineteenth century that of Hipp, described in 1884 by Hirsch, is the only one that could compete in precision with existing weight-driven precision pendulum regulators. The Hipp regulator can maintain a rate of 0.03 seconds per day during several decades. The present clock was purchased in 1887 for Dfl. 2000.- by the Verificatie van Zee-Instrumenten, the division of the Dutch Navy for the verification of chronometers.
The clock's main part is again the pendulum. It consists of two metal bars with a number of cross bars that carry from top to bottom: the pendulum suspension spring, a contact mechanism, an electromagnet and a mercury compensated pendulum bob. The mechanism is in principle the same as designed by Hipp in *c.* 1860 (see nr. 406), but its technical realization is much improved. The low pressure tube is added to prevent disturbances of the motion of the pendulum by variations in atmosphere pressure.
Traditiekamer, Marine Electronisch and Optisch Bedrijf, Oegstgeest, 0003.
Bibliography: Ambronn, 174-5; Erbrich, 21, 22, 236; Hirsch.; Ambronn, 174-5.

409 Masterclock System of 1923 Made by Siemens and Halske in Berlin
One of the two pendulum clocks has an invar pendulum signed: 'Riefler München DRP 100870 23 66'.
2 m 100 x 3 m 400 x 700 mm.
This clock system was used by a large banking company in The Hague. Similar systems were used for municipal time services. The main regulator has a precision pendulum as developed by Riefler. The system is provided with several control mechanisms. One of these governs the synchronous motion of the two pendulums, another controls the minute pulses that are transmitted to the impulse dials that were connected with this system. It also includes a clock for presence registration and a number of electrical control-dials.
Formerly in the building of the Rotterdamse Bankvereniging (now AMRO Bank) on the Kneuterdijk in The Hague. Since 1980 in Museum voor het Onderwijs, Den Haag, Inv. nr. 60283.
Bibliography: Electrische tijdaanwijzing, 32; Erbrich, 10, 253.

410 Impulse Dial Clock, 1855
Signed: 'Froment à Paris'.
380 mm, diameter of the dial plate 225 mm.
Every time an impulse is received (every minute) from the master clock the armature of an electromagnet operates on a crank that turns the wheelwork. The instrument was purchased in 1858 for Dfl. 116.20.
Teyler's Museum, Haarlem, Inv. nr. 674.

396

397

399

399

411 Impulse Dial Clock, c. 1870.
Signed : 'M. Hipp Neuchatel Suisse 3517' 3517'.
170 mm x 90 mm.
The first system of public time was installed in 1864 by Hipp in Switzerland. It used a master clock and a series of *polarized* impulse dials. The present clock is designed to receive an impulse every minute. Because the direction of the current then changes, an armature is brought alternatively into one or another position with the help of an electromagnet. This motion is subsequently transferred to the wheel that counts the minutes. The permanent magnet that magnetizes the armature is here in the form of an S, but Hipp also used the more common horse shoe magnets.
Nederlands Goud-, Zilver- en Klokkenmuseum, Schoonhoven, eau 142.
Bibliography : Electrische tijdaanwijzing, 32 ; Hoogeveen, 122-126.

412 Impulse Dial Clock, c. 1880
Signed on the dial plate: 'M. Hipp Neuchatel'. On the mechanism: 'M. Hipp Neuchatel Suisse no. 12801'.
Diameter 340 mm.
This clock receives an impulse every second. The corresponding gear work, brought into motion by a polarized armature, is adapted to indicate the time in seconds.This impulse dial was used with the Hipp regulator described above (nr. 408).
Traditiekamer, Marine Electronisch and Optisch Bedrijf, Oegstgeest, 002.
Bibliography : Electrische tijdaanwijzing, 32; Hoogeveen, 154.

413 Impulse clock mechanism of c. 1925
Signed: 'Patent Bohmeyer 553'.
220 mm x 95 mm.
The firm Bohmeyer of Halle made polarized impulse dials that operated with an armature that, like Hipp's, moves back and forth each time the direction of the current is changed. This movement is transferred to the wheel by two pallets.
Nederlands Goud-, Zilver- en Klokkenmuseum, Schoonhoven, eau 126.
Bibliography : Electrische tijdaanwijzing, 32.

414 Impulse dial mechanism of c. 1900
Signed : 'C. Theod. Wagner Wiesbaden' and 'H. Grau Cassel DPR 1093'.
1093'.
220 mm x 90 mm.
The system Grau-Wagner was patented in 1880. It employs a double rotating armature in order to transfer the impulses, received every minute, into the motion of the minute dial hand.
Nederlands Goud-, Zilver- en Klokkenmuseum, Schoonhoven, eau 125.
Bibliography : Electrische tijdaanwijzing, 32 ; Hoogeveen, 117-121.

415 Impulse Dial Clock
Signed : 'Electrique Brillé'.
Diameter 300 mm.
The essential element of the Brillié system is an electromagnet in the form of double T. It is enclosed between the poles of an horseshoe magnet. Pulses, now in one, then another direction, cause small oscillations of the electromagnet. With the help of an anchor escapement (of the type Brocot) this motion is converted into that of the minute hand. Since a pulse is received every half minute, the wheel is then turned by half a tooth (see *figure*).
Nederlands Goud-, Zilver- en Klokkenmuseum, Schoonhoven, eau 163.
Bibliography : Electrische tijdaanwijzing 32; Hoogeveen, 137-141.

416 Impulse Dial Mechanism by the E.N.E.M. (Eerste Nederlandsche Electro-Magnetische Uurwerken Fabriek)
Signed : 'Ned. Octr. no 4604' and 'no 2996'.
90 mm x 90 mm.
The armature in the mechanism, used here to translate a pulse into a movement of the minute hand, is executed as a permanent magnet.
Nederlands Goud-, Zilver- en Klokkenmuseum, Schoonhoven, eau 127.
Bibliography : Electrische tijdaanwijzing 32; Hoogeveen, 152-3.

417 Model of the Mechanism of the Impulse Dial System Synchronome
65 mm x 70 mm.
The mechanism employed here is based on the synchronome switch patented in 1897 by Hope-Jones. Through a current pulse the electromagnet is activated and an arm is released. When the current drops a spring pushes the arm back and the wheel advances by one step.
Nederlands Goud-, Zilver- en Klokkenmuseum, Schoonhoven, eau 216.
Bibliography : Electrische tijdaanwijzing, 34; Hope-Jones, 101.

418 The Prototype of W.H. Shortt's Free-Pendulum Regulator with a 'Synchronome' Regulator as its 'Slave', 1921
Brass and copper with oak case for the 'Synchronome' slave, overall height 1 m 530.
To exploit his synchronome switch, Frank Hope-Jones set up the Synchronome Company which successfully installed a large number of time systems. In 1910 he began a collaboration with the Edinburgh civil engineer William Hamilton Shortt to develop an even more accurate regulator. In 1921 Shortt developed a novel method of linking two regulators to provide a near perfect free pendulum clock. Each clock received half minute impulses so that one of them was freed from all influences that would affect its natural swing. The pendulum of the second clock is synchronised to the master pendulum and does all the work of operating the system. The prototype shown here differs in some respects from the production model in particular in being impulsed at the base of the pendulum. The 'slave' clock is a synchronome regulator which has been modified for this special purpose.
Royal Museum of Scotland, T 1950.
10a & 10b.

419 Endless Chain with Electromotor from Hope-Jones, 59
A very useful application of electricity in time-keepers has been the automatic winding of the weight of pendulum clocks. This was and is often applied to church clocks and to precision clocks situated in low pressure housing where the winding-up presents special problems. The system shown here uses the endless chain designed by Christiaan Huygens.

400

403

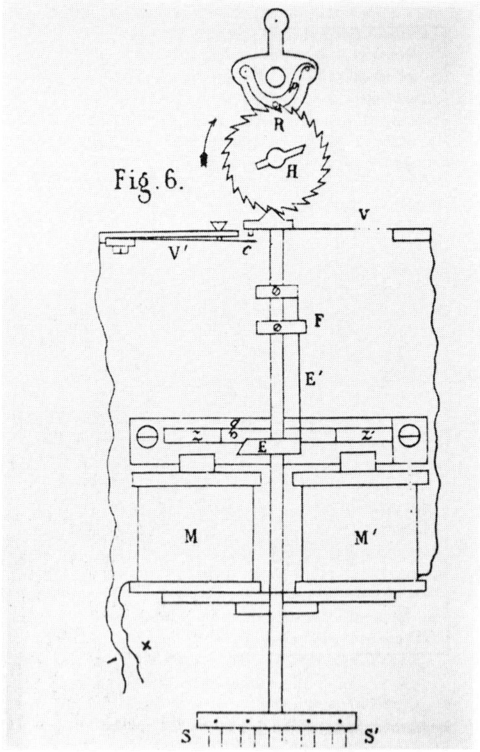

403

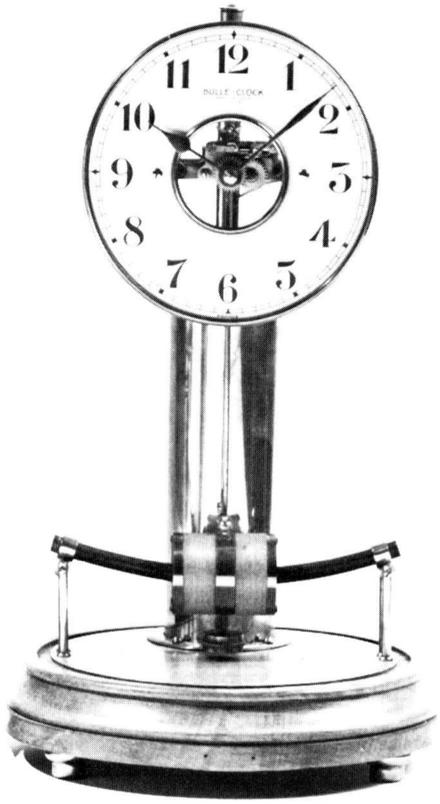

404

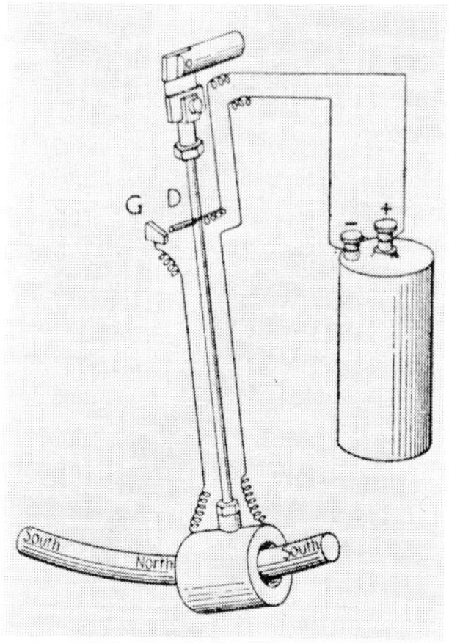

404

406

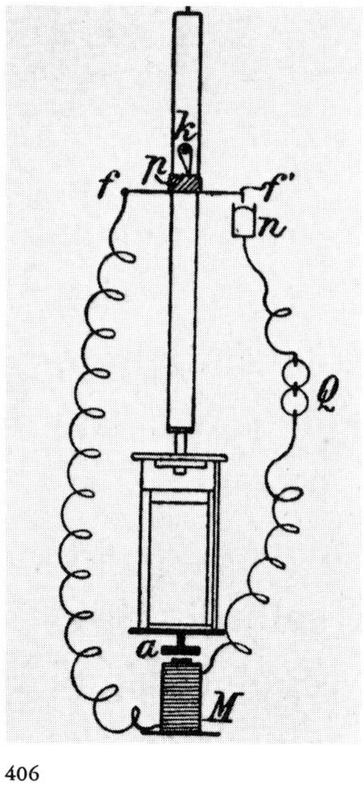

406

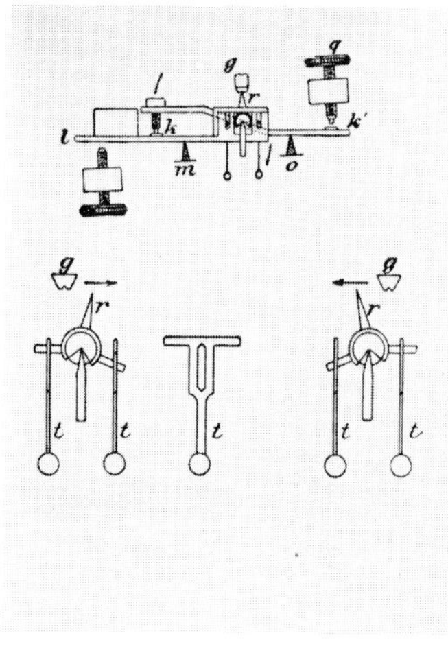

406

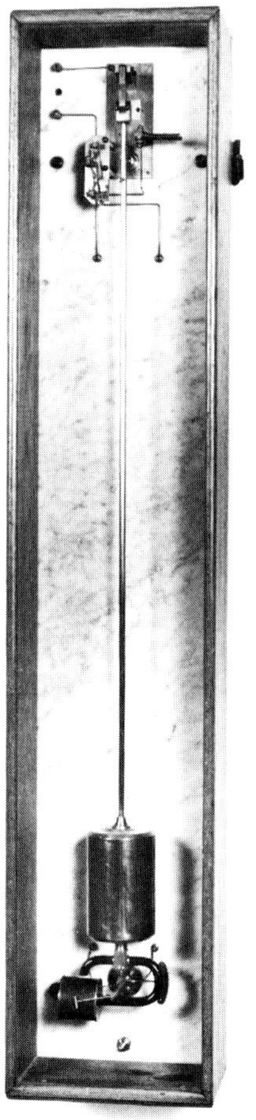

407

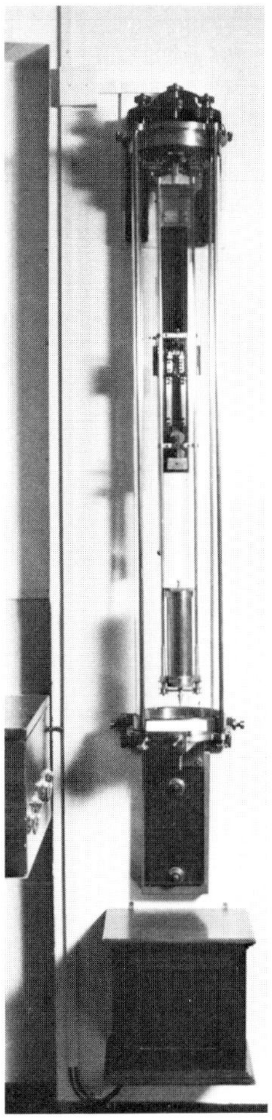

408

409

410

410

411

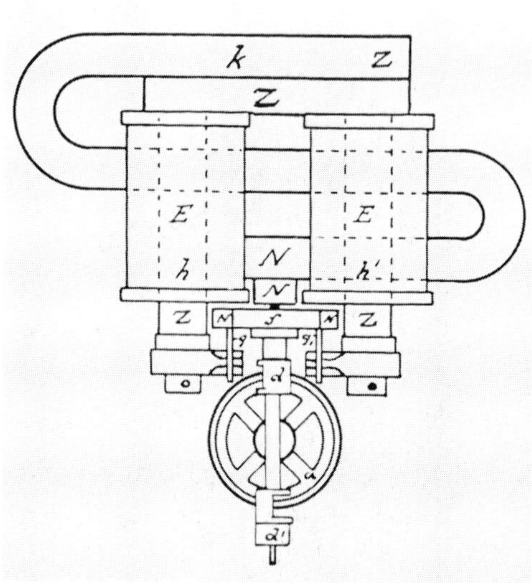

411

412

414 a

413

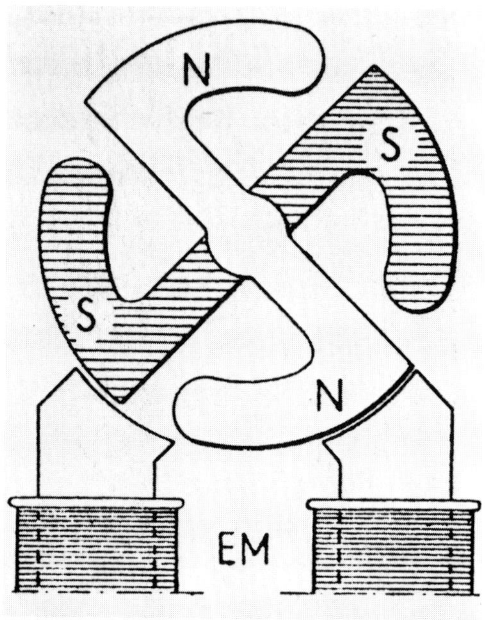

414 b

416

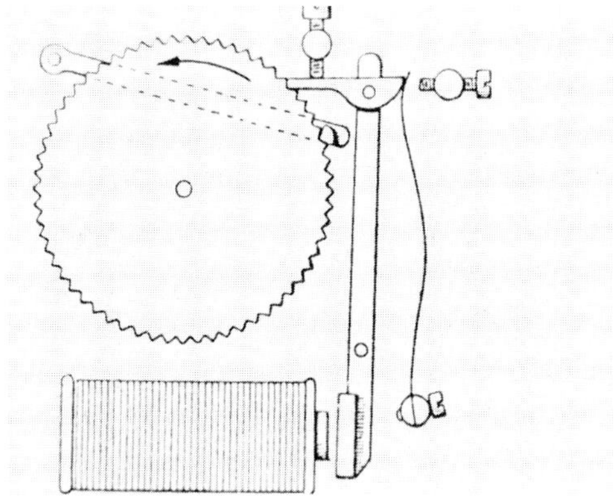

417

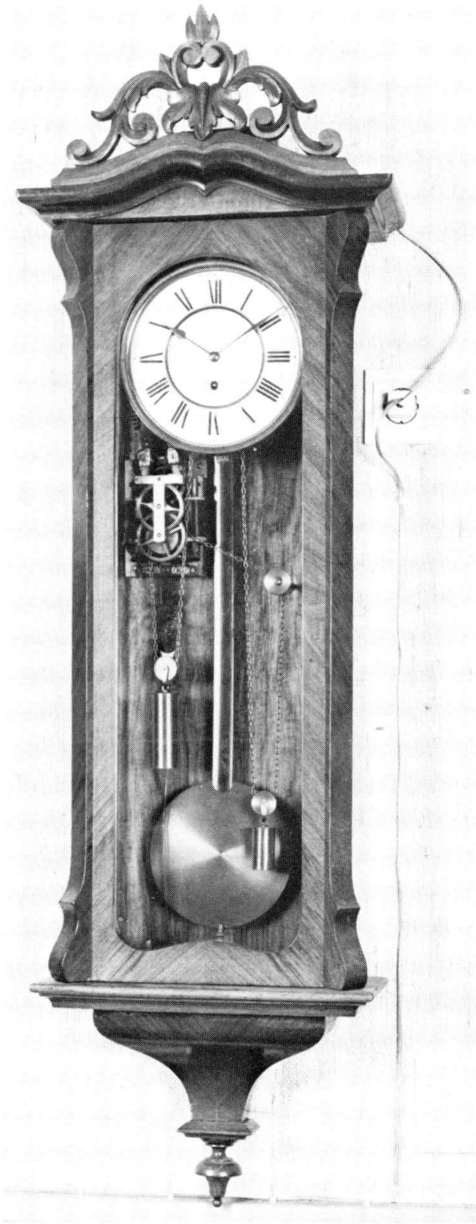

420

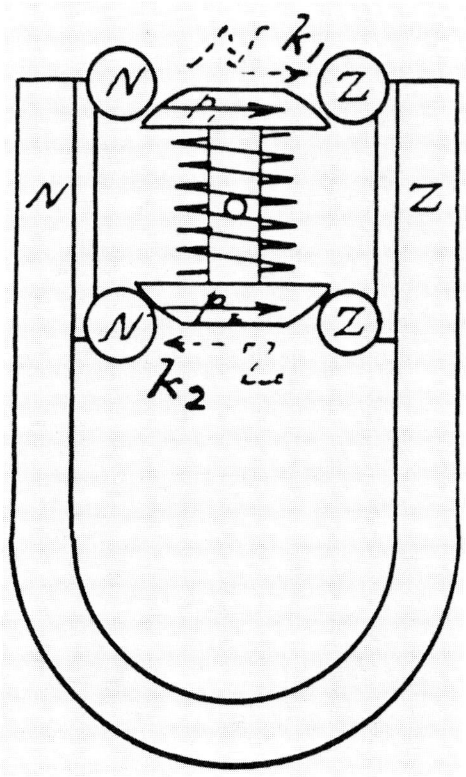

415

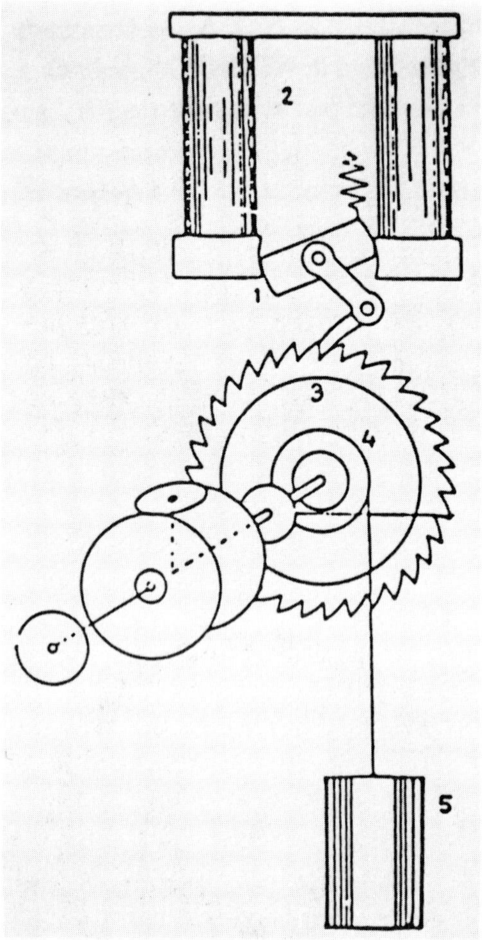

421

420 Regulator with 3/4 Seconds Pendulum and Electrically Wound Weights Made by Sardemann, Amsterdam around 1900
1 m 240 x 380 mm.
The Vereniging Christiaan Huygens proposed, around 1900, a competition for the electrical winding of clocks. The present clock won the first prize. When the weight is below a contact is closed and an electromotor started to wind the clock. When the weight has been sufficiently raised its pulley activates a lever which stops the motor.
Nederlands Goud-, Zilver- en Klokkenmuseum, Schoonhoven, eau 107.
Bibliography: Electrische tijdaanwijzing 29.

421 Master Pendulum with Electrical Winding Made by Siemens and Halske, c. 1920
1 m 150 mm x 28 mm.
The winding mechanism here applied is triggered by a minute pulse. This energizes two coils which turn an armature. By this latter movement a wheel is turned by one tooth and the weight is lifted by precisely amount it was lowered during the preceding the minute.
Museum voor Onderwijs, Den Haag, 48956.
Bibliography: Hoogeveen, 178-179.

422 Synchronous Motor Clock with an Alarm, c. 1930
Signed: 'Telechron USA'.
115 mm x 115 mm.
In 1918 H.E. Warren designed the first self-starting synchronous motor. It served as a frequency meter in generator stations and as the core of clocks produced by the firm Telechron in America. The hands of this kind of clock are driven through suitable gearing by a very small electric motor, the speed of which is regulated by the frequency of the alternating current mains feeding it. In Europe this is 50, in the USA 60 cycles per second.
Museum voor het Onderwijs, Den Haag, 59509.
Bibliography: Hope-Jones, 227.

423 Synchronous Motor Clock, c. 1925
Signed: 'Mauthe made in Germany nr. 291350'.
170 mm x 175 mm.
In the course of the twentieth century many types of synchronous motor clocks were designed. Like Warren's the present clock is self starting. Whether the clock runs can be checked by observing the black and white disk rotating through a little window in the dial plate.
Nederlands Goud-, Zilver- en Klokkenmuseum, Schoonhoven, eau 137.
Bibliography: Electrische tijdaanwijzing, 35; Hoogeveen, 227-8, 233.

424 Synchronous Motor Clock, c. 1925.
Signed: 'Philips'
140 mm x 195 mm.
Once the principle of the synchronous motor clocks was understood, simpler clocks were also designed that had to be started by hand. The present clock is an example of this type available on the market from 1925. The hour scale in the outer ring is continued from 13 tot 24 to make it easy for use with timetables.
Nederlands Goud-, Zilver- en Klokkenmuseum, Schoonhoven, eau 152.
Bibliography: Electrische tijdaanwijzing, 35; Hoogeveen, 223.

425 Synchronous Motor Clock with Calendar, ca. 1940
Signed: 'paauwe's patent'.
390 mm x 375 mm.
In 1937 Pieter Johannes Paauwe was granted a patent for a mechanical calendar which he incorporated in a synchronous clock. In 1937, he founded 'Paauwe's Volautomatische Kalenderuurwerken N.V' in Amsterdam'. Next to the time, the day of the month and the day of the week are indicated. From 1947 on this firm worked under the name NUFA, from 1975 under that of NEDKLOK.
Nederlands Goud-, Zilver- en Klokkenmuseum, Schoonhoven, eau 150.
Bibliography: Electrische tijdaanwijzing, 15, 35.

426 Synchronous Motor Clock Designed by the Nederlandse Uurwerkfabriek Nufa N.V., Amsterdam
The motor used in this clock is signed: 'Made in USA'.
184 mm x 145 mm.
Synchronous motor clocks were designed in many forms to compete on the domestic market. This model, named 'Golden Helmet', was especially designed for export of the United States of America.
Nederlandse Klokkenindustrie B.V. (Nedklok), Amsterdam.
Electrische tijdaanwijzing, 15, 35.

Standard Times

427 Map of the American Time Zone System
The rapid development of railroads in the nineteenth century was the primary incentive for introducing standard time. In small countries like The Netherlands the difference in local time between the most easterly and westerly cities is a matter of minutes. In bigger countries like the United States of America those differences amount to several hours and one standard time for the whole of the country is confusing. In 1870, therefore, the American professor Charles Ferdinand Dowd proposed *A System of National Time for Railroads.* America should

be divided into a set of zones, each of which would have its own standard time. This system was introduced in 1883. Railroad men now only had to change their watches by 1 hour when moving from one zone into another.

428 Map of the International Time Zone System
In 1884 the meridian passing through the centre of the transit instrument at the Observatory of Greenwich was adopted as the initial meridian for the geographical longitude. As a result a standard time for the world, Greenwich Mean Time (GMT), could be defined and used for all kinds of purposes. This time is the basis from which all standard times in the international zone system are derived. The world has been divided into 24 zones with the zero zone centred on the meridian of Greenwich. The time zones east of Greenwich have time later than GMT, and those to the west have earlier time -one hour difference for each zone. Travelling eastward, time becomes later and later until the International Date Line at *c.* 180° from Greenwich is passed, whence the date has become one day later. Travelling towards the west, time is gained and on passing the International Date Line the date changes to the preceding one.

429 World Clock Made by Geochron, ca. 1980
560 mm x 870 mm.
The dial plate of this clock is a map of the world with the international zones marked on it. For each zone the standard time is indicated by a number from 1 to 12. Moreover, the parts in the world where it is day or night are shown. In the course of the day the brightened areas moves to the left. The international date line is also indicated.
Private Collection.
Bibliography: Electrische tijdaanwijzing, 36.

430 World Clock with Day and Night Indication
Signed: 'Smiths Supply the World with TIME'.
Diameter 350 mm.
This synchronous motor clock indicates the standard time for 24 towns throughout the world. One can read directly from the clock whether it is day or night in these towns.
Nederlands Goud-, Zilver- en Klokkenmuseum, Schoonhoven, eau 245.

431 A Universal Geographical Clock, c. 1875
Signed 'César Pascal Breveté S.G.D.G.'
Blackened wood, brass and steel, printed paper dials.
Two cardboard dials turn on either side of the chapter ring which separates them. They

423

424

425

430

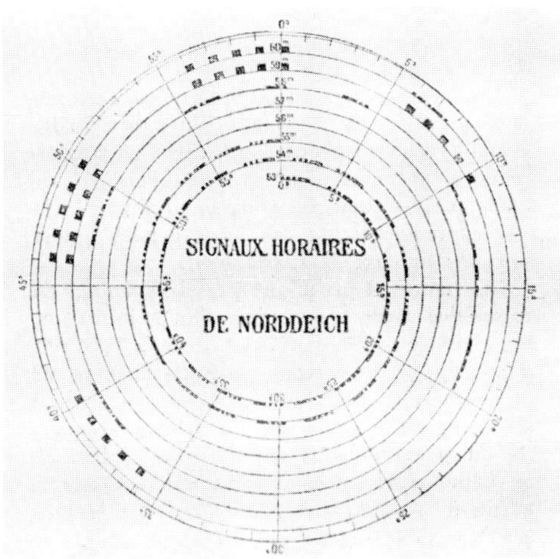

432

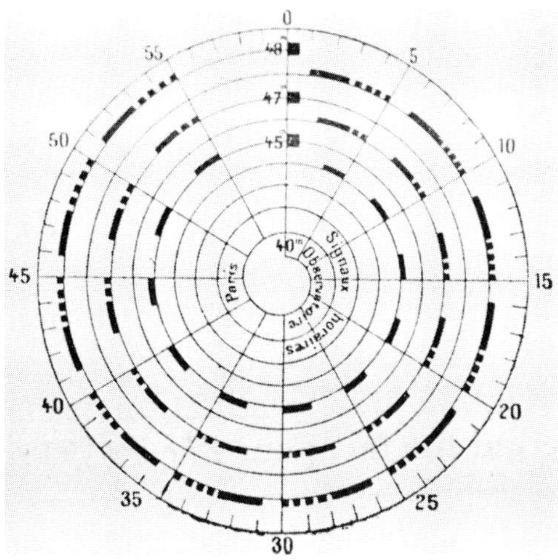
432

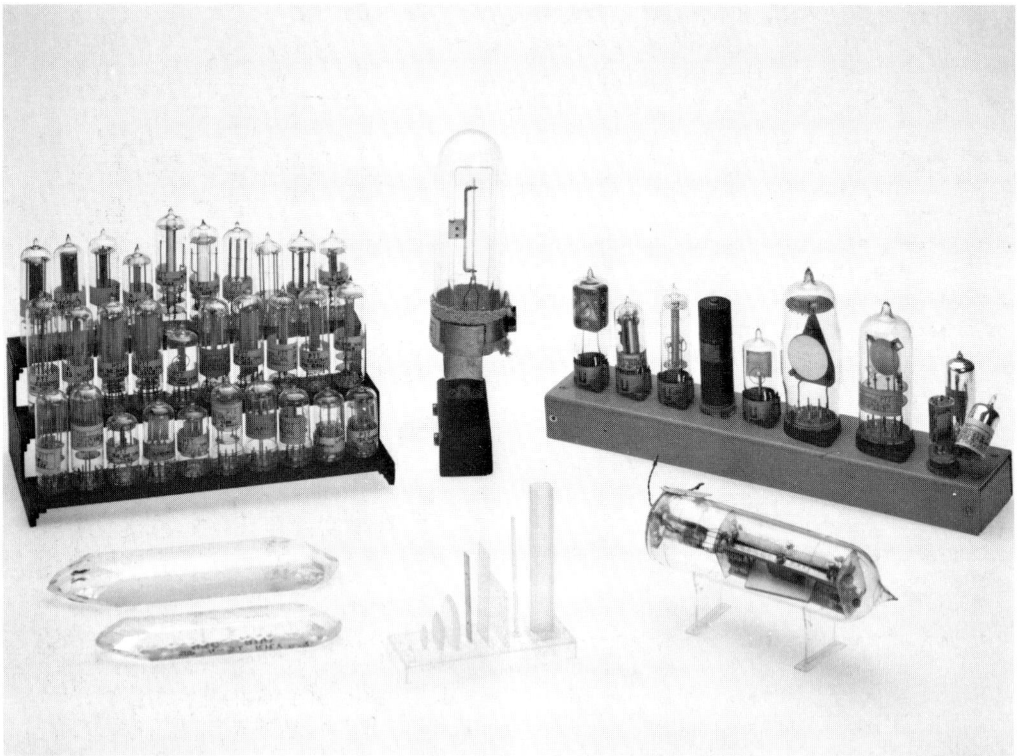

438-40

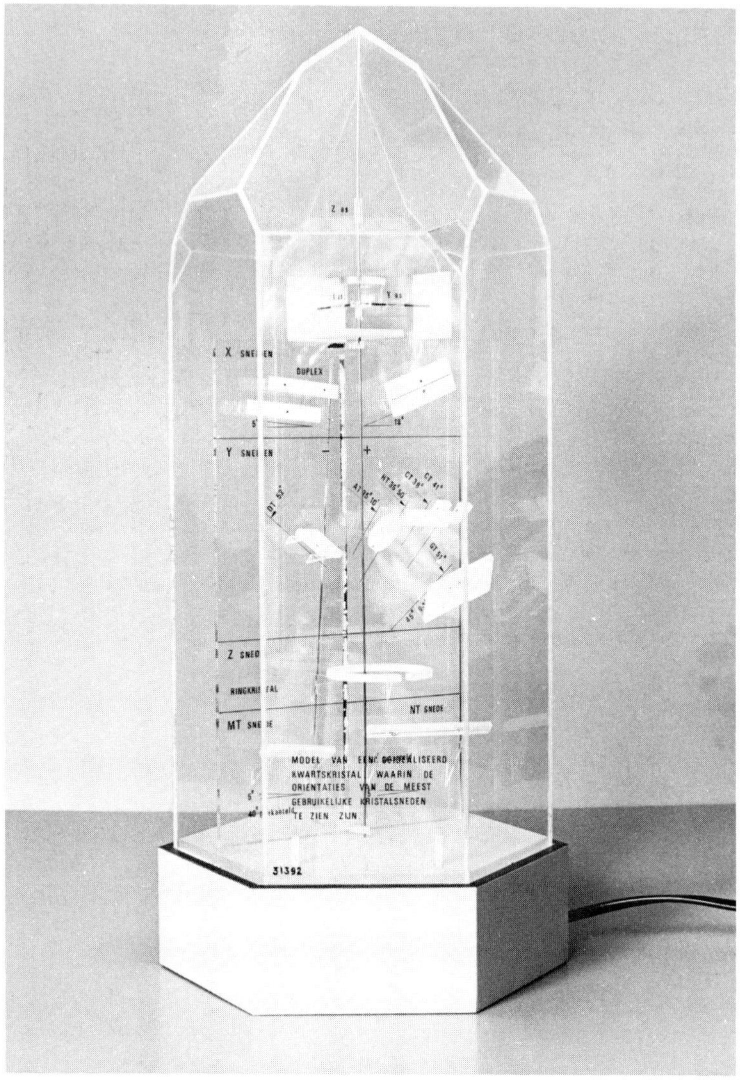

436

indicate time in principal cities throughout the world. Movement adapted from a 'Paris' movement with platform cylinder escapement, circular case.
Pascal took out a patent for this device in 1873
Private collection
Bibliography : Chayette-Sabrier 4 June 1981, lot 57

432 Time Schedule of Transmission of Radio Time Signals from the Eiffel Tower in Paris and from Norddeich, Germany
from Bigourdan, B78.80.
The radio time signal was a fundamental step in the development of the dissemination of time, particularly for navigational purposes. The pioneering work of Guglielmo Marconi in wireless telegraphy very rapidly led to the broadcasting of wireless time signals for navigational purposes. The Naval Observatory in Washington began a service of radio time signals in 1905. Soon radio time signals could be received worldwide. In 1912 an organisation called Bureau International de l'Heure was founded : to Nowadays such resonators form the core of unification of radio time signals, and to prepare plans for an international time service to suit the needs of all.

433 Marine Radio Receiver, type M 0.13, made by the Nederlandse Seintoestellen Fabriek, 1937
Signed : 'N.S.F. Holland'.
Serial nr. 4538/5.
460 mm x 740 mm x 350 mm.
Marine navigators were among the very first customers to use radio signals for their work. At pre-set times the ships' radio receiver was connected with the chronometer room to verify the time-keepers. The present receiver was used in Her Majesty's vessels Kinsbergen, Van Heemskerk and Tromp.
Traditiekamer, Marine Electronisch en Optisch Bedrijf (MEOB), Oegstgeest, 000580.

434 Page from the Time Signal Recording Book, Observatory University of Utrecht, c. 1925
330 mm x 200 mm.
The regular service of radio time signals was soon to be used by all astronomical observatories to verify their clocks. The astronomer in charge of the clocks recorded their reception in order to calculate the rate of his clocks.
Utrecht Universiteitsmuseum, Utrecht.

Quartz Clocks and Standard Frequency Generators

435 Quartz Crystal.
Quartz crystal (rock crystal, single crystal of SiO) is an almost ideal material for the construction of mechanical resonators. Nowadays such resonators form the core of all modern precision time-pieces.
PTT Research, Leidschendam.

436 Model of Quartz Crystal in Which the Orientations of the Most Commonly Applied Crystal Cuts are Indicated
Made in the PTT research laboratory at Leidschendam.
308 mm x 180 mm.
Fifty years of research have established the optimal way for cutting resonator crystals to be used in various applications. In the model a number of suitable orientations are shown. Each cut results in a resonator with a particular frequency (within a band depending as a whole on the size of the cuts).
PTT Museum, Den Haag, 31392.
Bibliography : Schneiders, 45-7.

437 Demonstration Model of Oscillating Quartz Ring of 1972, made in the PTT research laboratory at Leidschendam
180 x 260 x 290 mm.
In 1929 the American engineer W.A. Marrison developed quartz crystal resonators in the form of a ring for the benefit of a frequency standard of the Bell Telephone System in New York. Mechanical vibrations are excited by means of an oscillating electric field on electrodes deposited on the surface of the crystal. The present large sized model oscillates at the extremely low frequency of 583 Hz.
PTT Research, Leidschendam.

438 Two Crystal Oscillators Made by the PTT Research Laboratory, Dr. Neher Laboratorium, around 1950
Sizes : 100 x 70 x 210 mm and 150 x 20 x 20 mm (wxdxh).
For a long time the PTT Research Laboratory had a special division for the construction of crystal resonators. After cutting and polishing, the electrodes were soldered on the surface of the cut crystals. The whole construction was then housed in a vacuum for protection from external influences.
PTT Research, Leidschendam.

439 Series of 6 Crystal Oscillators Made by the PTT Research Laboratory, Dr. Neher Laboratorium, around 1950
140 mm x 300 mm 60 mm.
PTT Research, Leidschendam.

440 Series of 30 Crystal Oscillators Made by the PTT Research Laboratory, c. 1950
130 mm x 250 mm x 130 mm.
PTT Research, Leidschendam.

441 Diagram of a Quartz Crystal Frequency Generator from the PTT Research Laboratory
The mechanical oscillations of a quartz crystal have a high frequency (5,000 to many million oscillations per second) and for that reason are very suitable to stabilize and control radio waves. Moreover, crystals are much easier to protect from disturbing outside influences. In a frequency generator the quartz crystal is placed in a tuned oscillating circuit, such that the oscillations at the natural resonance frequency of the crystal are reinforced. The Americans Horton and Marrison were the first to describe such a construction in 1928.

442 Quartz Crystal Frequency Generator of 100 kHz Made by the PTT Research Laboratory around 1950
This generator was used in the period 1951-1972 as a frequency standard in combination with a synchronometer (see below nr. 443).
300 mm x 360 mm x 500 mm.
PTT Research, Leidschendam.

443 Synchronometer, type no 693-B of 1952
Signed: 'General Radio Company Cambridge Mass. U.S.A.
Serial no 306'.
220 mm x 270 mm x 190 mm.
This 1000 Hz synchronous motor clock used by the PTT in the 1950's for the synchronization of their time service. The 1 kHz signal was obtained in the PTT research laboratory from a 100 kHz quartz crystal generator (see nr. 7 above) with the help of frequency dividers. The clock itself closes a contact every second, thus generating the second impulses for the synchronization of a series of master clocks throughout the country. These in turn provide the minute pulses which regulate various impulse dials and also the speaking (telephone) clocks then in operation.
PTT Research, Leidschendam.
Bibliography : Schell, 466.

442

443

444

444 Speaking Clock, c. 1935
Signed : 'Constructeur F.H. Leeuwrik' and 'A. de Jong, Vlaardingen'.
380 mm x 980 mm.
This speaking clock was designed in 1934 and used until 1970 by the PTT for its telephonic time service. It was operated by the minute pulses received from a master clock. The core of the instrument is a metal cylinder with 84 strips of sound track of a movie film. On 24 of these strips in hours 0, 1, 2,...23, and on the remaining 60 strips the minutes 0, 1, 2,...59 have been recorded by voice. The photoelectrical 'reading' of these strips is guided by the motion of two electromagnets along a rail. This motion is regulated by the minute impulses received from a master clock in such a way that the 'minute' strips are always read after the 'hour' strips. By calling 002 the reading was put through.
Nederlands Goud-, Zilver- en Klokkenmuseum, Schoonhoven, eau 200.
Bibliography : Electrische tijdaanwijzing, 15, 35 ; Hoogeveen, 156-163.

445 Demonstration Model of Oscillating Quartz Ring of 1970, made in the PTT Research Laboratory, Leidschendam
120 mm x 260 mm x 200 mm.
The quartz crystal in this large sized model oscillates at 627 Hz. In clocks and watches tiny quartz crystals vibrate in general at very high frequencies of 2^{15} = 32786 Hz.
PTT Museum, Den Haag, Inv. nr. 31232.

446 Quartz Clock, c. 1967, made by Junghans, Schramberg
114 mm x 200 mm.
To be able to use quartz crystals in normal clocks and watches during the last decades the clock-making industry had to change very rapidly from mechanical to electronic time-keeping design and construction. The high frequencies of quartz crystal resonators have to be reduced (by frequency dividers) to much lower ones in order to drive the slow motion of the hands of a clock. Electronics made it possible to do so on a scale appropriate to domestic clocks and watches.
The first quartz clocks for domestic use were introduced in 1967 by the West-German firm of Junghans. The quartz crystal oscillator itself is enclosed in the horizontal tube. The so-called flip-flop circuits divide the frequency repeatedly in subsequently lower and lower values. The current generated by the last step in the series is used to synchronize the balance of the clock.
MTS Vakschool, Schoonhoven.
Bibliography : Electrische tijdaanwijzing, 37.

447 Quartz Crystal Master Clock, c. 1980
Signed : 'Siemens HU104'.
The progress of time measurement is well illustrated by the variety of clocks produced by the firm of Siemens in Berlin during several generations. The present master clock is not only much more accurate than the one of 1923 described above (409), it is much smaller too. The first difference is effected by the use of a quartz crystal, the second by the possibilities for miniaturization opened up by the electronic devices incorporated in such clocks.
Siemens Nederland N.V., Den Haag.
Bibliography : Electrische tijdaanwijzing, 37.

448 Quartz Crystal Clock Marine Chronometer, c. 1964
Signed, 'Lip No 08' and inscribed 'Marine service hydrographique'.
Mahogany case with silvered dial, 180 mm x 140 mm
Quartz crystal clocks when properly constructed can nowadays easily match, even surpass, mechanical chronometers in accuracy. Such quartz crystal time-keepers came to be referred to as chronometers. Yet for lack of long term reliability (battery exhaustion and electronic contacts that easily brake down at sea by oxidation) mechanical chronometers continued to be preferred at sea. Only in the 1970's when the problems of long term reliability were solved, were these replaced by quartz crystal chronometers.
The present example was used in the French hydrographic service.
Musée du Temps et de l'Histoire, Besançon.

449 Marine Quartz Crystal Chronometer QM-20, 1986
Signed : 'Seiko' and 'Seikosha Co Limited Japan, serial nr. 10871'. Together with certificate and observation booklet.
The present chronometer was used during several transatlantic voyages by the Dutch Royal Navy and will be used again during future voyages.
Magazijn voor Electrotechnisch Materieel, Oegstgeest.
Bibliography : Higginson, 20-3.

450 Radio-Controlled Quartz Crystal Clock, 1988
Junghans, Schramberg.
The latest revolutionary development in domestic quartz clocks is the radio-controlled quartz clock. It includes a long wave radio receiver which, following the instruction of an integrated circuit incorporated in the clock, listens at regular times (each hour) to radio time signals transmitted at a frequency of 77.5 kHz by a German radio station near Frankfurt am Mainz. Through their reception the clock setting is verified and when needed automatically corrected. One can not set such a clock wrong!
Junghans Nederland B.V.

Modern Time Measurement

451 Atomic Caesium Frequency Standard
Anno 1990 time measurement is determined by the atomic clock. Radiative oscillations from atoms provide an extremely stable frequency standard which, moreover, is completely independent of external factors like temperature, atmospheric pressure and so on. The atomic frequency generated is very high (around 10,000,000,000 oscillations per second) but with the help of a suitable electronic feedback circuit it can still be used to control a quartz crystal oscillator which thereby increases about a thousand times in stability and thus in precision.
In this apparatus a beam of caesium atoms is produced from a heated oven containing liquid caesium. This beam passes through a magnetic field (state-selector) in order to select atoms in a definite energy state. In a microwave cavity - excited by a quartz resonator - transitions between two (hyperfine) levels of the ground state of the atoms of the isotope Caesium 133 are triggered. In another state-selector that follows, those atoms that experienced such a transition are separated from the others and their number is subsequently counted by a detector. This information is fed back to the quartz oscillator (also called slave oscillator). By a kind of tuning process the frequency is found for which the output from the detector is optimal. If that is the case the frequency of the oscillator settles down at 9.2 GHz. with an overall accuracy of 0.01 Hz.
Van Swinden Laboratorium bv. NMI, Delft.
Bibliography : Kartaschoff, 57.

452 Atomic Caesium Clock
The progress in the development of frequency standards has resulted in a new definition of the second. In 1967 the second was defined as the duration of 9 192 631 770 periods of oscillations of the radiation generated in an atomic caesium clock, that is corresponding to the transition between the two hyperfine levels of the ground state of the Caesium 133 atom.
Bibliography : Kartaschoff, 56.

453 Rubidium Frequency Generator Type 304 D, 1970
Signed : 'Tracor, Austin, USA'.
The instrument consists of 1) a tube signed : 'Optical microwave unit, serial nr. 369' (130 mm x 280 mm x 130 mm) and 2) a quartz crystal resonator, type 5LW, serial nr. 268 (60 mm x 90 mm 150 mm).

In the rubidium standard the far red *optical* radiation (7800 A) produced in a rubidium lamp is used to excite Rubidium 87 vapour in a cavity. The absorption of this radiation by the rubidium atoms continues as long as sufficient atoms are present in the lowest level of the ground state. This is promoted by a microwave field of 6843 MHz radiation, which favours transitions from the upper (hyperfine) level to that of the lowest level. The optical absorption is greatest if the microwave frequency causes an optimal number of such transitions. A photo-detector behind the cavity is used to detect the absorption and to stabilize the frequency at the maximum value. The optical set up is simpler and cheaper than the atomic beam standard but it is 10 to 100 times less stable.
Used for the generation of second pulses from 1970-1981.
PTT Research, Leidschendam.
Bibliography : Kartaschoff, 58.

454 Rubidium Frequency Standard, type R-20, 1967
Signed : 'Varian Associates, Palo Alto, Calif. USA, serial nr. 117'.
135 mm x 200 mm x 540 mm.
The rubidium frequency generator shown here is the first atomic standard of the Dutch PTT. It was used from 1967 till 1979 for the determination of the second pulses that control the speaking Assam telephones operative in Holland.
PTT Research, Leidschendam.

455 Ammonia Maser Frequency standard, 1962
Brass, nickeled steel, and steel, 1 m 02 length, diameter 125 mm
This standard oscillator uses the vibrations of ammonia molecules at 24 gigahertz (24 milliards per second). A maser (=Microwave amplification of stimulated emission of radiation), is the low frequency equivalent of lasers. This example was built in Besançon at the Laboratoire de Physique et de metrologie des Oscillateurs and was the first clock of its kind to be built in France.
Musée du Temps et de l'Histoire, Besançon.

445

446

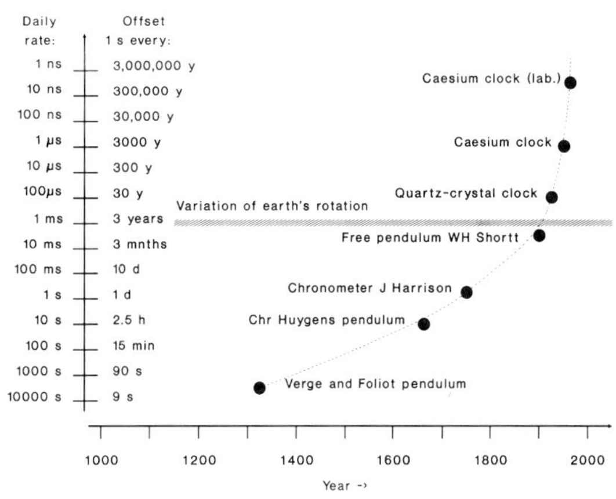

456

456 Clock Precision from 1650 to 2000

Graph.

The precision of a clock is usually described by its daily rate. This number indicates how many seconds or parts thereof the clock gains or loses during a day. This rate tells one therefore also after how many days or years the clock might have gained or lost 1 second.

During the last 5 centuries the precision of clocks has increased tremendously. Pendulum clocks of Huygens' design (1655) gained or lost 1 second per *3 hours*. Harrison's chronometer (c. 1735), did so after only *3 days*; the first Shortt free pendulum, 1921, after *3 months*. A further step forward in precision was taken when in 1928 Marrison invented the quartz crystal clock which gains or loses 1 second after *3 years*. In other words, the quartz clock is about 10,000 times more precise than that invented by Huygens.

During the *last fifty years* the performance of clocks improved even faster: the first atomic frequency generators will have gained or lost 1 second after *300 years* only, whereas the most recent ones should have done so only after *300,000 years*.

457 Variations in the Position of the Polar Axis Measured During the last decades

Small changes in the rotation of the Earth could only be measured after quartz crystal clocks became available. Since then these changes are being constantly recorded. Here the change in the position of the polar axis during a number of years is shown. The shift of the pole is yearly of the order of 5-15 metres, indeed very small, considering the circumference of the Earth of 40,000 kilometers.

458 The International Organization of Time

Display showing International Atomic Time (TAI) and Universal Coordinated Time (UTC).

Today our basic time scale is the *International Atomic Time Scale* (TAI). In physical laboratories all over the world the oscillations of their atomic clocks are counted. The results of these individual counts are sent to Sèvres, Paris. There all data are compared by the Bureau International des Poids et Mesures (BIPM). From all data received a kind of average is worked out which constitutes the International Atomic Time Scale.

Next to the International Atomic Time Scale we keep several others. The most important of these is Coordinated Universal Time (UTC). It is kept artificially as close as possible to the astronomical time scale based on the rotation of the Earth, although it runs as fast as TAI, the atomic scale. Differences between UTC and rotating Earth time are regularly measured and if necessary corrected by the addition of 'leap seconds'. Because of the regular addition of leap seconds UTC always differs from TAI by an integral number of seconds. Since 1972, 15 seconds have been added, which shows that since then the rotation of the Earth has slowed down by an averaged amount of 0.83 seconds a year.

459 The National Organization of Time

Display showing the local dissemination of the International Time.

From the Bureau International des Poids et Mesures (BIPM) in France reports are distributed to national bureau of standards. In Holland this is the Van Swinden Laboratorium bv, a division of the Nederlandse Meetinstituut (NMi). These national bureaus take care of the adjustment of their own clocks to UTC in case a leap second must be added, and of the dissemination UTC on a national scale. Local standard times (MET here) are of course directly obtained by adding the required number of hours to UTC. If one needs the precise time one can compare one's clock with that in the Van Swinden laboratorium. This is done by the Dutch PTT for its telephonic time service, and by several scientific and industrial institutions as well as by anyone who has an interest in the precise time. One can also use radio time signal services from abroad, as is done by the broadcasting companies in Holland.

460 Radio-controlled Clock by Junghans, 1990

Junghans, Schramberg, Germany.

These radio-controlled quartz clocks come in the form of ordinary clocks and watches. Through a built-in longwave radio receiver standard time (MET) is maintained by these devices with atomic precision, since the transmitted radio signals are generated by an atomic caesium standard. The accuracy of these *domestic* clocks is therefore unrivalled.

Junghans Nederland B.V.

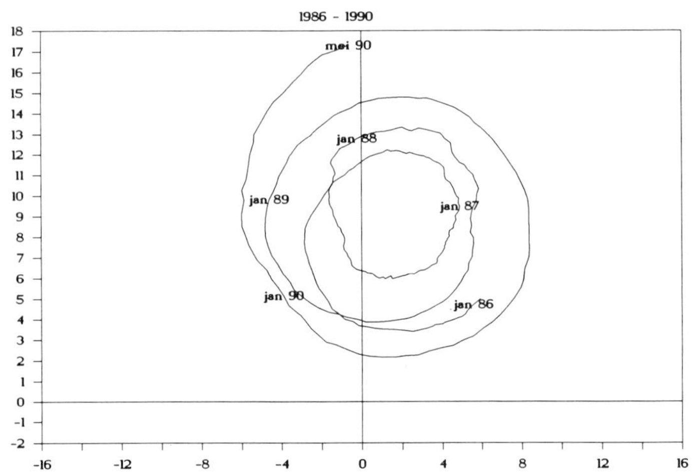

457

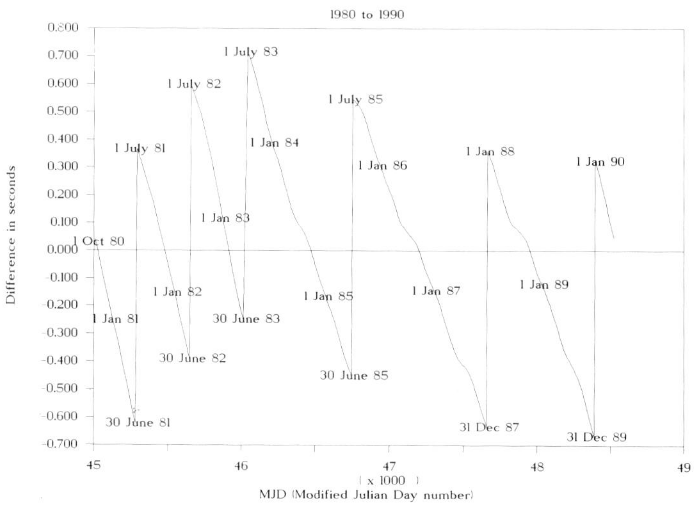

458

Horological Production for a Mass Market'

'Tis with our judgements as our watches, none
Go just alike, yet each believes his own.

Pope, *Essay on Criticism,* 1.6.

The 18th century had seen a vast expansion in demand for clocks and watches. Already in the mid-18th century Dutch and Swiss watches were being made in English styles, and, supplied with English maker's names, were undercutting their rivals. By the end of the century Swiss manufacture had virtually destroyed French competition and was posing a serious threat to the English.

In 1776, Fréderic Japy set up a watch manufacture equipped with tools to produce a standard *ébauche.* For this other standardised parts could be produced, and the production of such watches increased as dramatically as their price declined. In the early 19th century Swiss watches, made in any and every style desired, swept all competition before them to totally dominate the world market.

463

461 A French Watch with an Unusual Dial, *c.* **1800**
Signed 'Les Frs. Raiguel à Paris' (dial & movement), 'Leton Mag' (case).
Brass and silvered brass, steel and blued steel, enamel dial.
Diameter 59 mm.
Dial with centre seconds hand making one revolution in 75 seconds which quantity is marked on the dial. Subsidiary dials for hours and minutes, and calendar. Virgule escapement, fusee with chain, stop lever in side of case. Nederlands Goud-, Zilver- en Klokkenmuseum, Schoonhoven Di 2.

461

462 A Swiss Repeating Watch, c. 1800
Signed 'Dufour, Fol & Compe, Genéve', case-maker's initials 'A.R.'.
Brass and steel with enamelled dial, silver case.
Diameter 46.5 mm.
Full engraved back plate with English type clock, engraved and pierced case, verge escapement, dust cap.
Musée Internationale d'Horlogerie I-55.

463 Watch Movement, c. 1800 and later
a. Signed 'Degoumoi Franeker', diameter 43 mm.
Enamelled dial, verge escapement, silver regulator disc.
b. Signed 'Fres Viss & Menu àu Genéve', diameter 45 mm.
c. Very thin cylinder movement with silvered engineturned dial, subsidiary seconds, plain steel balance and going barrel.
d. Signed 'Monnier A Paris' No 7650, diameter 46 mm. Full plate movement with silver regulator disc, bridge cock, verge escapement, enamelled dial.
e. Signed 'Moulinie Genéve', diameter 49 mm. Semi-Gothic roman hour numerals, subsidiary counter-sunk seconds, three quarter plate movement, bi-metallic compensated balance, straight line lever escapement with auxiliary weight to one end, balance spring regulator.
Zaans Uurwerkmuseum.

464 Contrasting Movements of the early 19th century
a. Movement of a watch for a ring
Brass, gilt brass, and steel, enamelled dial, 30 mm x 15 mm.
Oval dial with two dials, hours and minutes on one, seconds on the upper. Cylinder escapement.Nederlands Goud-, Zilver- en Klokkenmuseum, Schoonhoven HCW 438.
b. Movement of a Pocket Chronometer
Signed 'Tobias London'.
Brass and steel, Diameter 47 mm.
Unusually shaped back plate with spring detent escapement and compensated balance. Nederlands Goud-, Zilver- en Klokkenmuseum, Schoonhoven HDI 444.

465 The Layout of the Lever Watch in the 19th century
Working model.
Brass and steel 690 mm x 350 mm
Musée du Temps et de l'Histoire, Besançon

466 A Stop-Watch by Robert Roskell, early 19th century
Signed 'Robert Roskell No 41986'.
Brass and steel with enamelled dial.
Diameter 39 mm.
Dial with counter sunk subsidiary seconds, plain steel balance with balance spring regulator, verge escapement. The stop click works directly on the rim of the contrate wheel.
Zaans Uurwerkmuseum.

467

467

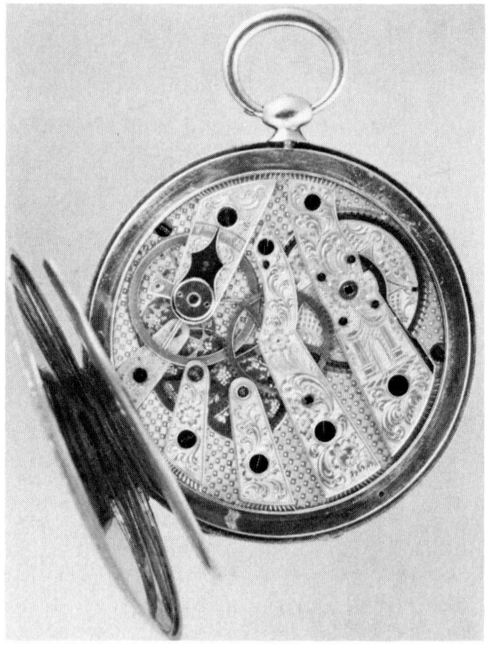

469a

467 A Souscription Watch, 1808
Signed 'Breguet & Fils' 1590.
Brass, steel and blued steel, silver & gold dial, engine-turned gold case.
Diameter 56 mm.
Bridge movement, ruby cylinder escapement, parachutes on balance-arbor endstones, balance spring regulated by sliding index. The certificate accompanying the watch states that it was sold to Recordon, probably Louis Recordon (1778-1824), a well-known Soho watch-maker.
Museum of the History of Science, Oxford.

468 A Miniature Masterpiece, early 19th century
Signed 'P M- (=Piguet & Meylan)
Brass and steel with engine-turned gold case, enamell dial diameter 9 mm
Lever escapement, seventeen jewel movement with an extra wheel in the train. Doubtless one of the smallest lever watches ever made.
The Science Museum, London, 1952-1.

469 Two Thin Hunting-cased Watches, c. 1840 & 1875
a. Signed 'Ami Sandoz & Fils, La Chaux-de-Fonds'.
Brass and steel with gold case.
Diameter 40 mm.
Engraved and *guilloché*-worked case, cylinder escapement with 8 jewels.
b. A Decorative Hunting-cased Watch, with Lever Escapement, 1875.
Signed 'Haas & Privat,
Brass and steel with enamelled, *guilloché*-worked and engraved gold case, lever escapement.
Diameter 39 mm.
Musée Internationale d'Horlogerie, la Chaux-de-Fonds, I-174 & I-2115.
La Chaux-de-Fonds'.

470 Key Wound Watches 1830-1880
a. Bridge movement decorated with coloured flowers, cylinder escapement.
b. Cylinder escapement and enamelled dial.
c. Lever escapement and enamel dial.
d. Skeletonised movement with lever escapement, enamel dial No 14617.
e. Signed 'Vacheron & Constantin'.
Zaans Uurwerkmuseum.
Of the mass of innovations, changes and ameliorations introduced in 19th century watch-making, one only must stand for all - the introduction of keyless-winding.Watches can be made without keys in various ways, but the system that became universal in the second half of the 19th century was essentially that devised in 1842 by Adrien Philippe (of Patek & Philippe), and published by him in 1856.

471 The Classic Treatise on Keyless Winding, 1863
Adrien Philippe, *Les montres sans clef ou se montant et se mettant à l'heure sans clef,* Paris, 1863.
80, pp. 308 with 3 lithographed plates.

Adrien Philippe devised his system of keyless winding for watches while struggling to establish a watch manufactory at Paris in 1842. Although aware of the existence of earlier systems - in particular those of Breguet and Audemars - Philippe claimed to have no detailed knowledge of them while devising his own. This could be applied easily to all kinds of watches. A winding rod is rigidly attached, through a rod passing through the pendant, to a pinion which meshes at right angles with the crown wheel itself geared to the spring barrel. Turning pendant in a single direction therefore causes the barrel to turn and wind up the spring. A clickwork system prevents winding in the wrong direction.
That the various elements of this system already existed Philippe acknowledges. The merit he claims is 'to have made the most practical and long-lasting application' (p. 88). The value of his system was first recognised at the 1844 exhibition where he won a medal, and where it also brought him to the attention of the Comte de Patek. Their subsequent partnership was the beginning of a legend.
Private Collection.

472 A Half-plate Keyless Wound Watch, c. 1860-1870
English with cylinder escapement.
Zaans Uurwerkmuseum.

473 Three Stem Wound Watches, 1870-1900
a. A half plate and bridge movement, Swiss made for the English market with cylinder escapement.
b. Marked 'Globia', with cylinder escapement.
c. Half plate and bridge movement with cylinder escapement.
Zaans Uurwerkmuseum.

469b

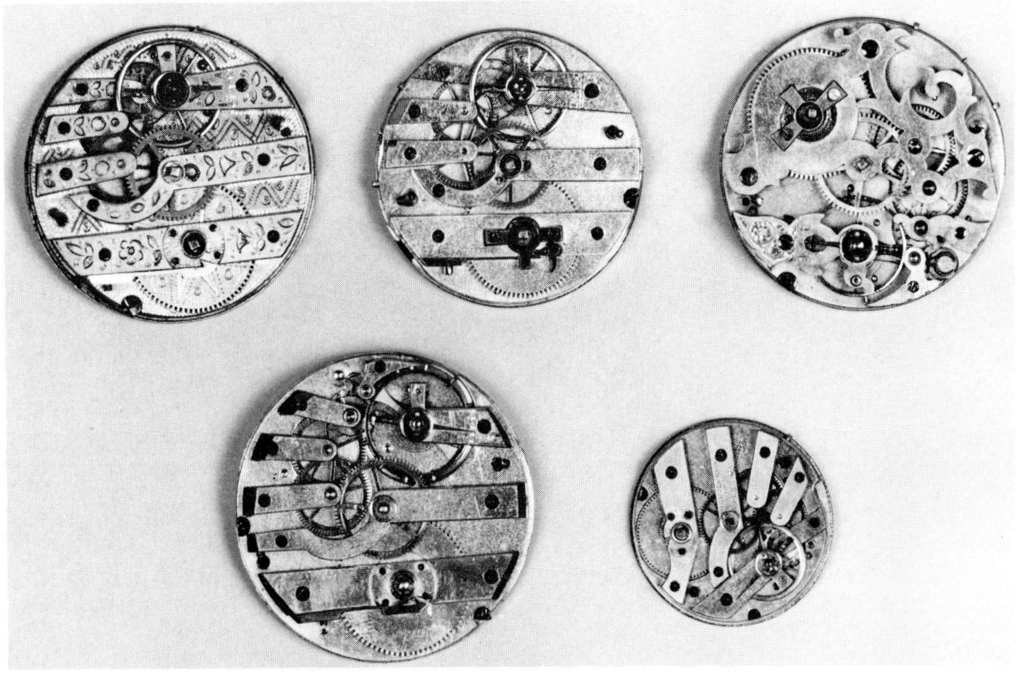

470

474 **A Chronograph Watch with Facetted Case, 1894**
Inscribed 'Philémon Makeef 1894'.
Brass and steel with gold case.
Diameter 54 mm.
Lever escapement.
Musée Internationale d'Horlogerie, La Chaux-de-Fonds.

475 **A Simple System of Keyless Winding, c. 1895**
Not signed, Swiss, market '660 5019/14528' and 'SC 0,800' on the case.
Brass with silver case.
Diameter 49 mm.
Enamelled dial with arabic numerals, gold hands and subsidiary counter-sunk seconds. Glazed inner cover showing the simplified jewelled movement with lever escapement and cut bimetallic compensated balance. There is no spring barrel or fusee, the normal spiral main-spring being replaced by a helical spring attached at one end to the winding stem and at the other to a crown pinion meshing with the centre pinion. When the watch is wound the spring is coiled round the winding arbor.
This system was the subject of British patent No 23,671 taken out by I. Grasset and A. Meylan on 22 December 1892.
Goud-, Zilver- en Klokkenmuseum, Schoonhoven, Inv. AAN22.
Bibliography : *Abridgements 1889-1892,* 82 ; Jaquet & Chapuis, pl. 114.

476 **Three Swiss Watches, late 19th/early 20th century**
a. Signed 'Vacheron & Constantin Genéve'.
b. Signed 'International Watch Co Schaffhausen'.
c. Signed 'La Génerale Succr de Louis Brandt & Frére Helvetia'.
Nederlands Goud-, Zilver- en Klokkenmuseum, Schoonhoven.

477 **A Minute Repeating Watch, end 19th century**
Inscribed 'Alice Borel Noel 1897' by Lecoultre, Le Sentier.
Brass & steel with gold case.
Diameter 37 mm.
Anchor escapement.
Musée Internationale du Temps, La Chaux-de-Fonds.

478 **A Self-winding Watch, post 1878**
Signed 'Loehr Breveté S.G.D.G.' 'No 1882 Perpetuelle Breveté SG DG'.
Brass, gilt brass, steel, nickel and enamel.
Diameter 49 mm.
Automatic winding on the principle of a pedometer. An eccentrically placed, crescent-shaped weight on a lever is mounted at the back of the case. A pawl fixed to the lever works on the ratchet wheel connected to the barrel arbor. A soft spring keeps the weight lifted. If the watch is worn upright by means of the straight clip, the weight will rock up and down at each step taken thus winding the watch. Cylinder escapement, provision for key-winding, time-setting by the rippled knob.
Nederlands Goud-, Zilver- en Klokkenmuseum, Schoonhoven HCK 442.
Bibliography : Meiss, 41.

479 **A Stop-watch with Reversed Motion Seconds Hand, end 19th century**
Brass, steel, blued steel, nickel and enamel.
Diameter 53 mm.
Roman hour numerals with subsidiary seconds reading to 15 in the reverse direction. The seconds hand mounted on a large cylinder wheel makes four revolutions per minute. Cylinder escapement, stop lever in the side of the case. Nederlands Goud- Zilver- en Klokkenmuseum, Schoonhoven HCK 441.

480 **A Watch by Patek Philippe, c. 1896-1899**
Signed 'Patek Philippe & Cie Genéve No 110742' and inscribed 'Louise Golay'.
Brass and steel with gold case.
Diameter 31 mm.
Lever escapement. The case is engraved with the owner's initials 'LG' in a monogram.
Musée Internationale d'Horlogerie, La Chaux-de-Fonds I-358.

But too much confidence could be placed in the performance of a watch.
'Above all', he said, 'I must take care not to miss the steamer'!
'You have the time', replied Fix, 'it is only noon!'
'Noon, Pshaw! It is eight minutes to ten!'
'Your watch is slow!' replied Fix.
'My watch! A family watch that has come down from my great-grandfather! It don't vary five minutes in the year. It is a genuine chronometer.'
'I see what is the matter,' replied Fix. 'You have kept London time, which is about two hours slower

474

than Suez. You must be careful to set your watch at noon in each country.'
'What! I touch my watch!' cried Passepartout. 'Never.'
'Well, then, it will not agree with the sun.'
'So much the worse for the sun, monsieur! The sun will be wrong then!'
And the good fellow put his watch back in his fob with a magnificent gesture.
Jules Verne, *Round the World in Eighty Days,* Ch 8.

Watch and Clock-making in America

Traditional methods of watch- and clock-making and repairing were imported from Europe into North America during the 17th and 18th centuries. At the beginning of the 19th century they were transformed first by Eli Terry's application of water-power to drive a series of machines and so manufacture wooden clocks in quantity, secondly by Chauncey Jerome's mass production of very cheap clocks with brass movements.
New forms of clocks appeared in the wake of these innovations. The most famous was Eli Terry's scroll and pillar shelf clock, but many different forms were produced.

481 Wall Clock, c. 1860
Signed on paper label in back of case 'Brass Clocks made for A. Colemen Savannah, G.A. Warranted if well used' with instructions. On a watch paper 'Reeves Watch and Clockmaker Cra..ey' assortment of clocks and watches'.
Pine with palissander veneer, glass, steel and brass, 650 x 390 mm x 105 mm.
Rectangular case with central glass panel back painted with above the dial and below a town scene. Anchor escapement with separated pendulum set immediately behind the door, weight drive over two pullies at the sides of the case. Hour striking with locking plate on going. 24 hour duration.
Nederlands Goud-, Zilver- en Klokkenmuseum, Schoonhoven.

482 A Wall Clock of the Later 19th Century
Paper label with the 'Time is Money Trade Mark'
662 mm x 398 mm x 111 mm

483 Bracket or Table Clock, 3rd quarter 19th century
Trade mark of Atlas carrying the earth with motto *Diligens vincit omnia* (Diligence overcomes everything).
Veneered pine, painted glass, painted zinc, glass, brass and steel, 355 mm x 200 mm x 100 mm.
Rectangular case with triangular top and two turned and pointed finials. Centre of the door of glass back painted with the dial carrying a brass alarm setting disc at the centre. Anchor escapement with spring suspended pendulum, spring drive. alarm striking on a bell.
Nederlands Goud-, Zilver- en Klokkenmuseum, Schoonhoven, Pen 42.

484 Wall Clock, c. 1890
Signed 'New Haven Trademark'.
Veneered pine with inlay, glass, steel, brass and zinc, 950 mm x 400 mm x 150 mm.
Turned two pillar frame with leaf carved pediment and bottom scroll surrounding a rectangular case with circular dial. Anchor escapement, imitation mercury compensated spring suspended pendulum, spring drive, hour striking on bell.
Nederlands Goud-, Zilver- en Klokkenmuseum, Schoonhoven.

485 A Universal Time Watch, 1860-1861
Signed 'Auguste Mathey New York'.
Silver case guilloché, brass and steel.
Diameter 55 mm.
The twice XII hour dial and concentric minutes are mobile, time being indicated against a fixed pointer and against the names of 72 towns throughout the world marked on an outer concentric fixed scale.
Cylinder escapement.
Musée Internationale d'Horlogerie, La Chaux-de-Fonds I-214.

486 A Lower Price Watch, c. 1900
Signed 'Ingersoll Yankee' ; 'R.H. Ingersoll &

473

481

483

484

Bro New York Pat Dec 28 90 Jan 13.91 Made in USA 562113'.
Blackened & gilt brass, brass and steel, diameter 52 mm.
Pin pallet anchor escapement, winding by folding fixed key.
Nederlands Goud-, Zilver- en Klokken-museum, Schoonhoven MAK 44.

487 **A Connecticut Watch Factory in the mid-19th century**
Photograph
By 1850 there were said to be 1436 clock-makers in Connecticut, the centre of American horology. The American technique of mass production was winning them a place in the world market. When the Gebrüder Junghans' factory at Schramberg, Würtemburg was set up in 1864, it was organised on American principles.
Meanwhile in America and elsewhere attempts were being made to mechanise watch production.
In the 1840's, Pierre Frédéric Ingold had attempted to manufacture watches using machines that would cut interchangeable parts. He failed.
Only after years of experiment was the American Watch Co at Waltham able to make watches by machines successfully and their parts were not truly interchangeable. Even so between 1858 and 1864 their production passed from 14,000 to 118,000. Competitors quickly appeared. Between about 1850 and 1925, some 70 American watch factories produced about 120,000,000 jewelled watches, and about 250,000,000 cheap 'dollar' watches.

American Watches

The industrialisation of watch-making, pioneered in America but quickly adopted in Switzerland and Germany, represents a complete democratisation of time. Anyone and everyone could now own a watch accurate to within seconds per day. Between 1885 and 1913 204,181,000 watches were exported from Switzerland alone. Such figures reflect the fact that the industrial societies of Europe and America were now entirely time-structured.
The clockwork metaphor of the 18th century had become social reality in the 19th.

488 **Watches 1900-1940**
a. Signed 'Vacheron & Constantin' with ruby lever and compensated balance.
b. Signed 'Antoine fréres' with lever escapement and compensated balance.
c. Cylinder escapement.
d. 'Fritz Roskoff' & 'Cu Patent'.
Nederlands Goud-, Zilver- en Klokken-museum, Schoonhoven (a & c); Zaans Uurwerkenmuseum (d).

489 **A Swiss-made Watch for the English Market, c. 1930**
Signed 'Tavannes Watch Co'.
Brass and steel with composition case.
Diameter 50 mm.
Half-hunter type Dennison case with roman numerals. Enamelled dial with roman hour numerals reading to minutes. Subsidiary counter-sunk seconds dial with arabic numerals. Quarter-plate and bridge engine-turned 17 jewel movement with four adjustments. Lever escapement with ruby pallets and bimetallic compensated balance with blued steel spring. Pendant wind with pull set.
The Dennison case is inscribed on the inner cover 'English made. Guaranteed to be Made of Two Plates of 10 ct gold with plate of Composition between and to Wear 20 Years 166762/32/0' and on the inside of the outer back cover 'Moon Trade Mark A L D Dennison Watch Case Co Ltd 166762' and with several repair dates.
With a quite high quality movement placed in a cheap, mass-produced case, this watch is an excellent example of marketing. The Dennison watch case retained all the allure of an expensive gold case while being, as the inscription makes clear, quite the opposite. Combined with a good quality movement however it could be sold in England at a price to appeal to the upper working/lower middle classes - sufficiently expensive as to

488a-c

be an effort and so worthwhile, but not so expensive as to be out of reach.The Dennison Watch Case Company was originally founded c. 1874 by Aaron L. Dennison (1812-1895), a pioneer with Edward Howard (1813-1904) of American factory produced watches (see Milham 394-398).
Private Collection.

490 A Skeleton Watch, 20th century
Glass plated movement, pendant wound, subsidiary seconds.
Lever escapement with compensated balance.
Nederlands Goud-, Zilver- en Klokkenmuseum, Schoonhoven.
Reacting to the American challenge, Swiss manufacturers began themselves to mechanise and by 1900 once more dominated watch-making.
Despite short-term difficulties, the Swiss industry improved its position. In the late 1940's nearly 80% of world production was Swiss. Quickly reduced, throughout the 1950's and 1960's Swiss manufacturers continued to hold a 50% share.

The Wrist-Watch

Although watches had been attached to bracelets from the 18th century onwards, the idea of manufacturing and retailing a watch and strap as *a unit* to be worn on the wrist, was not exploited until the later 19th century. Patek Philippe produced a bracelet watch as early as 1868, but the mounting of watches on wrist-straps only became common in the last decades of the century. A considerable boost to the habit was given by the first World War when the convenience of the arrangement was strongly felt. Since then production has been continuous, especially as the wrist-watch lends itself to treatment as piece of jewelry.

491 Early Wrist-Watches and Bracelets, 1889
Advertisement by J.W. Benson for his 'Bond Street Novelties'.

492 An Early Wrist-Watch, c. 1900
The small ladies' pendant wound watch has been mounted in a double clip and ring which carries the black silk strip.
Zaans Uurwerkmuseum.

493 A Group of Four Ladies' Wrist-Watches, c. 1920
Zaans Uurwerkmuseum.

494 Three Wrist-watch Movements, c. 1930
a. Omega
b. Incabloc
c. Swiss
Zaans Uurwerkmuseum.

Variety in the wrist-watch is enormous. Artists' responses to its double function of utility and adornment across seven decades is shown in a representative series of watches by Cartier (numbers 495 - 514).

495 Man's Gold-cased 'Santos' Wrist-Watch with Leather Strap and Sapphire Crown Cabochon, 1908
The watch was named after aviation pioneer Alberto Dumont Santos (1873-1932).

496 Lady's Bracelet Watch with Cushion-Shaped Gold and Platinum Case, Diamond bezel and acanthus scrolled lugs, silver-plated dial with roman chapters, meshwork bracelet of five rows of pearls attached by four roads of onyx beads, gold buckle, 1912
Provenance : The Princess Orlov.

486

489

490

492

494

497 Lady's Platinum Cased 'Tortoise' Wrist-Watch, diamond bezel, silver dial, black moiré strap with gold buckle, 1914
Provenance : Sir Philip Sassoon.

498 Lady's Platinum and Gold Mounted Wrist-Watch with barrell-shaped diamond and onyx case, silver-plated dial, platinum Milanese bracelet with rose gold buckle, c. 1915

499 Lady's Rectangular Platinum and Gold-cased Wrist-Watch with diamond bezel and attaches, silver dial and white gold Milanese strap, c. 1919

500 Lady's Gold-cased Wrist-Watch with double black cord bracelet, 1924

501 Man's Gold Hunter Case Wrist-Watch with jump cover, lizard skin strap, 1926

502 Lady's Rectangular Wrist-Watch Platinum Case and Cover Pavé Set with cut diamonds, diamond attaches, double cord bracelet and clasp, c. 1927

503 Man's Hunter Cased 'Tank' Digital Wrist-Watch, 1928
Provenance : The Maharajah of Patiala.

504 Man's Rectangular Platinum Cased Wrist-Watch with Patek Philippe movement and leather strap, c. 1930

505 Small Rectangular Platinum Cased Wrist-Watch with black double leather rope strap, c. 1930
Provenance : Presented by Count Marzotto to the Italian aviation pioneer Italo Balbo (1896-1940).

506 Lady's Small Platinum Cased Wrist-Watch with Flexible link platinum strap, 1932
Provenance : Prince Tikkar of Kapurthala.

507 Lady's Watch Clip with Matching Bracelet, rectangular gold case with sliding cover set in step-cut bracelet, *c.* 1938

508 Lady's Triangular Wrist-Watch, gold case with black enamel zig-zag and point decoration, leather strap, c. 1940

509 Man's Gold Cased 'Tank' Wrist-Watch with flexible link pavé strap, 1944

510 Lady's Gold Wrist-Watch with double plated strap around caged movement with cabochon ruby embellishments, 1956

511 Man's Double Winder Gold Cased Wrist-Watch, c. 1959

512 Man's Calendar Gold Cased Wrist-Watch Indicating Date, day of the week, month date and phase of the moon. Movement by Audemars-Piguet, 1962
Provenance : Countese Volpi di Misurata.

513 Man's Elliptical Gold Case Wrist-Watch with movement by Audemars-Piguet, c. 1970

518

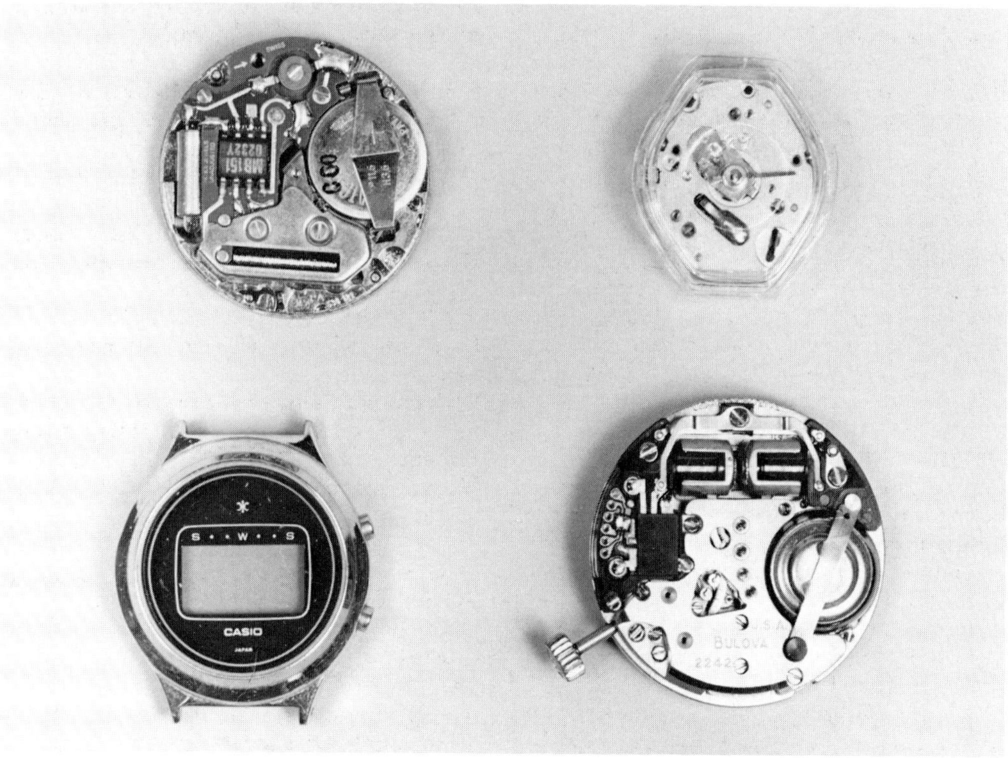

525

514 Man's White Gold Cased 'Duo Plan' Wrist-Watch for 'Local' and 'Home' Time, 1972
Cartier Geneva numbers 495 - 514

515 A Wrist-Watch Giving Time in Britain and India, c. 1925.
Gold case with crocodile strap, overall length 45 mm, width 20 mm
Private Collection

516 A Very Small Mechanical Watch Movement, c. 1960
Diameter 13 mm.
Zaans Uurwerkmuseum.

By the 1930's, fully interchangeable parts had become fairly common. The cheap 'Timex' watch exploited this fully. It depended entirely upon automated production. It is the culmination of two centuries effort.

517 Examples of Timex Watches, c. 1950
Private Collection.

518 Automatic Watches and Stem Wound Movements, c. 1955
a. A 25 jewel Incabloc 'Ultramatic'.
b. The 'Baumatic' by Baume & Mercier also with 25 jewels.
c. The 'Dugena'.
d. Stem wound 17 jewel movement signed 'Gérard Perregaux'.
e. Ladies' watch movement, 'Gebo'.
Zaans Uurwerkmuseum.

519 The Years of the 'Timex' were also the Years in which Electrically-Driven Watches were Developed
The first commercial electric watch was made by the Hamilton Watch Co in 1957.

520 Launching the Hamilton Electric Watch, c. 1957
Photographs of Hamilton models, promotional material and a copy of the essential handbook.
W.O. Smith Jr., *The Hamilton Electric Watch Repair Manual,* Pittsburgh, 1958.
This is devoted to the Hamilton model 500.
Private Collection.

Much development however had taken place elsewhere

521 An Electrically-powered Experimental Movement, c. 1900
Brass and steel, diameter 49 mm.
Two magnetic bobbins move the central stator which has fixed lever. The lever gives impulse to a pin on the balance wheel and the circuit is completed at the appropriate instant by contacts fixed to the balance staff.

Nederlands Goud-, Zilver- en Klokken-museum, Schoonhoven, Eau 149.
Bibliography : *Elektrische,* 114.

522 The First Electric Watch in France, Lip, c. 1960
A selection of materials relating to the development of the Lip electric watch.
a. Seven experimental movements used in laboratory research.
b. Three experimental batteries and a plate.
c. Record of the performance of batteries and a memo from Lip announcing the success of the research to Tortey, a member of the laboratory staff and announcing the award of a prime of 10,000 francs.
d. Fred Lip's own electric watch movement mounted in a triangular gilt bronze case. Inscribed 'electronic Incabloc Patents Pending 3913'. This is the first model incorporating two batteries and is numbered '6'. Of these first six made one belonged to de Gaulle (see no 523 below), one belonged to Nikita Khrushtshev and the present example was used by Lip throughout his career.
Musée de Temps et de l'Histoire, Besançon.

523 Man's Gold Cased Electronic Watch, Signal 'Cartier', movement by Lip, 1960.
Provenance : One of the first six electronic watches made by Lip. It belonged to the Général Charles de Gaulle.
Cartier Genève WG11C60.

But new ideas followed swiftly. Probably the first electronically maintained watch was the Bulova Accutron.

524 The Tuning-fork Watch with a Model of its Escapement, 1965
Signed 'Swiss made 2141 Bulova Brevet - Patent'.
Brass, steel and plastics, 40 mm diameter (watch) ; 250 x 190 mm (escapement model).
A pair of electro magnets linked to a circuit maintain a tuning-fork in vibration. The amplitude is kept constant by an auto-regulating system, and the oscillations of the tuning-fork are transmitted to the gear train by a clock system. Developed in the early 1960's, the 'Accutron' watch was free of the problems created for precision watches by friction, wear, position and temperature change. Despite its success however the 'Accutron' would have only a short life.
Cité des Sciences et de l'Industrie, Paris.

In 1968-1969, the first quartz watches appeared.Two decades later they have totally supplanted the mechanical watch in everyday use.
But they still retain its appearance.

525 Examples of Quartz Watches and a Piece of Quartz, c. 1980
a. 'Bulova Accuquartz 2242' with calendar.
b. 'Caroline Quartz'.
c. Not signed.
d. 'Casio Japan'.
Zaans Uurwerkmuseum.

Examples of contemporary watch production

526 - 549 Baume & Mercier
The seigniorial family of Baume originated in the Franche Comté.
In 1834 the 'comptoir Baume Frères' was officially registered in Les Bois.
In 1860 serial production was introduced to the Swiss manufactory, and in 1878 the firm changed its name to Baume & Co.
In the last decades of the 19th century the firm won several awards in international competitions, and expanded its retail-organisation in the English colonies and throughout the Far East. In 1918 William Baume was joined in partnership by Paul Tcheréditchenko, a Russian emigré who took the name Mercier on acquiring the Swiss nationality.

550 - 573 Bertolucci
Launched in 1987 in Basel, Switzerland, Bertolucci is a new name among prestige watch manufacturers. Typical of their products is the original 'pebble'-design, in which the most recent horological technology is presented in modern cases of refined decoration.

574 - 597 Cartier
The maison Cartier was founded by the jeweller François Cartier (1819-1904). During four generations his descendants developed watches as jewellery creating such classic models as the 'Santos'.

598 - 621 Chopard
Léon Ulysse Chopard established a manufactory of high precision pocket chronometers at Soinvillier (Jura) in 1860, quickly obtaining dominance in the supply of such watches to the Swiss railways. In 1920 the business was transferred to Geneva, continuing as a family business, until taken over by the Scheufele family, watchmakers of Pforzheim, Germany. Further expansion followed : a new factory in Geneva in 1974 and new branches in Paris and New York in 1975 and 1976.

622 - 645 Ebel
Founded in 1911 in La Chaux-de-Fonds by Eugène Blum, Ebel remains a family business, being now directed by Blum's grandson. Watches for sporting use and appropriate in style have become characteristic of the house.

646 - 669 Rolex
In 1905, Hans Wilsdorf founded the Rolex Company to manufacture high quality precision wrist watches. In 1910 a Rolex wrist watch was awarded a Swiss chronometer certificate, the first wrist watch ever to be so successful. Four years later, a Class A certificate was obtained in Kew Observatory Trials, an unrivaled break-through in precision for the new time-keeper. In 1926 an improved waterproofed case, the 'Oyster', was released, and in 1931 a new self-winding rotor watch, the 'Perpetual' was designed. The development of watches for deep sea diving and long range exploration have become a special tradition of the company.

670 A Mantle Regulator Showing Local and Mean Time, c. 1840
Signed 'C. van Spanje te Tiel No 15'.
Gilt bronze, brass silvered brass and steel on wood base.
Height 491 x 425 mm.
The clock shows hours and minutes in the two systems, beats seconds on the large outer dial and half seconds on the upper subsidiary dial. The local time train is linked to the hour and half hour striking. Maker's own system of temperature compensation to the pendulum.
Nederlands Goud-, Zilver- en Klokken-museum, Schoonhoven, Pen 17.

671 A Typical French Decorative Gilt Bronze Clock with Sentimental Scene, mid-19th century
Musée Internationale du Temps, IV404.

672 An Automaton Picture of a Clock Repairer and his Family, late 19th century
Signed 'Adalbert Hawsy Leipzig'.
Pine, carton brass and iron, 330 mm x 410 mm.
Spring driven movement manipulating the figures through a cord and pulley arrangement.
Nederlands Goud-, Zilver- en Klokken-museum, Schoonhoven, Di 16.

673 A Late Diptych Dial, 1908
Belgian. Inscribed 'Breveté S.G.D.G. Marque deposée J.M.'.
Printed paper on wood,
103 mm x 66 mm.
A typical diptych dial has on the back instructions for converting the true solar time of the dial into Belgian mean time.
Private Collection.

674 A Large Geographical Horizontal Sun-dial, c. 1900
Signed 'F. Barker & Son Scientific Instrument Makers London E.C.'.
Brass, diameter 510 mm.
The dial is drawn for latitude 29° or 30°, region of Delhi, reads to 2 minutes and gives

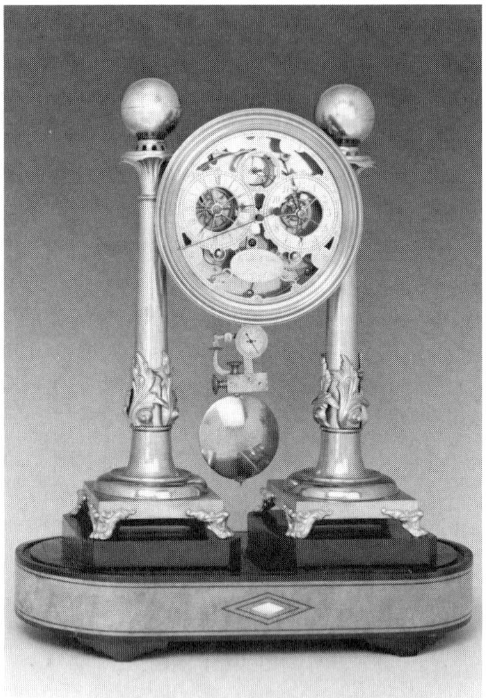

670

682

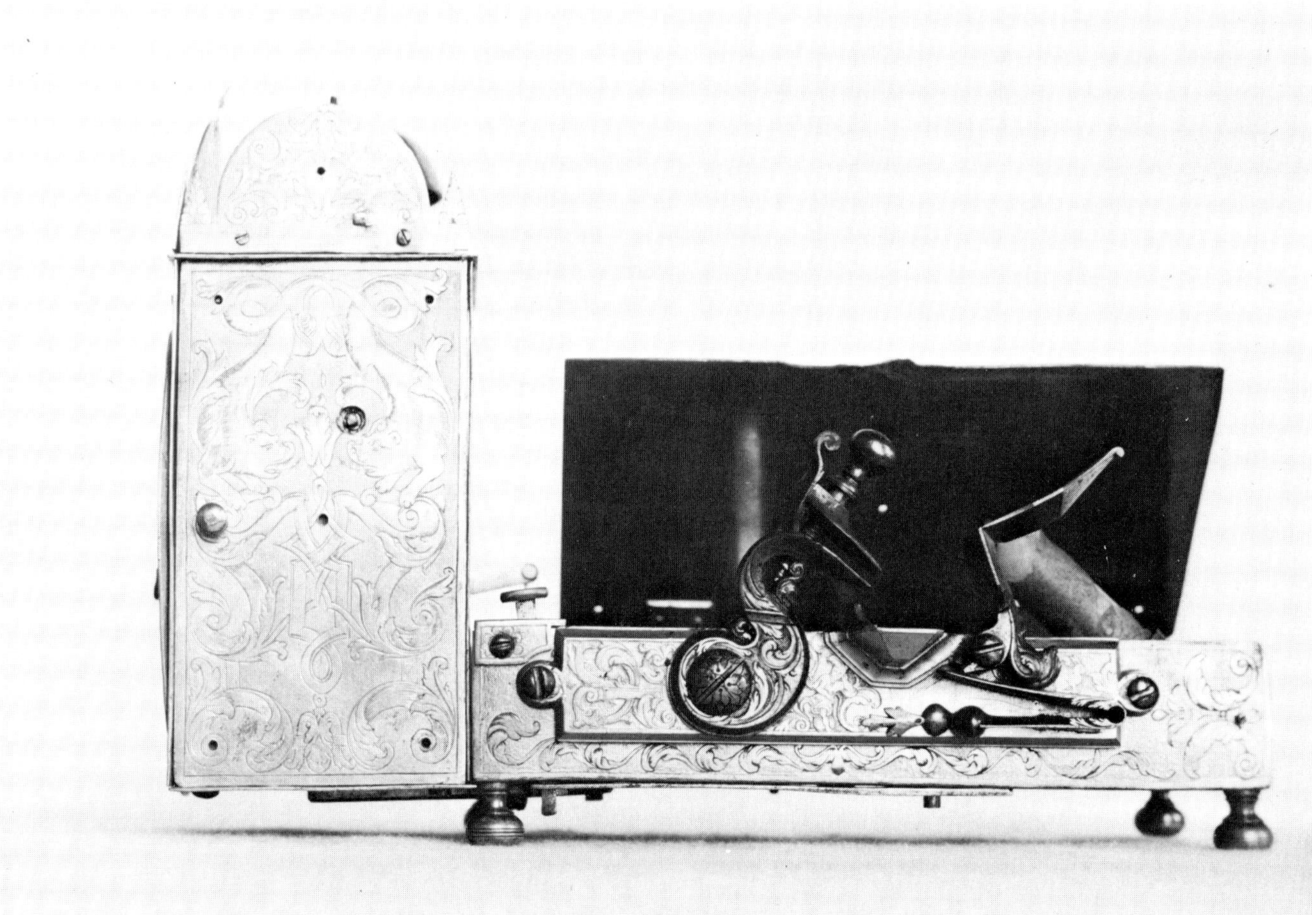

682

the latitudes, longitudes and local time of thirty towns throughout the world.
Private Collection.

675 **A Precision Sun-dial, post 1861**
Signed 'Chronometre Solaire A. Molteni. 44 Rue du Chateau d'Eau Paris'.
Black lacquered brass.
Height approx. 200 mm.
The curved hour table is adjustable for latitude on a circular base. A sighting lens is mounted to the south of the meridian line opposite a vertical scale marking noon and with the analemma of the equation of time. Sunlight focussed through the lens onto the hour scale thus provides readings accurate to a minute. Dials of this kind patented by Fléchet in 1861 were regularly used in midcentury France for regulating railway clocks.
Musée du Temps de l'Histoire, Besançon, 953.7.1790.

676 **An English High Precision Mechanical Equinoctial Dial, early 20th century**
Hemispherical bronze bowl containing the geared mechanism. Reads to 1 minute.
Private Collection.

677 **A Water-powered Clock, 1867**
Signed 'Idrocronometro P. Embriaco'.
Sheet iron.
Height 1 m 250.
A simple pendulum clock in which impulse is given to the pendulum by two weak springs attached to two shallow troughs which seesaw as first one then the other fills with water. Invented by the Dominican G.B. Embriaco (18291903) several such clocks were manufactured, some large scale examples being set up in public places in Italy and elsewhere.
Musée Internationale d'Horlogerie.
Bibliography : Ungerer 372-3.

678 **A Swiss Travelling Clock, early 20th century**
Signed 'Omega 3341654'.
Nickeled and silvered brass and steel, 82 mm x 25 mm.
Nederlands Goud, Zilver en Klokkenmuseum, Schoonhoven, HAK 445.

679 **An Atmospheric Clock Using Air Pressure Variation to Automatically Rewind Itself, c. 1930.**
Signed 'Le Coultre Le Sentier', 240 mm x 200 mm x 168 mm.
The clock uses the principle patented by Jean Léon Reutter in 1926.
Musée Internationale du Temps, La Chaux-de-Fonds, IV-595.

680 **A Speaking Clock, c. 1910**
Marked with a crowned 'GB' on the movement and 'DRP' on the dial.
Wood and metal, 240 x 260 mm x 430 mm.
The movement with going barrel and platform lever escapement carries a vertical gramophone system and loudspeaker functioning with a tape carrying twelve time-signals.
Musée Internationale du Temps, La Chaux-de-Fonds, IV-350.

681 **A Quartz Speaking Clock, c. 1985**
Signed 'Seiko quartz'.
Bronzed metal and steel.
The clock has no dial, when the top of the pyramid is pressed a woman's voice announces the time to seconds.
Private Collection.

683

684

685

The Uses of Time

Alarm Mechanisms and Transport

As the precision with which time can be kept or measured increases, so does the range of uses to which it can be put.
Its use for structuring the life of a monastery illustrated earlier depended less upon precision in measurement than on determining periods of time that were proportionately congruent. The successful use of time in nineteenth century navigation however depended upon a high regularity of performance in the time-keeper.
The growth of precision in counting, and determining, smaller units of time, and of higher reliability in the keeping of time, make an increasing number of applications possible. The most obvious use of keeping time is to receive a warning when to do something -most commonly when to wake up.
Alarms have a history almost as long as clocks. A water-clock alarm was used by Aristotle, and their history thereafter is continuous.
Like clocks and watches, alarms illustrate the democratisation of timemeasurement as they evolve from luxury objects into the most banal items of everyday life.

682 A Percussion System Alarm, late 17th/early 18th century
Gilt brass & steel with silvered dial, 250 mm x 80 mm.
Table clock with vertical verge movement with short bob pendulum in front of the dial with central alarm disc and going barrel. The percussion, pistol-type, alarm is set off by a spring transmission release, the resulting spark igniting a powder trail which runs to a small cannon. At the sounding of the cannon the lid of the clock springs open and a candle is ignited.
Musée du Temps et de l'Histoire, Besançon.
Bibliography: (For a similar percussion system) Welcomme, 436.

A Sounding of Alarms 1800 - 1970

In the following sequence only the signature if any, one key dimension, and unusual methods of operation are indicated.

683 Signed 'Invt. Lareche No 830' 'J.L. Galerie de Valois Palais Royal No. 164' and stamped under the alarm lever 'Japy'.
Width 110 mm.
A watch with handsetting on the square through the dial is placed on a disc and clamped in place. The hour is selected on the star wheel. If the toothed key is placed on the centre arbor corresponding to the minute hand, the alarm can be wound. Sounds on a bell.

684 Signed 'Henry Robert Paris' '5610', mid-19th century.
Signed on the movement 'Foi a Paris No 106'.
Length 65 mm.
Operation as for number 683.

685 Signed 'R à Paris', mid-19th century
Height 150 mm.
A simplified version of nos 683 - 684. The watch is placed on the plate and the key placed on the winding arbor. The spring-suspended bell is locked in place and a cord with a weight at one end is coiled round the shank of the key, one turn equalling one hour. The operation of the watch causes the cord to unwind until the weight presses on a spring set below it, thus releasing the bell to sound.

686 An Alarm Watch, 2nd half 19th century
Diameter 51 mm.
Single spring for the alarm and the cylinder movement sounds on a gong.

687 Marked with an Anchor in a Shield with a Five Pointed Star Above and 'Fot Rte 17675'
Horizontal drum, time being shown by a fixed hand alarmsetting by means of the cockerel above.

688 Signed 'Chronoreveil A.V. Hors concours Exposition Paris 1889 Breveté S.G.D.G. Deposé', c. 1890
Diameter 115 mm.
Watch form alarm with central setting dial set through the pendant.

689 German, c. 1890
Height 190 mm.
Pin pallet escapement, fixed key winders, timesetting by a pin through the main wheel arbor.

690 Signed 'Van Galen & Terlinden', c. 1890

691 Signed 'Lenzkirck A G U 1 million 715098', c. 1890
77 mm square.
Pin pallet operation.

692 Signed 'Reveil Avertisseur Breveté S.G.D.G.'
On movement 'Medaille de Bronze' and 'Fot Breveté SGDG 29519'.
Height 106 mm.
Pendulum, alarm sounds by hammering on the side of the case.

693 French, late 19th century
Diameter 58 mm.

686

687

688

689

690

691

694 **Marked 'A V Déposé Breveté' and mark of a walking lion (Duverdry & Bloquel for whom see Allix 439). End 19th century.**
Height 155 mm.

695 a **French, marked 'AG Breveté 2217'**
Diameter 59 mm.

695 b **Signed 'Repetition Junghans' with their mark**
Diameter 125 mm.

695 c **Signed 'Lenzkirch A G U' and on backplate '00974 Lenzkirch A G U 1 million', c. 1920**

696 **Signed on movement 'Junghans B 23 with their mark, c. 1925**
Diameter 125 mm.

697 **Signed 'Uhrfabrik Muhlheim, Muller & Co', c. 1930**
Diameter 126 mm.

698 **Signed 'Kienzle Piano-Forte', c. 1935**

699 **Signed 'Kienzle', c. 1935**
Height 50 mm.

700 **Signed 'Junghans', c. 1950**
Height 74 mm.
Travelling alarm.

701 **Signed 'Junghans', c. 1950**
Height 50 mm.

702 **Signed 'Mauthe', c. 1955**
Height 77 mm.

703 **Chinese, c. 1960**
Height 150 mm.

704 **Signed, 'Mauthe Made in Germany, c. 1960**
Height 100 mm.

705 **Signed 'Zetka 7 jewels Germany', c. 1965**
Width 100 mm.
In form of book.

706 **Marked 'Made in China 6705' and Chinese Characters, c. 1970**
Diameter 115 mm.
Nederlands Goud-, Zilver- en Klokkenmuseum, Schoonhoven.

As the structure of European societies became more complex, so time became essential. The sophisticated, interconnected coach and rail transport systems of the late 18th and 19th centuries would have been unworkable without reliable timekeepers.

707 **A Clock for Use in a Sedan Chair or a Coach, late 18th century**
Signed 'Berthoud à Paris', '4646'
Diameter of case 350 mm,
of dial 125 mm.
Turned mahogany case with suspension mount and ring with cord. Horizontal verge escapement with fixed pendulum probably replacing an original silk thread suspension. Rack striking hours and half hours at will by pull cord.
Nederlands Goud-, Zilver- en Klokkenmuseum, Schoonhoven, WA 83.

708 **A Dial for Indicating the Departure Time of Coaches, c. 1800**
Inscribed 'M'.
Mahogany, brass and glass.
Diameter 255 mm.
A manually set dial which would have been set up in the hall of a coaching inn to indicate the time of the next scheduled coach departure. 'M' indicates mail coach as opposed to passengers, and it is interesting to note the early use of a 24 hour dial.
Museum of the History of Science, Oxford.

709 **A Mail Guard's Watch c. 1830**
Signed 'John Walker 490 Princes Street, Leicester Square' , 100 mm x 80 mm x 40 mm.
The Science Museum, London.

710 **A Railway Watch, c. 1865**
Signed 'Lancashire Watch Co.' Prescot
Full plate movement with keyless winding.
The Science Museum, London 1975-223

711 **A Motor-car Dashboard Clock, early 20th century**
Signed 'Omega'.
Diameter 85 mm.
Eight day duration.
Nederlands Goud-, Zilver- en Klokkenmuseum, Schoonhoven, HAK 445.

712 **A Chronograph-counter for Automobiles, c. 1895**
Inscribed 'Automobile'.
Brass, steel with nickeled case.
Diameter 55 mm.
The instrument uses the system patented by

692

693

694

695a

346

695 a

695 b

696

697

698

699

700

701

702

703

704

705

706

707

717

708

711

719

Alb. Jeanneret & Frères of St Imier in 1891 (Swiss patent No 3364).
Musée Internationale du Temps. La Chaux-de-Fonds, lnv 720.

713 The 'Siderograph', a Sidereal Time-piece Graduated Directly in Degrees of Arc for Aerial Navigation, c. 1955
Signed 'Longines Breveté' '6865964'.
Aluminium, brass and steel.
Diameter 105 mm.

714 A Recording Speed Indicator for Trains, c. 1960
Signed 'Indicateur Flaman vitesses en Kilométres à l'heure Ateliers Vaucanson Paris'.
295 mm x 320 mm.
Musée du Temps et de l'Histoire, Besançon.

Time that is the measure of speed also determines resting periods.

715 A Parking Meter and a Park Card, 1990
Eltra Parkeergroep.

716 The Most Famous of Timed and Scheduled Journeys, 1874
Jules Verne, *Round the World in Eighty Days,* Paris, 1874.
Early editions in French and English.

Timing in Warfare

...there is nothing so conducive to an interest in time as military service, unless it be factory employment. I myself spent most of my first army pay for a Swiss-made wrist-watch with luminescent hands... Reveille in my camp came at 04.45 hours, and we had carefully measured times to wash, dress, and do everything else we did. With such a time-table, the only way to free oneself for other things was to know the hours and minutes. Looking back, I can see that I was like one of those medieval monks for whom lateness could be a grievous sin. He risked loss of Salvation ; I risked kitchen police. He had his bells ; I had my trusty watch.
'David S. Landes, *Revolution in Time,* 1983.

717 An Aid to Keeping Time in Warfare, c. 1812
Signed 'Adie Edinburgh'
Brass, with red straightgrained morocco fitted case.
Diameter 76 mm.
A typical box sextant, of which Adie was probably the retailer rather than the maker, reading to one minute. At a later date (?1880/90), it has been inscribed with the information that 'This sextant was used during the Peninsular War by General Sir Thos. MakDougall Brisbane Bart, with which he kept the time of the British Army'. The last phrase, which quotes a remark by the Duke of Wellington, presumably means that Brisbane used the sextant to make meridian altitude observations to check his time-piece with which those of his officers were synchronised.
Thomas MakDougall Brisbane (1793-1860), soldier, colonial governor and astronomer served throughout the Napoleonic Wars. Governor of New South Wales 1821-1825, the city of Brisbane is named after him.
Royal Museum of Scotland, T1956-87.

361

497

499

501

502

503

510

722

725 a

725 b

724

718 A Terrorist Watch, late 19th century
Melchior metal.
Diameter 75 mm.
The watch has a bridge movement with lever escapement and a very large going barrel. An electric contact is positioned at 12 o'clock in the path of the minute hand. The watch is contained in a red leather box together with three dry batteries and has two screw contacts on the outside. By a series of contacts and resistances the box acts as a conductor for electricity to a bomb. The minute hand contact acts as the switch.
Musée du Temps et de l'Histoire, Besançon. Collection : Maitzner.

719 A Stopwatch Timer with 'Flyback' mechanism for Projectiles, c. 1930
Not signed but probably Swiss made. Inscribed 'Patt. 6 47376'.
Brass and steel with plated case.
Diameter 56 mm.
The white painted dial with red and black numerals, records distance travelled in yards up to 5000 against time - 6 seconds - by means of a single brass hand and with subdivisions for each 100 yards. Keyless wound half-plate movement with ruby endstones and ruby lever escapement. Pump operation through the pendant with stop-arm acting directly on the shallowly toothed outer edge of the balance wheel. Flat steel balance spring and regulator. Plain case with milled decoration on the inner face of the inner cover.
Private Collection.

720 A Clock for Controlling a Convoy, 1940-45
Signed Lilley & Reynolds Ltd London E.C. 3' and on the movement 'Coventry Astral Made in England 123'.
Brass.
Diameter 153 mm.
Ships' clock showing hours and minutes with subsidiary seconds used as a control clock for executing convoy zigzag and other manoeuvres.
Rijksmuseum 'Nederlands Scheepvaartmuseum', Amsterdam, S 4642/12.

721 Northern Irish Terrorist Bombs with their Timing Mechanisms
Photographs.
Royal Museum of Scotland, Edinburgh.

Timing in sport

Since time is the measure of speed, all speed-based sporting competitions are ultimately controlled by time.
In Byzantine Constantinople, chariot-races were controlled by water-clocks and in late 17th and early 18th century England, the rise in popularity of horse-racing can be linked with the development of the seconds-indicating watch and with the stop-watch.

722 A Stop-watch for Horse-race Timing, mid 18th century
Dutch.
Gilt and silvered brass, brass and steel.
Diameter 73 mm.
Centre seconds and minute hand both

indicating on a scale of 60. An ordinary verge watch movement was adapted by the addition of a second motion work of the seconds hand, and a toothed wheel was mounted on the crown wheel to intercept the stop work lever operated from the side.
Nederlands Goud-, Zilver- en Klokkenmuseum, Schoonhoven, Di 439.

723 A Recording Chronograph, c. 1825
Signed 'Chronograph Rieussec'.
Brass and steel.
Diameter 57 mm.
The dial divided to 60 rotates making a complete revolution in 1 minute beneath a fixed arm which is linked to an ink reservoir. A button in the side of the case allows a spot of ink to be deposited on the dial when desired. A second button controls the start/stop of the movement. Patented by Rieussec in 1822, this instrument was first tested timing horseraces 1 September 1821 at St Mandé for which purpose it seems to have been devised.
Musée Internationale d'Horlogerie,
La Chaux-de-Fonds, I-932.
Bibliography : Grunenwald.

724 A Marker for Skating Competitions, 1897
Signed 'G. Tasma Gorredijk'
Oak, brass and steel, 265 mm x 150 mm x 45 mm
The simple mechanism contains two levers attached to cords set in the eyes on either side of the case, and internally to the 'pendulum' of which the bob is visible in the curved aperture. The skater's aim in Frisian competitions, is to pull the cord on his side of the box before his opponent, the locking of the pendulum in place on his side signalling the victory.
Nederlands Goud-, Zilver- en Klokkenmuseum, Schoonhoven, Di 1.

725 Three Clocks for Timing Pigeon Races, 1907, c. 1920 and c. 1925
a. Hattersly & Turner's patent.
b. Toulet (Excelsior) Brussels.
c. Benzing-Strobbe Brussels.
The Science Museum, London (a), Nederlands Goud-, Zilver- en Klokkenmuseum, Schoonhoven (b & c).

726 A Sports Stadium Time early 20th century
German.
Sheet iron and brass, diameter 332 mm.
The dial carries two sets of graduations 0 - 9 minutes and seconds. The wheel train and hands are turned by a synchronous motor which can be arrested manually or by electrical impulse.
Musée du Temps et de l'Histoire, Besançon, Inv Jules Haag, 16.

727 A Timer for Motor Races, c. 1920
Signed 'T & F Mercer'
Time recorder developed by General Sir Capel Holden for the Royal Automobile Club. In a convenient carrying case a chronometer drives and supplies the time input to a chronograph on the paper of which two pens mark by electrical impulse the moment when a car's wheels pass over a specific point, while a third marks each half second.
The Science Museum, London 1988-429
Bibliography : *The Autocar* 28 October 1927, 956.

Timing in science

Time is part of the content of science, and the crucial role played by time-measurement in astronomy and navigation has been illustrated above.

The ability to time short durations and speed with precision enables a further variety of phenomena to be studied.

728 The Apparatus of Gerard Moll for Measuring the Speed of Sound, c. 1822
Speed and distance travelled are measured in terms of time. But to measure accurately the speed of a phenomenon such as sound, very great experimental care and high precision instruments are required. During the late 18th and early 19th centuries the measurement of the speed of sound excited considerable scholarly curiosity. In 1822 a commission of the Bureau des Longitudes in France made a series of experimental determinations of the velocity of sound arriving at a value of 330.6 metres per second. Such determinations however are only of value when they can be compared with results obtained by other observers in different but comparable conditions. Moll repeated measurements using a similar method to the French. An observer was positioned at a known distance from a cannon and recorded the time interval between his seeing the flash of discharge (virtually instantaneous) and hearing the sound of the explosion. An artillery battery was placed at the disposal of Moll and Albert van Beek. Cannon were fired simultaneously at night from positions 9 miles apart and the interval between flash and sound was recorded at each end, the results being averaged. Moll's result was 332.05 metres per second. The accepted value today is 331.45 metres per second.
Apparatus used by Moll shown is
a. A late 18th century English binocular telescope by Dolland on a tripod stand.
b. A repeating circle signed Lenoir of the type developed by Borda and Lenoir for the 1787 triangulation between the Paris and Greenwich observatories.
c. Two clockwork decimal timers with heavy pendulum equipped with stop work. The dials carry three hands operating on three scales of 100 parts.

728

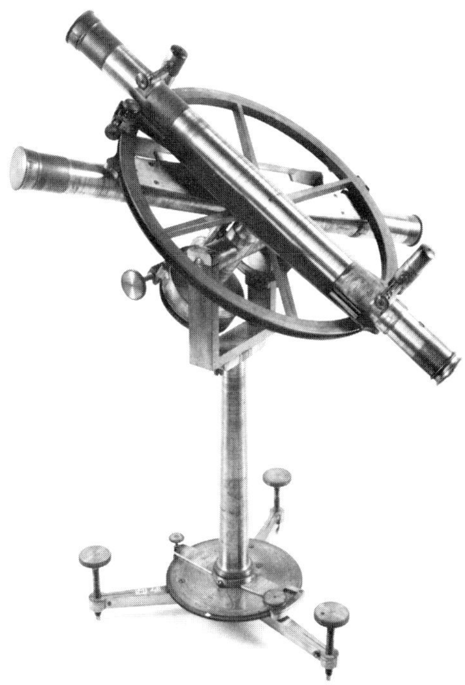

728

527 530 538 544

598 600 602 604

554 562 568 570

623 629 632 644

574 576 586 594

646 648 650 654

731

732

729

Gerard Moll (1785-1838) was Director of Utrecht University Observatory (from 1812) and concurrently Professor of Physics from 1815 in the same university.
Universiteitsmuseum Utrecht, M411.
Bibliography : Moll.

729 **An Apparatus for Measuring Tenths and Thousandths of Seconds, c. 1880**
Signed 'M. Hipp Neuchartel No 7741'.
Brass, silvered brass, boxwood and black lacquered pine.
Overall height approx. 600 mm.
The chronoscope consists of a clockwork movement started and stopped electrically, either automatically at a distance or manually. Once started the fall of the weight which drives the wheelwork and so the indexes is controlled by the vibrations of a thin metal oscillator. Such instruments were commonly used in late 19th/early 20th century research laboratories concerned with such diverse research as the measurement of reaction times or the speed of nervous and muscular action in experimental psychology laboratories, or the movement of projectiles mainly for military purposes.
Musée International d'Horlogerie,
La Chaux-de-Fonds, IV-298.
Bibliography : Favarger 5047; Toulouse & Pieron ii, 26-9.

730 **Heliostat and Accessories, late 18th Century**
Brass and steel, clock 125 mm x 71 mm, mirror pillar 224 mm.
Rectangular clock case with arched top. The case is set at 38°, the co-latitude of the place of use, on the top of a three foot stand with levelling screws and plumb-line. Side-mounted pendulum with silkthread suspension around cycloidal cheeks. The dial shows hours, minutes and seconds and from the centre arbor extended through the dial-plate, a drive arm in the same plane and therefore parallel with the polar axis connects through a U-shaped bracket with a sleeve sliding on a rod fixed to the back of a mirror pivoted in a 'U'-bracket at the top of a separate stand.
As the clock drives the linked arms around the angle of the mirror will be altered. If the heliostat is correctly set up a steady beam of sunlight can be obtained for use through a solar microscope or for experiments with lenses and prisms, for the clock drives the mirror round to follow the sun.
This instrument closely resembles one described by 's Gravesande in 1748. It was purchased for Teyler's Museum in either 1789 or 1791.
Teyler's Museum, Haarlem.

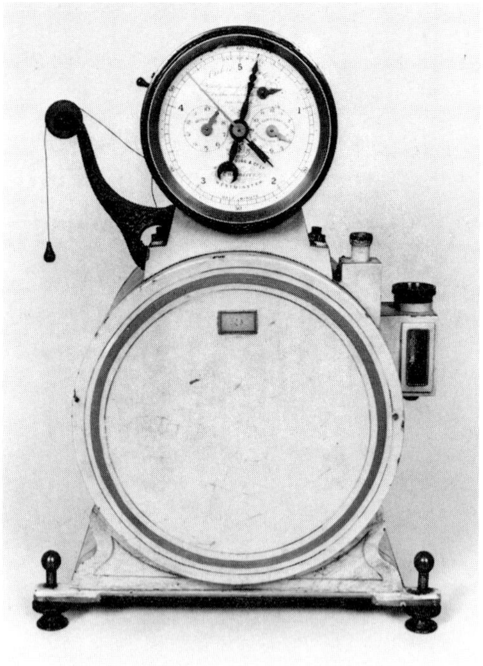

733

734

731 A Drive Clock for a Telelscope, mid to late 19th century.
By Merz.
Robust four post movement with air-braked speed governor using a pair of weighted vanes linked by a pair of parallel coiled cylindrical springs.
Universiteitsmuseum, Utrecht, WM3254.

732 A Chronometer Used During the Transit of Venus Observation, 1874
Signed 'Knoblich Krille Altona Nachfolger No 1700'.
Brass and steel in mahogany case 195 mm x 195 mm x 100 mm.
Teyler's Museum, Haarlem.
Bibliography : Kaiser (2).

733 A Timer for the Combustion of Gas, c. 1890 and an automatically timed gas Switch, c. 1900.
a. Signed 'William Sugg & Co Ltd Engineers Westminster' and inscribed '5 Cubic Feet Hourly rate of Consumption by the observation of one Minute'.
Cast iron, silvered brass, brass and steel, approx. 480 mm x 270 mm.
The main dial is divided into 60 seconds and marked at the 60 point 'one minute'. Subsidiary dials for minutes 1 - 10 and 'one cubic foot of gas divided into twelve'.
Circular half plate movement contained in a massive case, plain balance, warning alarm on a bell.
Teyler's Museum, Haarlem.
b. Patented by Gunning in 1899 this is the first fully automatic on/off gas switch for a street lamp.
Double arm and star wheel timing mechanism controlled by an 8-day clock movement.
The Science Museum, London, 1938-463

734 A Timer for An Unknown Purpose, c. 1930
Signed 'Smiths English Clocks Ltd Cricklewood London'.
Electric dial-clock with handset at bottom of the case. The dials is marked in minutes 0 to 20 by two minute intervals.
Museum, The Hague 94203.

Memento Mori

Ever since the fact of death first produced anxiety about the future in primitive man, death and time have been fundamentally linked. Whether time is identified with death, or merely seen as its instrument, time the destroyer is one of the most potent and persistent themes in art, literature, and religion.

Time, Money & Productivity

For time is the measure of business, as money is of wars; and business is bought at a dear hand, where there is small despatch.
Francis Bacon, 'Of Despatch', *Essays,* xxv.

In contrast with time's traditional role as destroyer, in many aspects of commercial life, time is the factor that increases value. In banking and life insurance, time is an essential element of the commercial equation.

736 A Selection of Interest Tables, 16th to 20th centuries
Economisch - Historische Bibliotheek, Amsterdam.

737 The passage of time which brings to perfection and then destroys the flavour of wine, is also the cause of the monetary value of this - undrinkable - bottle.

But time may serve in other ways. In order to wind up mantle clocks supplied by insurance companies, the owner must deposit his weekly contribution which will later be collected.

738 Insurance Clocks
a. French, Signed 'La Pendastrava. Dispositif Bte S.G.D.G. Marque et modèle déposés'.
Gilt brass with veneered wooden case with exotic wood inlay, 225 mm x 330 mm.
The movement has a lever escapement of which the balance may be blocked in position by a lever. This lever is lifted by the fall of coins introduced into the machine, which are collected in a drawer below the movement. This drawer is locked, being emptied once a year by the insurance company collector.
b. Signed 'Groot Noordhollandse van 1845'.
180 mm x 220 mm x 70 mm.
c. Danish. Inscribed on the front 'Spaarkanven for slagelse og' and on the back 'N K s 22/1 54'.
Green painted metal, 101 mm x 95 mm x 50 mm.
a. Musée du Temps et de l'Histoire, Besançon; b & c) Spaarpot Museum, Amsterdam.

739 Early Inculcation of the Time/ Money Relation: Moneyboxes late 19th/ early 20th century
a. In the form of a white clown inscribed 'Time is money' in red letters.
400 mm x 80 mm x 5 mm.
b. In the form of an earthenware clock.
160 mm x 90 mm x 40 mm.
c. A green tin-plate box in the form of a bracket clock, German.
75 mm x 55 mm x 45 mm.
d. In the form of a stoneware kitten with a clock, Dutch.
85 mm x 95 mm x 60 mm.
e. Circular blue tin-plate box with japanese dolls indicating the day and the week. Signed 'Fujimo...Japanese dolls'.
f. Tin-plate box in the form of a clock.
Inscribed around the rim:
'Tickety tock
This elfin clock's
Really a money box'.
When a coin is inserted the dial of the clock moves round.
Diameter of dial 150 mm.
Spaarpot Museum, Amsterdam.

740 The Inexorability of Time and Money, c. 1840
L'Horloge de Credit, coloured lithograph from Epinal.
Lithographed by 'J.P. Delmain C M', published by 'Pellerin & Cie, 397 mm x 295 mm.

741 The Time Difference in Banking Today
Reconstruction of a dealing room.
Rabobank.

Time and Productivity

Time which is used to order systems and coordinate activities can also be used to determine rational work patterns and so increase productivity and profits.
From early in their history the sand-glass and the weight-driven clock have been used for this purpose.
Already in the mid-14th century the periods to be worked by stone-masons and quarry workers in north Italy, and by weavers and wool-combers in N.W. France and Flanders were being determined by specially designated clocks, and break-periods were timed by sand-glasses.
Wage payments by the hour, which begin c. 1400, were a direct result of the clock, for they depended upon the equal hours of the clock to be acceptable.
Industrial operations were also timed by sand-glasses.

742 A Sand-glass Timing the Operation of a Stamping Mill, c. 1450
Illustration from a manuscript *Das Feurwerkbuch....*
Photograph: The Science Museum, London.

743 A Sand-glass Used for Timing a Potter's Furnace, mid-16th century
From Cipriano Picolpasso, *Li tre Libri dell' arte del Vasaio.*
As the scale of industrial enterprises grew and the capital at risk was greater, careful owners and managers created elaborate systems of time management.
Already in c. 1700 the Crowley Iron Mills in Yorkshire were operated and costed entirely on a basis of time. Sounding the bell and noting attendance times were the responsibility of the factory 'monitor'.

'Wm. Wright's time paper for [Account] 892
Hrs.
Mins.
Saturday, begun at 4, entred not till 5, and went to Bladen to ship a keel of goods for Ralph Jackson at 7, so employed till 11
5 hours 0 mins
Monday, entred 'Run' at 7 at night, continued at work till 9
2 hours 0 mins
Tuesday, went at 8 to trye Thomas Meanly's stock,and came [back] at 1/2 hour past 8
0 hours 30 mins
Wednesday, begun at 4, entred 'Come' not till 6
2 hours 0 mins
Thursday, the Monitor absent, 'Come' at 5, 'Run'at 8, 'Come' at 9. 'Run' 12. 'Come' at 1, 'Run' 4.
9 hours 0 mins
Friday, entered 'Run' at 7 at night, 'Come' continued till 10
3 hours 0 mins
The time control of large enterprises became increasingly current during the 18th and early 19th centuries. Such commercial practices underlie the rational principles for organisating industry that Charles Babbage introduced. Particularly intrigued by the relationship between time and force, Babbage also advocated the use of mechanisms to carry out repetitive tasks, and of timing and recording mechanisms to check upon people doing repetitive tasks.

Perhaps the most useful contrivance of this kind, is one for ascertaining the vigilance of a watchman. It is a piece of mechanism connected with a clock placed in an apartment to which the Watchman has not access; but he is ordered to pull a string situated in a certain part of his round once in every hour. The instrument, aptly called a tell-tale, informs the owner whether the man has missed any and what, hours during the night.
Charles Babbage, 1836.

744 The Truth of Babbage's Assertion a Watchman's 'tell-tale', c. 1920
Signed 'Dent's portable Registering Tell-Tale no 44416'.
100 mm x 100 mm x 50 mm
The Science Museum, London.

728

730

745 **Two Punching-in Clocks, early 20th century**
The ring on the front of the clocks has numbered holes. Each number represents a worker and by rotating the arm and pushing a pin on the end of it through the appropriate hole, the time of the worker's attendance and departure is recorded on a time chart held on a drum revolving inside the case.
(b) Signed 'Wurthembergische Uhrenfabrik Börk & Söhne'.
Royal Museum of Scotland, Edinburgh (a); Musée Internationale d'Horlogerie, La Chaux-de-Fonds (b).

746 **A Master Clock and Slave Dial for a Factory Time-system, c. 1920**
Brass and steel, overall height approx. 1.20 m.
The synchronome clock was developed by Frank Hope-Jones between 1895 and 1910. A first patent of 1897 described an impulse system which produced a good electrical contact without interfering with the time-keeping functions of the clock (see no 417 above). In 1905 a second patent described an improved form of contact system in which the gravity arm was 'hung up' on a catch while the pendulum counted out 30 one second beats. The arm was then released firmly onto the contact.
The synchronome system was that most widely used in Britain for controlling step by step dials in circuits during the early decades of the twentieth century, but it never gained much popularity elsewhere.
Private Collection.
Bibliography : Hope-Jones, 55-7; 88-91.

With the establishment of rationally time-structured factories, it was a natural step to a rational analysis of the time-structure of work.

747 **The Pioneer of Time and Motion Study, F.W. Taylor and some of his works**
Photograph
Modern time and motion study is identified with the name of Taylor (literally in some non-English speaking countries where the subject is known as Taylorism), who was the pioneer of its application in America at the beginning of the 20th century. Initially adopted with some enthusiasm by management, his ideas quickly provoked labour opposition, which feared that they would lead to unemployment, and were denounced as un-American. Ironically at the same time as they were being attacked in the United States, Taylor's ideas were being enthusiastically adopted in Soviet Russia.
Books by F.W. Taylor
The Principles of Scientific Management, New York & London, 1913
Etudes sur l'organisation du travail dans les usines (translated by L. Descroix) Angers, 1907.
Over arbeidsprestatie en loon regeling (translated by H.J. Hendrikse), Amsterdam, 1909.
The Royal Library of the Hague.

748 **The Work of Frank B. Gilbreth**
Books and Video film
With his wife Lilian, a psychologist by profession, Gilbreth extended Taylor's work. His particular contribution was in the quantitative analysis of work for which he invested special measuring machines and a new unit - the Therblig an anagram of his name - equal to one ten thousandth of a second. All Gilbreth's time-measurements were decimally based.

Books shown

Das ABC des Wissenschaftlichen Betreibsführung, Berlin, 1917, a translation of *The Principles of Scientific Management.*
W.R. Spriegel & Clark K. Myers (eds), *The Writings of F.B. Gilbreth & Lilian B. Gilbreth,* 2 vols., 1953.
The video film presented is composed of extracts from a longer film made shortly after Gilbreth's death from archival films in the possession of his widow.
The Royal Library of the Hague.
Department of Sociology, University of Leiden.

749

751

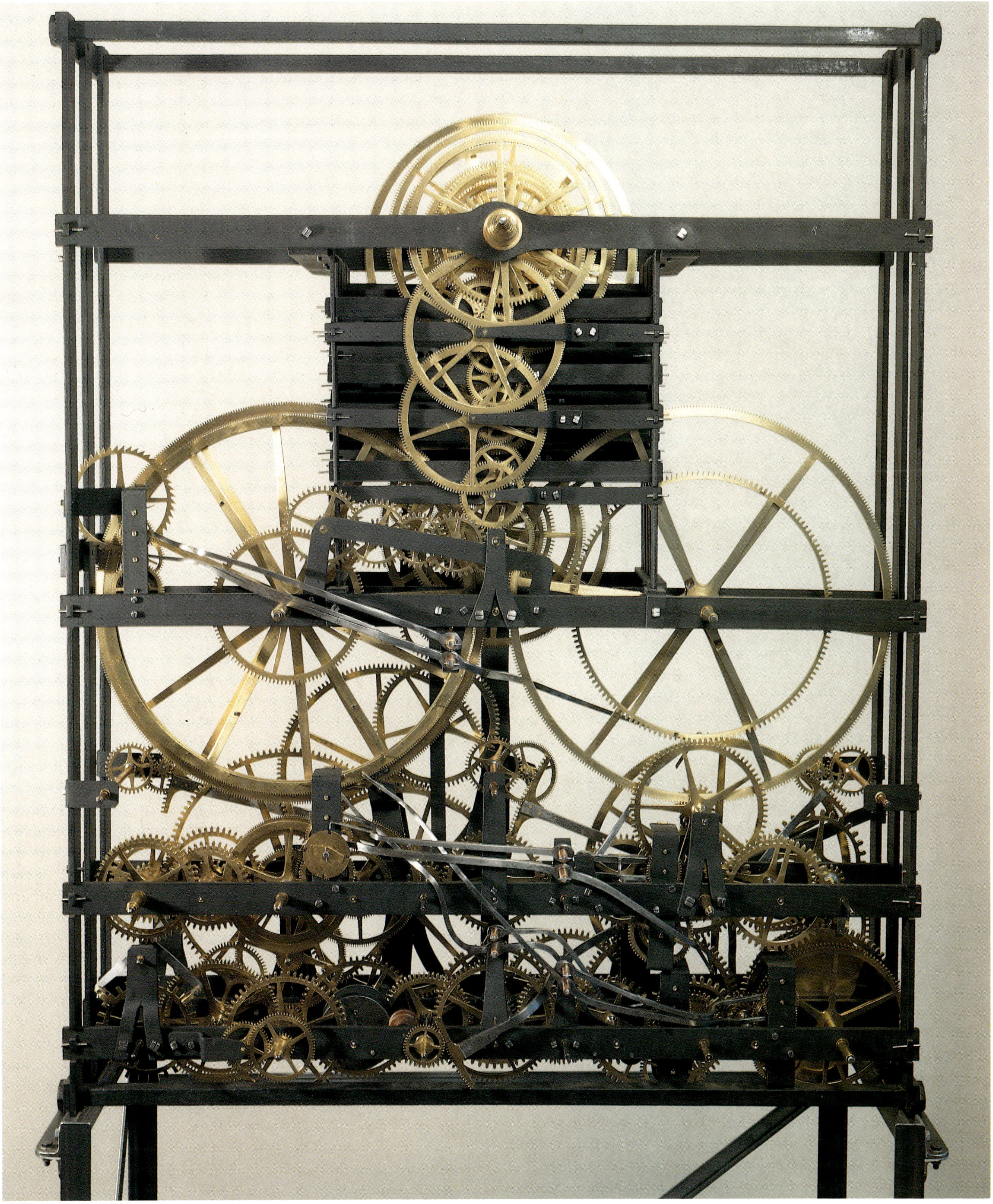

809

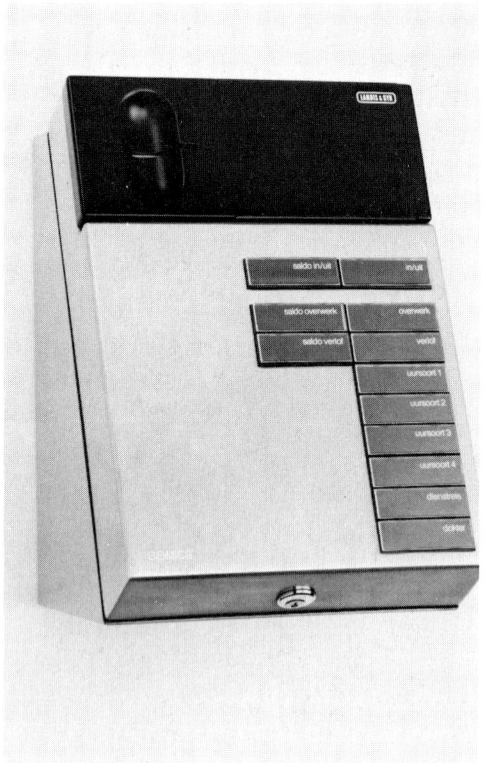

753

749 A Decimal Stopwatch Counter with Flyback Mechanism, c. 1900
Signed 'Elka Watch Cy Amsterdam'.
Nickel, brass and steel.
Diameter 51 mm.
The dial graduation runs from .001 to .01 subdivided to ten. It is thus theoretically possible to time to one tenth thousandth unit. The small centre dial counts the number of revolutions made and the flyback mechanism enables two actions to be timed at once. Swiss lever escapement with seven jewels.
Watches graduated in Terbligs (one ten thousandth of a second) must have resembled that shown here although since on the present watch the hands make one revolution in 36 seconds it provides decimal division of angles and was probably intended for astronomical or navigational purposes.
Nederlands Goud-, Zilver- en Klokkenmuseum, Schoonhoven, Di 437.

750 The Chronograph 'Michelin', c. 1931
From 1920 onwards, the Michelin tyre company set out to popularise the ideas of Taylor in France setting up a special committee to do this and publishing numerous books and pamphlets.
The registering chronograph shown here was intended to replace other forms of timemeasurer which had the disadvantage that they obliged the operator to momentarily stop observing his subject while recording a time thus introducing errors into his time profile for a given operation.
Société Michelin.
Bibliography : *Prospérité,* April June 1931.

751 An Alarm Diary for the Office, c. 1930
Signed 'Horo-Memo '.
Metal alloy with marbled finish, 230 mm x 160 mm x 100 mm.
The alarm mechanism controlled by the clock within the box can be set for any hour between 6 a.m. and 5 p.m. by setting the levers on the side of the box. Reminder calls for appointments can be programmed for a week at a time.
Cité des Sciences et de l'Industrie, Paris.

752 The Time Stamp 'Cosmo', c. 1915
90 mm x 70 mm x 50 mm
Model Patented in 1913
The Science Museum, London.

753 A Modern System of Time Management, 1990
Bomics of Landis & Gyr.
An integrated system for access control, time recording, planning and calculation, Bomics measures and controls time, provides a complete record of the use of time, and signals deviations from the plan, even those as small as a minute.
Bomics is an instrument for planning before and control after, as precise or as global as required, for work time and project time. It provides automatic warning signals whenever needed, and may be used for other purposes such as company purse and time controlled access keys.
Bomics is user-friendly for both the card holders and management, and can provide all requisite information concerning time and money.

Time in Music

Words move, music moves
Only in time
T.S. Eliot, *Burnt Norton,* V.1-2.
The nature of a piece of music can be varied dramatically by changing its measure, or the time in which it is played.

754 Changing Speeds
Recording of a well known piece of music played at several different speeds, each slower than the last.

Increasing mastery of time-measurement led to the development of the idea of time in music as something that could be prescribed.

Stand by a large Chamber-Clock and beat your Hand or Foot... to the slow Motions of the Pendulum, telling one, two, *with your Hand down as your hear it strike and* three, four, *with your Hand up; which I would have you observe in this* slow *sort of* Common-Time. *Also you must observe to have your Hand or Foot down at the beginning of every Bar.*
The second sort of Common-Time *is a little faster which is known by the* Mood, *having a stroke drawn through it. The third sort of* Common-Time *is quickest of all and then the* Mood *is retorted; you may tell* one, two, three, four, *in a* Bar, *almost as fast as the regular Motions of a Watch.*
Henry Purcell, *A Choice Collection of Lessons for the Harpsichord or Spinnet*

Musical time was perhaps the first time-system encountered by man which was completely independent both of himself and of the natural world. The time-structure of a musical work is entirely determined by the composer.
Investigations of the pendulum as a timer for music parallel its development as a timer for time.
Early suggestions by Mersenne of the use of a simple pendulum moved by hand were to have a long history.
More complex devices however were soon imagined and investigation of them throughout the 17th and 18th centuries eventually led to the model used with only minor changes since.

755 The Earliest Metronome with Centre Pivoted Pendulum, 1814
Signed '... D.N. Winkel in 1814...zu Amsterdam'.
Mahogany box enclosing a centre pivoted pendulum with two equal weights. The lever of the two weights is fixed, the upper may be slid along the rod to change the speed of beat. The interest of Winkel's invention was that it made possible a compact instrument, previous 'music chronometers' having used the full length of a long pendulum to attain the slower tempi. Gemeente Museum, The Hague.
Bibliography: Harding, 24.

755

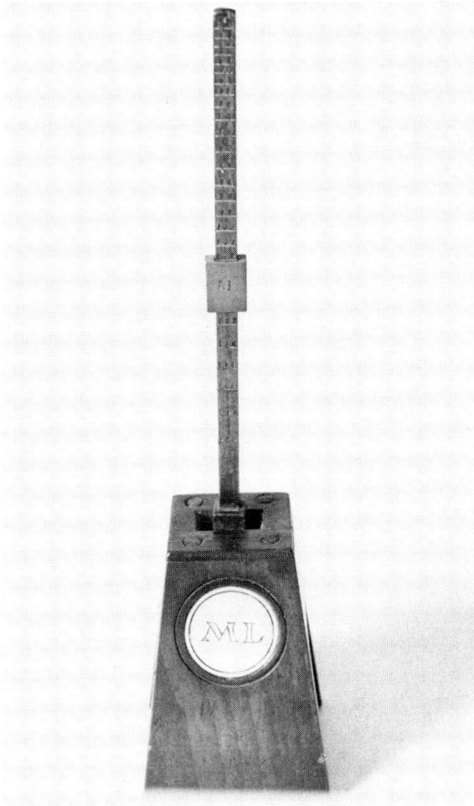

756

756 An Early Simple Metronome, c. 1820/30
Inscribed 'M' on the morable weight and on a later inlaid circular ivory plaque, 'ML'.
Mahogany, overall height 265 mm.
An open pyramidal stand carries the pendulum pivoted at its centre. The pendulum rod is graduated for number of beats per minute from 50 to 160.
From 50 - 60 the intervals are 2.
From 60 - 76 the intervals are 3.
From 76 - 120 the intervals are 4.
From 120 - 152 the intervals are 6.
From 152 - 160 the interval is 8.
Adjustment is by a sliding weight on the upper part of the counter-weighted pendulum rod.
Early form of the Maezel pendulum or metronome after he had seen Winkel's instrument (1815) and adopted its design despite Winkel's refusal to part with his rights.
Private Collection.
Bibliography: Harding, 22-28.

757

757 **A Decorative Version of the Winkel/Maezel Metronome, c. 1870**
Signed 'Pinfold's'.
Gemeente Museum, The Hague.

The metronomic scale is not borrowed from the measures of any one length peculiar to any one country, but is founded on the division of time into minutes. The minute being thus...the element of the metronomic scale, its divisions are thereby rendered intelligible and applicable in every country : a *universal standard measure for musical time* is thus obtained, and its correctness may be proved at all times by comparison with a stopwatch.
Anon, *'Maezel's Metronome', The Quarterly Musical Magazine and Reivew,* III, 1821, 303.

758 **A Clockwork Maintained Metronome, c. 1900**
Signed on the movement 'Paquet à Paris' ; inscribed 'Metronome selon Maezel Londres, Paris, Bruxelles, Hollande, Paris Amerique and on the scale 'JT L 2301'.
Mahogany, with brass movement and paper scale, overall height 213 mm.
Pyramidal box containing the metronome arm with sliding triangular weight for setting against the weight scale calibrated 40 to 208 (largo to presto). The pendulum is maintained by a simple mechanism composed of a going barrel gearing through a pinion to an escape wheel controlled by a pair of parallel pallets on the pendulum pivot arbor.
Private Collection.

759 **The Difference Time Makes**
Audio-visual presentation.
A Film and music are played together at correct and incorrect speeds.

The Times of the Natural World

Once when I howled for my food half an hour before the right time, my Chinese amah [nurse] wanted to give it to me, but my mother pointed to the clock and refused, at which the amah said in her beautiful pidgin English, 'Maskee mississy clock he no savvy. Baby clock, he inside tummy, he savvy'.
C.M. Bowra, *Memories 1898 - 1939,* 1966.

760 Drill Core from a Sediment in Liberia, Africa
Formed in the Pre-Cambrian, more than 570 million years ago, the rock is layered iron ore consisting of $Fe3^{+}$ and $Fe2^{+}$ - oxides (magnetite and hematite).
Technical University, Delft.

761 The Ground Beneath the Nieuwe Kerk to a depth of 24 m
Below the sand which was put in during the building of the church, are layers of peat consisting of the remnants of (swamp)-plants. The layers of clay under these were formed by sedimentation in the sea, as shown by many shells. The thick layer of sand on which the poles of the Nieuwe Kerk are resting was formed by windsupply, during the last glacial period (between 100,000 and 10,000 years ago). From 21m downwards sand follows the sediment was formed in the then Eam Sea.
Data of the Public Services, Amsterdam.

762 The Sub-soil of western Holland
The deepest drilled layer (Carbon) is approx. 300 million years old. At a depth of 12 kilometres, between Leiden and Medemblik, a number of layers (partly Jura and Cretaceous period) are worn out by erosion. The layer of sand on top of this lacuna dates from the last 70 million years (tertiary and Quaternary). The layer from which the stones of the Nieuwe Kerk originate, such as the Cretaceous period (sand stone from Bentheim) and the Lower Carbon (blue stone from Hainault), are surfacing in Belgium (not yet Antwerp).

763 Sedimentation Demonstration Apparatus, 1990
The Sedimentator consists of two perspex barrels filled with water, through which a long perspex tube is fitted. The tube is open at the top but also has an opening on the side in the lower barrel. The lower barrel is filled with white sand, the other with red sand. By continuously agitating the water in the lower barrel, a grain of white sand will land in the tube at regular intervals and there white sediment is formed. The water in the upper barrel is only agitated for a couple of seconds when a beam of light from a photocell is interrupted. A thin layer of red sediment will then be formed on the white sediment.
The tube shows the growing history of the exhibition (small apertures in the tube, slowly agitating) or of a single day (large aperture, strong tube agitation).

764 Section from the Roman Aquaduct Which Led Water from the Eifel Mountains to Cologne
Photograph R. Stenvert, Utrecht.

765 Chalk Sinter from the Roman waterworks from the Eifel to Cologne
Preservation of Monuments and Historic Buildings Service, Zeist.

766 Dendrochronology
Going back in time by jumping from one tree to another : the four drill cores cover a cover a period of two hundred years. On the right the youngest drill core, with the cambium at the boundary of bark and wood on the edge of the last annual ring, from a tree which was cut down in 1980. The second drill core was taken from a beam of a house built around 1877. The third and fourth drill cores are old trunks which have been covered with earth through the years and have thus been preserved.
From Schweingruber.

767 Periodical Movements of Leaves on Bean Plants. Left : by day ; right : at night
From Bunning.

765

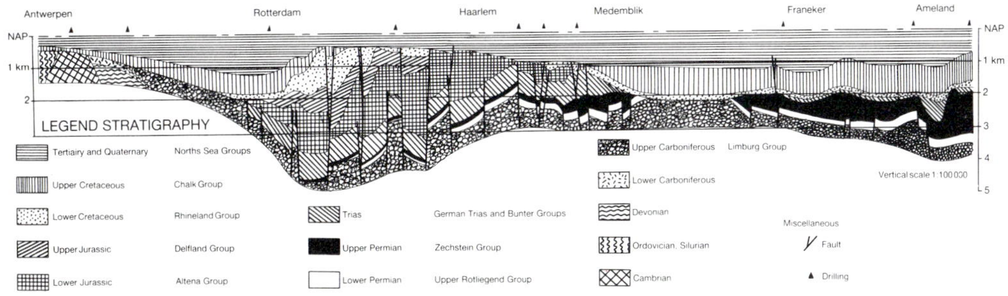

762

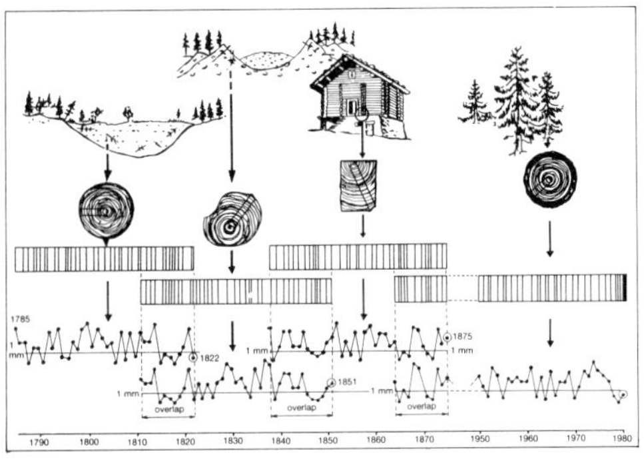

766

Phaseolus Multiflorus

767

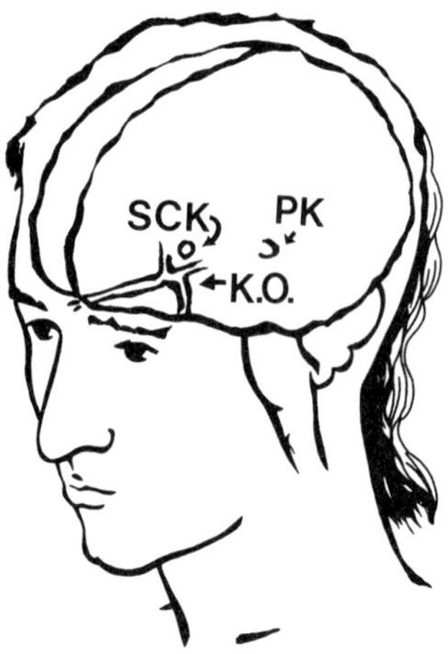

768

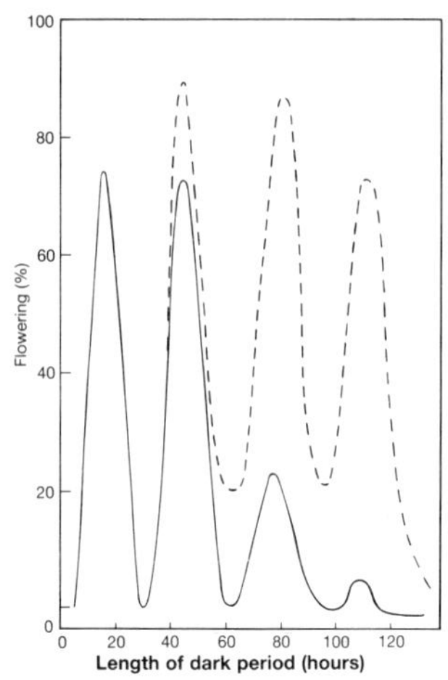

774

768 The Biological Clock in Man

The suprachiasmatic core (SCK) lies in the middle below the two large brain halves, immediatly above the crossing of the eye nerves (KO) and behind these the pineal gland (PK).

769 The Formation of Melatonine from Seratonine in the Pineal Gland

The transformation takes place in two steps, each of which is catalysed by an enzyme. The activity of the first enzyme is raised during the night and is then many times higher than by day.
From Arendt, Broadway, English, Kemp, Poulton and Symons.

770 The Scatter of Asthma Attacks During Twenty-four Hours Among a Large Group of Patients

771 The Field Mouse (Microtus arvalis).

772 A Bamboo with a Long Flowering Cycle

There are many kinds of bamboo, with various blossoming characteristics. The kind depicted here, *Phyllostachys bambusoides,* only blossoms once in 120 years and is planted for various purposes such as use in building, angling and in cooking (sprouts).
From Farrelly.

773 The Menstrual Cycle

774 Flowering Induction in the Chenopodium Rubrum

The plants are cultivated under continuous light. One week after sowing they can already be induced to flower by putting them in the dark for a number of hours and thereafter putting them again in continuous light. The percentage of plants which will blossom has been vertically stated. Continuous line : the plants are standing on wet filter paper. Dashed line : the plants are standing on filter paper with a solution of glucose. In the dark the plant cannot photosynthesize and hence can not produce any sugars. After a couple of days in the dark the plant's reserves become exhausted and the rhythm or the capacity to induce flowering, or both, are lost. The added glucose is absorbed via the roots.
From Evans.

771

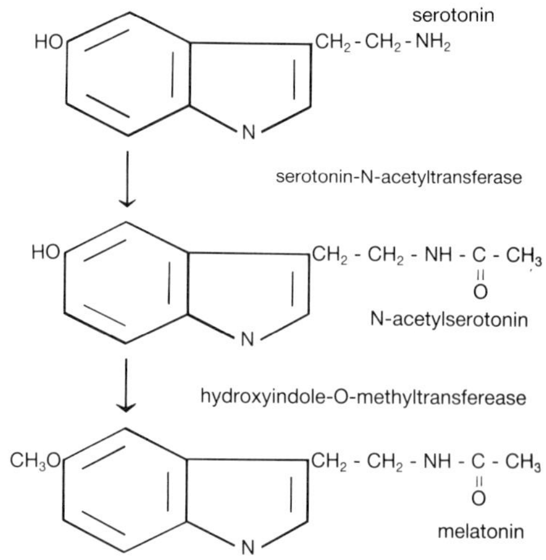

769

A flowering twig **of Phyllostachys bambusoides**

772

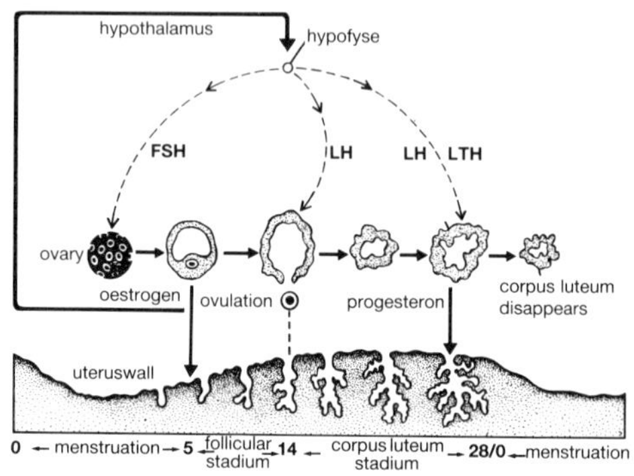

773

The Physics of Time

I can do a score of things that can't be done,... I can find a thing I cannot see and see a thing I cannot find. The first is time, the second is a spot before my eyes.
James Thurber, *The Thirteen Clocks,* 1951, 65.

The Arrow of Time

What is time? Such an open-ended question is difficult to answer. But physicists know the behaviour of time. In physics, there are two main aspects of time : the physical properties of time, and the way in which time is experienced by complex beings like people. Some of the physical properties of time are treated below ; let us look first at the question why time seems to be 'going on'.
Why does time pass? Why do we remember the past and not the future, and why can the past not be altered, while we can have at least some influence over things to come?

775 The Arrow of Time

Computer simulation of the irreversibility of time.
The user can choose from several simulations in which a few, or many, particles participate. It will be seen that any system which consists of more than a few particles have a vanishingly small chance of ever returning to a state of order. Thus, time seems to go in one direction only, whereas the individual particle processes impose no such direction at all.

776 The Ehrenfest Explanation

Video presentation discussing the irreversibility of time.

Try as you may, but you will never, never see anything like ink flowing back into a pen, or ten billiard balls spontaneously coming together in a perfect triangle.
Why is there an arrow of time?
How does nature know the difference between forward and reverse?
Let us ask a simple question : does nature in fact impose a direction on time?
Absolutely not! When we look in detail, we can not see a difference between forward and reverse.
In physics, we say that individual processes are reversible or symmetric in time.

Great scientists worried about the problem that there is no arrow of time for individual processes, and yet we see time pass.
Then Ludwig Boltzmann solved the problem, about a hundred years ago. Every single collision is reversible, but the result of many reversible collisions together gives the appearance of a direction in time. It is not impossible for ten billiard balls to come together in a perfect triangle ; it is just fantastically unlikely.
Simply because nature is made of infinitely many particles, we see that chaos is more likely than order.
This is what produces the Arrow of Time.
On a small scale, nature is time-symmetric, but on a large scale the time symmetry is lost.
A house built of bricks doesn't have to be shaped like a brick.

Atoms are the bricks from which our body is built.
The motion of atoms is reversible in time. But living things, made from an incredibly large number of atoms, have a direction in time.
Boltzmann's great achievement in explaining the Arrow of Time was not recognised by most other scientists, except a few. Paul Ehrenfest and his wife Tatiana invented the following demonstration of Boltzmann's theory. Suppose that you have two animals, and one flea. Then either one animal or the other has the flea.
The chance that the flea jumps from the panda to the tiger is exactly the same as the chance that the opposite occurs, namely that it jumps from the tiger to the panda. This we call a reversible process.
But if you have many fleas, let's say twenty, you never observe that they're all on the panda. On the average, each animal has about as many as the other.
Even if the fleas started all on the tiger, after a while they'd be evenly distributed by random jumping of the fleas, even if each individual flea jump is reversible. The order which was present in the beginning is lost because there are so many fleas.
With twenty jumping fleas, it would take eighty thousand million years (four times the age of the universe) until, by chance, they got back on one animal. Disorder is more likely than order. That is why there is an arrow of time.

It is not impossible in principle for time to be perceived as running backwards, but the odds against it are so small that they defy description. As always in physics, what matters is what happens in practice.

777 The Boltzmann Box

A box containing three white and three yellow ping pong balls has only one chance in sixty of showing three white balls in a row, followed by three yellow ones. In 59 out of 60 cases, the balls will be mixed up somehow. If you shake the box once every second, it should take you (on the average!) about a minute to shake them back into a three/three order. Try it! And don't cheat : shake without looking. If the box contained ten yellow and ten white balls, it would take you on the average 184756 seconds, i.e. just over two days, to shake them into a ten/ten order.

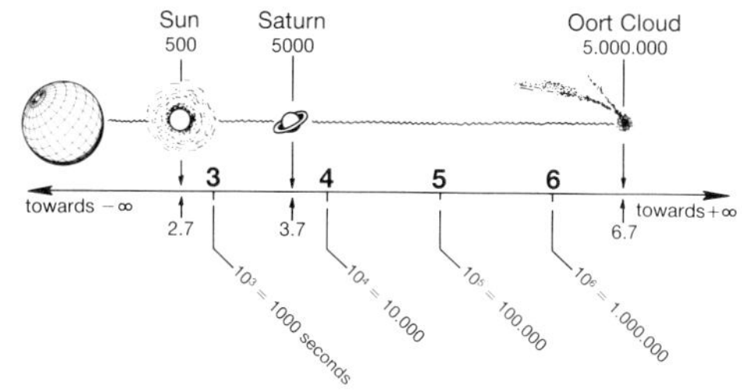

780

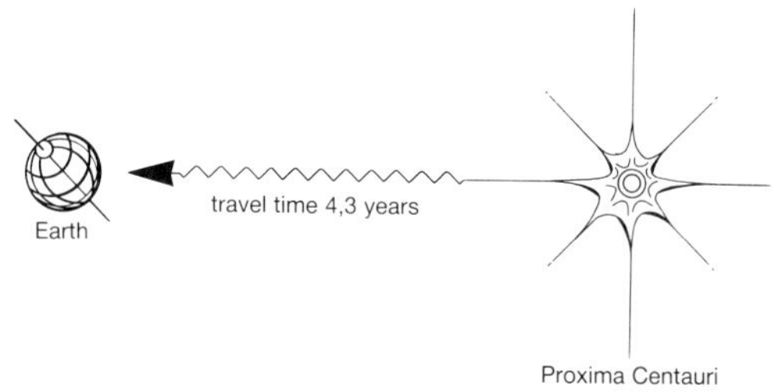

780

778 Sugar In Your Tea?

This is a test tube containing 50 grains of white and 50 grains of brown sugar. If you could shake them a thousand million times per second, it would take you 10^{60} years, or 10^{50} times the age of the universe, to shake them in order to have one chance that they became fully separated (just to remind you of number notation : that's a one with fifty zeroes).
On the other hand, if you started with a pinch of white sugar and added a pinch of brown to it, only a few seconds of shaking would suffice to mix them completely.

Time, the Fourth Dimension

In modern physics, time is a dimension similar to the more familiar three dimensions of space. Space and time together form a four-dimensional entity called spacetime. That means, that distances in space and intervals of time can be measured with the

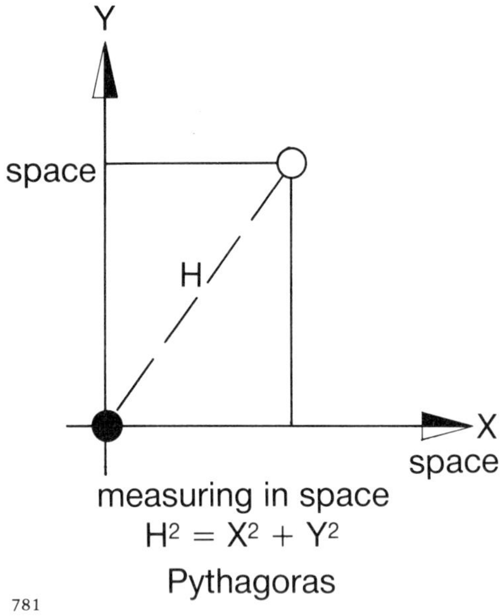

781

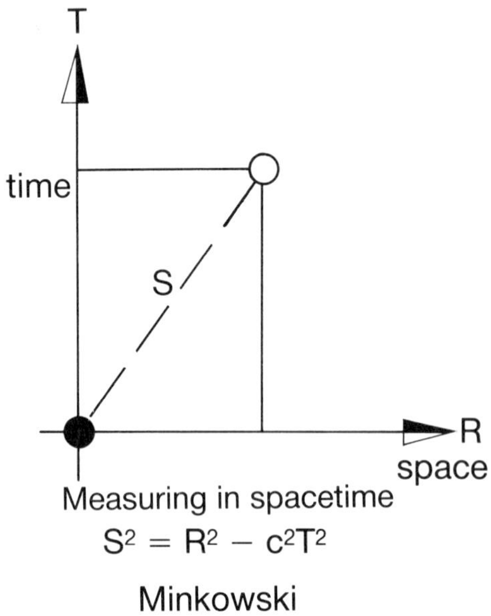

781

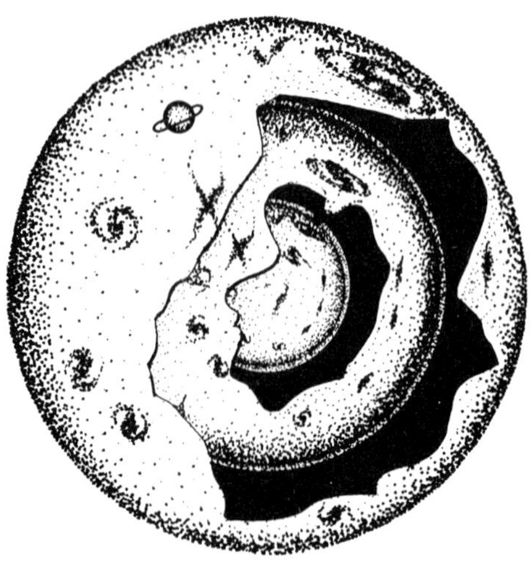
782

same yardstick. The yardstick we actually use is the second. Thus, we can say that the distance in time from here to the Big Bang is 20000 million years ; the distance in space from here to Jupiter is three quarters of an hour, while the spatial distance to the Andromeda galaxy is three million years.

779 Time, the Fourth Dimension
Video presentation of relativistic time.

That looks very strange : we can measure the universe in units of time! Time is the universal measure for distances in space as well as in time.
How can that be? The reason is as simple as it is enigmatic. Physicists have discovered that the speed of light is constant. No matter what you do, it is always the same. Surely you agree that that is very odd. When you drive a car at 150 km/h, the police can come alongside you to tell you to pull over, and receive a stiff fine. But if you travel with the speed of light, you always go with that speed, no matter how hard the Interstellar Police tries to catch up with you! Even a ray of light sees another ray of light go with that famous speed, called c, which is very close to 300,000 kilometers per second.
Thus, we have discovered that there is one special speed which nature always keeps constant. Now that is peculiar. A speed is a distance divided by a time interval : meters per second ; or kilometers per hour ; or miles per minute. Thus, nature must conspire to keep the ratio of space- and time-distances constant.

780 Measuring Time and Space with Light
Because the speed of light is always the same, we can use light signals to measure distances in units of time. Because the speed of light is constant, it is the perfect cosmic yardstick. If it takes light four years to get to the nearest star, we can say that the nearest star has a distance of four years. (Sometimes, this is emphasized by using the word light-years or lightseconds for a distance, but that is really superfluous).

We do not know why the speed of light is constant. It is one of the great mysteries of present-day science. Maybe if you design your own universe, you can build it differently. But in our universe, the speed of light is constant. And that is why we can measure space and time in time-units, such as seconds.

781 Measuring in Time and in Space
When measuring distances in space, we observe that the sides X and Y of a right triangle are related to the hypotenuse H by the famous $X^2+Y^2=H^2$, Pythagoras's Theorem. In a three-dimensional box with sides X, Y, and Z, this becomes $X^2+Y^2+Z^2=H^2$. When we measure in time (called T) as well as in space, Einstein has shown that our distance recipe should be different, namely $X^2+Y^2+Z^2-c^2T^2=S^2$. The quantity S is the 'spacetime hypotenuse', called proper time by physicists. The fixed number c is the speed of light. The expression $X^2+Y^2+Z^2-c^2T^2=S^2$ forges a permanent link between measurements in time and in space.

When we consider the expansion of the universe, there is a simple way in which we can illustrate Time as a fourth dimension. But we have to make one simplification. Because it is not easy to draw four-dimensional pictures, we pretend that space has two dimensions, rather than the three it actually has. Thus, we can draw our universe as the surface of a spherical balloon.

782 The Fourth Dimension
We can draw the motion of the universe as a nested sequence of expanding spheres. We, and all that is in the universe, live in such a sphere. Actually, the spheres are three-dimensional, and they expand in the direction of the fourth dimension : the direction of Time.
As the universe expands, the universal sphere moves outward in Time. Thus, the future universe surrounds the universe which is past.

What is outside the universe? Outside our universe lies our universe in the future. Inside lies the universe which is past. If we trace the motion of our world backwards in time, it shrinks ; ultimately, it vanishes to an expact point in the past. This is the point where Time and Space began. This 'infinity' point marks the beginning of time and space. We call it the Big Bang. There's no such thing as 'before' the Big Bang, because nothing existed, not even Time itself.

783 To Be or Not to Be : the Origin of Time
You cannot shrink something to a size smaller than zero ; thus, there is no 'before' this beginning. What is outside the universe? The universe tomorrow. What is inside the universe? The universe yesterday. What was there before the Big Bang? Nothing -- no space, and no time, not even the verb 'to be'.

The Lorentz Clock

You can synchronize your watch with that of someone else.
And if the watches are good, they will keep ticking the same time, and it makes no difference if you read your own watch or the other.
Right? Wrong! As it happens, nature is more subtle. But we do not notice this in our daily

lives, because the relativity of time is noticeable only if you move very fast. Fast, that is, with respect to the person whose watch you are reading.
How can that be? The basic reason is found in a rather peculiar property of nature, namely that the speed of light is always the same no matter what you do, where you are, or how you move.
This is a certified observed fact, and it is very, very odd. Many people think that the basic thing about the speed of light (called c) is, that it is the highest possible speed in nature. As it happens, the speed of light is really a maximum speed, but that is a secondary fact, not as fundamental as the constancy (physicists say invariance) of the speed of light. You cannot observe whether something is the highest attainable speed : it is always possible to make objections like 'You haven't tried hard enough'. In fact, the maximum-speed rule can easily be derived from the invariance.
The fact that c is invariant is totally bizarre, completely against our everyday experience. Light is an electromagnetic wave, but in daily life we do not see waves with constant speed at all. Suppose that a water wave rolls towards the shore. Then it has a certain speed with respect to the coastline. But a seagull can adjust its airspeed such that it sees the wave crests as if they are standing still underneath. With light, that is impossible. Everyone sees light move with the speed c. Even one light wave sees another one move with c!

784 Relativistic Time

Wall displays and computer simulation of the relativistic effects involving time, such as time dilatation and the measurement of spacetime.

By the way, this is the root of the proof that the speed of light is maximal. After all, if one light ray sees another one (say that it is coming on straight ahead) move at the speed of light, then that means that c+c=c. This may strike you as bizarre, but to a physicist, all this means is that because of the invariance of the speed of light, velocities do not add up like simple numbers. And why should they? If you think about it, you will realize that in fact very few physical quantities add up like simple numbers. For instance, distances do not. You can walk one kilometer, and then another one, and then a third kilometer, and end up at precisely the same place where you started. That means that distances can add up like 1+1+1=0 - of course, if you walk along a triangle! If distances added up like simple numbers, you could never go out to mail a letter and expect to come home afterwards, because then 1+1+1=3.

Thus, if velocities add like c+c=c (for light rays coming towards each other) and c-c=c (for rays travelling in the same direction), it is easy to see that c+c+c=c too : no matter how many lightspeeds you add, you can never exceed c. Therefore, the speed of light is a maximum.

A Theorist's Clockwork

The baffling consequences of the constancy of the speed of light can best be illustrated by means of a Lorentz Clock. This is a light ray caught between two mirrors. Anything periodic can measure time, and a light ray bouncing between two parallel perfect flat mirrors certainly moves periodically.

785 The Lorentz Clock : construction

We can see the relativity of time measurement by means of an idealized clock, the Lorentz Clock, a light ray caught between two mirrors. A light ray bouncing between two parallel perfect flat mirrors always moves periodically and does not escape. Therefore, it can be used to measure time.

786 The Lorentz Clock: moving the machine

Now what happens when the Lorentz clock moves? The light will have to move along a slanted line in order to catch up with each mirror.
The speed of light is fixed, and in a moving Lorentz clock it has to traverse a longer distance between the mirrors. Covering a longer distance with a fixed speed means that it takes more time. Thus, a moving clock seems to tick more slowly than a stationary one.

Precisely how much more slowly a Lorentz clock appears to tick, when observed as it flies by at a fixed velocity, can be calculated simply by following the path of the light rays in space and in time. We need nothing more exotic than good old Pythagoras's Theorem. Suppose that the distance between the mirrors is D meters. Then the time between ticks of the the Lorentz clock is t-L seconds, which is simply t-L=D/c. Travel time equals distance divided by speed! Now watch a light ray starting at mirror 1. As the light travels towards mirror 2, this 'target' mirror moves along with speed v. The light ray, as seen by some observer seeing the mirror fly past, reaches mirror 2 at time t rm obs. During the crossing between the mirrors, the target mirror moves over a distance v times t-rm obs.
Look at the figure : in order to hit the mirror just right, the light ray must travel along the hypotenuse of a right triangle with sides D and $v \times t_{obs}$.
By Pythagoras's Theorem, the length of the hypotenuse equals $\sqrt{D^2+v^2+t^2_{obs}}$. The crossing time t-rm obs for the light is equal to the distance travelled divided by the speed.

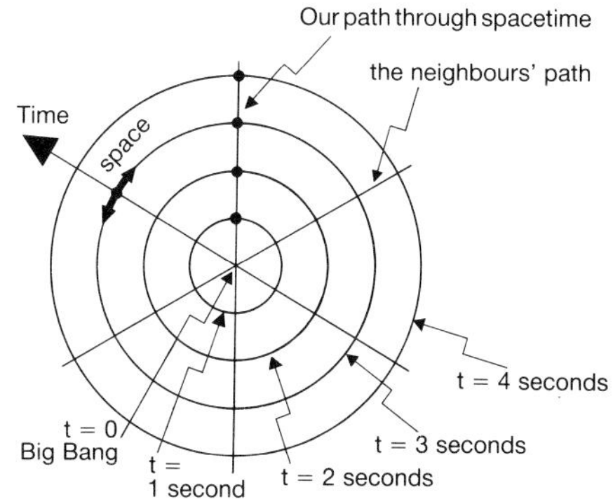

783

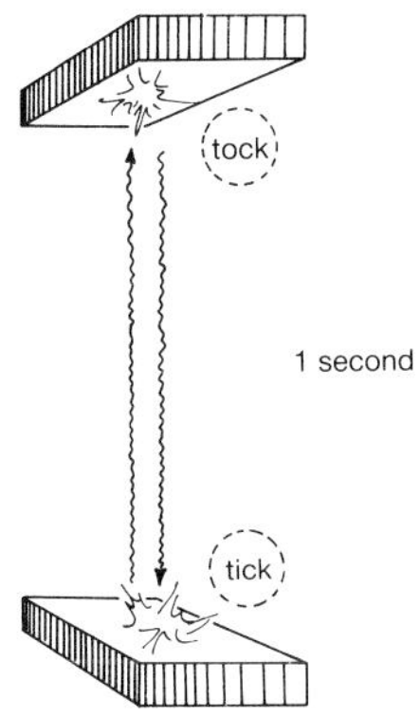

785

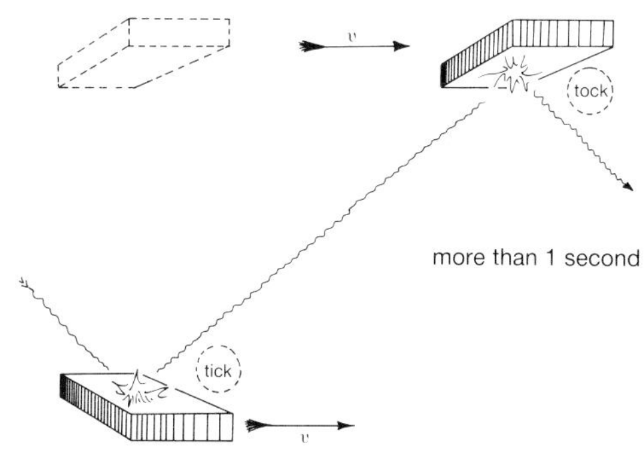

786

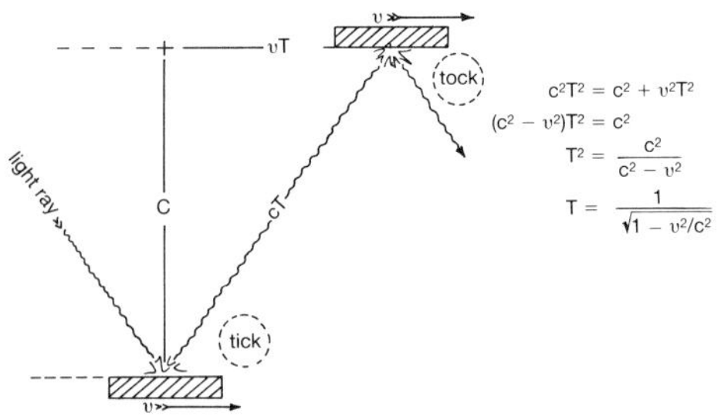

787

788

What speed? The speed of light. We can always use c for this speed, because c is invariant!

787 Time and Pythagoras

The ticking of a Lorentz clock can be calculated by measuring distances in spacetime by means of light rays. A simple calculation, involving no more than Pythagoras's Theorem, shows that a Lorentz clock ticking once every T seconds, moving with a speed v with respect to some observer, is seen (by that observer) to tick once every $T/\sqrt{1-v^2/c^2}$ seconds. As seen by the clock itself, it still goes 'tick' every T seconds.

Therefore, we get simply $t_{obs}=1/c\sqrt{D^2+v^2 x\; t^2_{obs}}$. Now we can extract the clock ticking time observed by someone seeing the clock fly by. Taking the square on both sides of the formula, we get by a little rearrangement $t_{obs}=tL/\sqrt{1-v^2/c^2}$.

Simple! If the clock is standing still, v=0, and we get t_{obs}=tL : the ticking time which is observed is the same as the time seen by the clock itself. But when v is not zero, the observed time is equal to the time seen by the clock, divided by a number smaller than unity.

A moving clock seems to tick more slowly. This is what physicists call time dilatation. No, you cannot synchronize watches with someone else and then expect the watches to tick the same seconds always!

By the way, you cannot make v larger than c without getting into deep trouble : if you do that, the time formula says you must take the square root of a negative number. The result is an imaginary number, and as far as we know there's no such thing as imaginary time! That is what would happen if the speed of light were exceeded.

Why Don't We Notice?

When you compare your watch with that of someone else, it does make a difference whether or not you are moving with respect to the other person. But this is only noticeable when your speed is close to the speed of light, which is 300,000 kilometers per second. If you traveled that fast, you could circle Earth seven times in one second. Even in a spacecraft, which goes at about 10 kilometers per second, the relativistic difference in clock rate is exceedingly small : an astronaut's watch ticks only one part in a thousand million slower than a watch at Mission Control on Earth.

Precisely how much more slowly a Lorentz clock appears to tick, can be calculated by measuring distances in spacetime by means of light rays, as introduced above. A Lorentz clock ticking once every T seconds, moving with a speed v with respect to some observer, is seen (by that observer) to tick once every $T/\sqrt{1-v^2/c^2}$ seconds. As seen by the clock itself, it still goes tick' every T seconds. Because of the constancy of the speed of light, Time is no longer universal. Instead, when observing things we must consider behaviour in space and in time.

An example : if T=0.000002 seconds is the decay time of a muon (a light subatomic particle), then a muon made by a cosmic ray hitting the upper layers of Earth's atmosphere would have decayed long before it hit Earth's surface, some hundred kilometers below. It could at most move c times T=600 meters before it disappears. But if the muon moves at 99.99% of the speed of light, then the decay time perceived by us would be $T/\sqrt{1-0.9999^2}$ = 0.03 seconds, long enough for the muon to travel to cosmic ray observatories on the ground : 0.03 seconds times its speed of 0.9999c equals 9000 kilometers, plenty of space. And indeed, plenty of muons are in fact observed to hit ground-based cosmic ray counters!

788 The Lorentz Clock : what difference does it make?

People never move close to the speed of light. But fast particles can go very near the speed of light. For them, it makes a lot of difference. A muon liberated by a cosmic ray smashing into Earth's upper atmosphere would decay long before it could reach the ground. But we do see cosmic muons at ground level. This is only possible due to the relativistic 'stretching' of time.

Powers of Ten in Time

Human experience of time is limited. A human heartbeat takes about one second, but by the time we celebrate our first birthday our heart has already beat thirty million times. When we consider the world outside our immediate experience, we find an even larger variety of characteristic durations (called time scales).

789 Time Scales : Powers of Ten in Time

A video presentation of the various time scales in the universe, from large (astrophysical) to small (sub-atomic).
In this overview we will take a series of steps through a large variety of known time scales in the physical sciences, from the furthest reaches of the universe down to sub-atomic particles. We begin with the longest time scales known in the cosmos, and descend -- via the time scales of human life -- to the shortest, ending at the very frontier of the unknown.

790 Powers of Ten in Time
By writing a very large number, such as 150,000,000, as something smaller times a 'clean' multiple of 10, we can write big numbers very compactly. In this example, 150000000=1.5 times 10^8.

791 Examples of time scales in nature
a. The age of Earth is 4500 million years.
b. The time it takes for the Sun to complete one orbit around the Galaxy is 250 million years.
c. Continental drift changes the landscape of Earth markedly in 25 million years.
d. The branche of evolution which led to Homo Sapiens, started to sprout about 4 million years ago.
e. The stars in the Orion Nebula have a characteristic evolution time of 400000 years. On a logarithmic scale we find the phenomena (a) through (e) at positions 17.2, 15.9, 14.1, and 13.1
universe
Brought within range by using the logarithmic 'Boeke telescope', these time scales stretch from 10**17.8 seconds, the age of the universe, to 10**-1 seconds, the human reaction time.

Pulsars, The Best Clocks in the universe

A pulsar is a dead star, which has been compressed to a size not much bigger than a large city. Such a star is called a neutron star by astronomers. The star has trapped a strong magnetic field, and spins around exceedingly fast: from about once per second in some cases, up to almost a thousand revolutions per second in the fastest pulsars. The spinning magnetic field acts as a huge dynamo, causing immense relativistic lightning storms above the polar caps of the pulsar. This makes the magnetic poles light up brightly throughout the electromagnetic spectrum. Each time the star whips around, a flash of radiation is sent into space. These flashes are the most accurate timekeepers known in the entire universe.

793 A Model of a Pulsar

This model of a pulsar was built to scale 1 :100000.
In the background you hear the signals of actual pulsars have been made audible as real sound. It was made at the Arecibo Observatory, which is run by the Astronomy Dept. of Cornell University.

794 A Pulsar in Amsterdam
Even though a typical pulsar contains up to 40% more mass than our Sun, it is only twenty kilometers across. That is to say, such a sphere, which contains more mass than our entire Solar System, is not much larger than Amsterdam.

795 Pulsars, the Best Clocks in the Universe
Because the pulsars send out flashes, they lose energy. This energy is extracted from the energy stored in their very rapid rotation. Thus, radiating pulsars must slow down. But the cosmic thunderstorms which make the pulses use up only a very small fraction of a neutron star's rotational energy. Therefore, neutron stars are the most accurate clocks in the universe.

The Future of Time

796 What Will our universe Do in the Future?
We know from observations that it will continue to expand forever. The future of time is truly infinite. And, in infinite time, a great many things can happen.

797 The Future Expansion of Our Universe
The only thing which can stop the expansion of the universe is the gravitational attraction between the matter and energy in it. Observations show that there is too little of those in the universe to ever stop it from expanding. Thus, our future lasts forever.

798 The Fate of the Solar System
About 4600 million years ago, an interstellar cloud of dust and gas started to collapse. A hundred million years later, the core of that cloud had formed a tight ball, in which hydrogen nuclei (protons) fused together, producing heat. Thus the Sun was born. But in another 5000 million years, the fusion reactions will run out of hydrogen fuel. The core of the Sun will contract within a few hundred million years, thereby driving off the outer layers of the Sun.

799 The Fate of Life on Earth
Some 4500 million years ago, the Solar System formed. About 500 million years after that beginning, Earth had cooled enough for the first complex molecules to form. Microfossils of primitive cellular life are known from about 1000 million years after the beginning of Earth. At present, the energy for life is provided by the Sun. But at the end of its life, the outer layers of the Sun will swell up dramatically. These layers will eventually engulf the entire Solar System, blasting the planets, and all that is on or in them. All that you see around you will evaporate in an incandescent blaze.

800 Even the Particles May Die
After the Solar System is gone, there will be many thousands of millions of years during which the process of star formation can repeat itself. But after a million million years, even these processes will end. There will then be no more stars in the universe, no more light. The universe will continue to

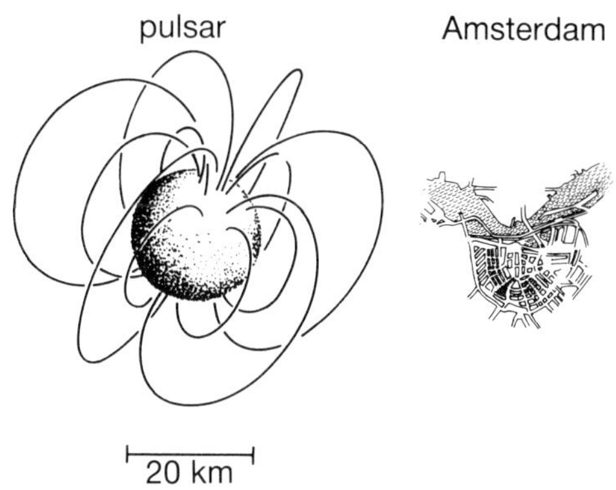

794

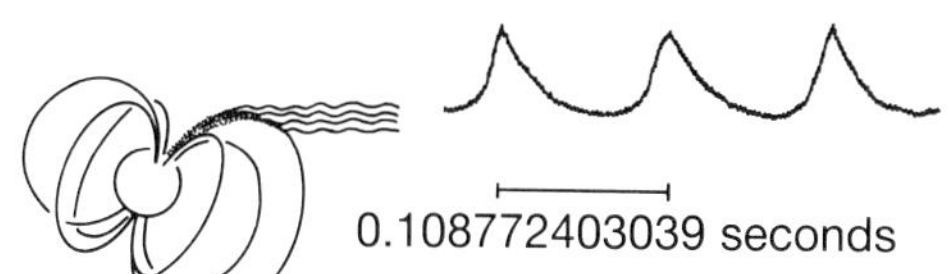

795

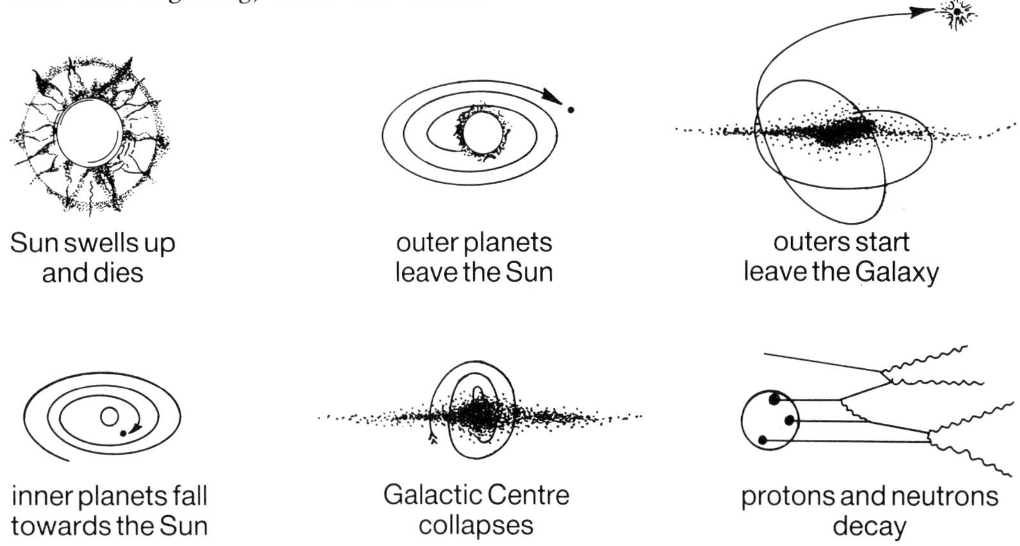

796

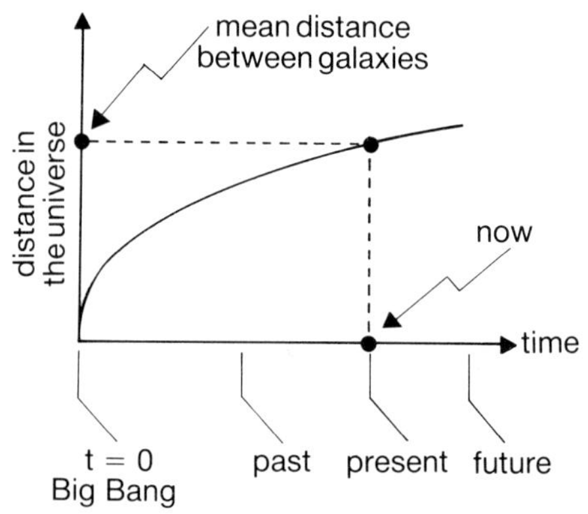

797

expand, containing only the dead cinders of former suns. But according to some physical theories, the very protons and neutrons of which these dead stars are made are very, very slightly radioactive. These particles may decay in about 10^{34} years, leaving nothing in the universe except the feeblest of light, coursing through otherwise empty space.

801 The Size of the Future
In an earlier illustration, we compressed the entire history of the cosmos into one week. Let us go further, and compress Time such that the whole past of the universe takes a single second. Then we would have to wait another 10^{18} of such seconds for the proton to have a fifty/fifty chance to decay. Thus, if we compress all of known Time into one second, we would have to wait ten thousand million years for the dead remains of the stars to vanish in miniature blazes of light. After that, nothing remains but speculation.

802 The Bird and the Mountain
The estimated decay time of the proton is very long. When trying to grasp such cosmic immensities, people often take refuge in parables. It is here that physics and myth occasionally meet.
In ancient tales, it is said that there is a mountain, ten kilometers high, to which a little bird comes every century. The bird scrapes its beak on the mountain and leaves, not to return until a century later. And when that bird has worn away the whole mountain, then one second of Eternity has passed.

One of the fun things about physics is, that one can actually calculate how much time something like that would in fact take. I have done just that. It is not too hard : just estimate the number of atoms in such a mountain, and the number of atoms dislocated by a single sweep of the bird's beak. Then multiply by the time between the bird's visits. I have found that one second of this Eternity, as defined in the tale, is about the time it takes for the proton to decay.

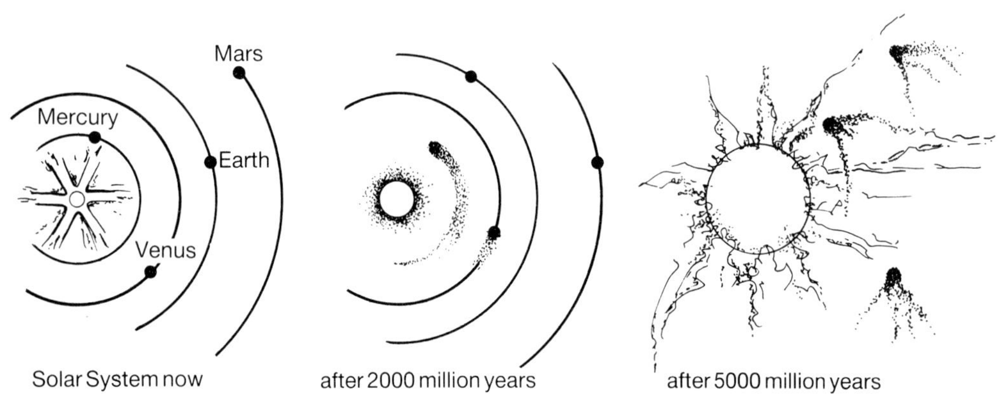

798

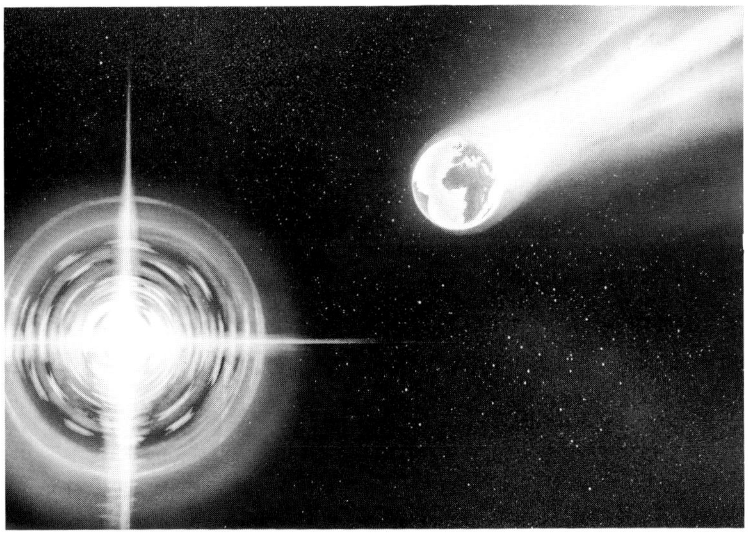

799

802

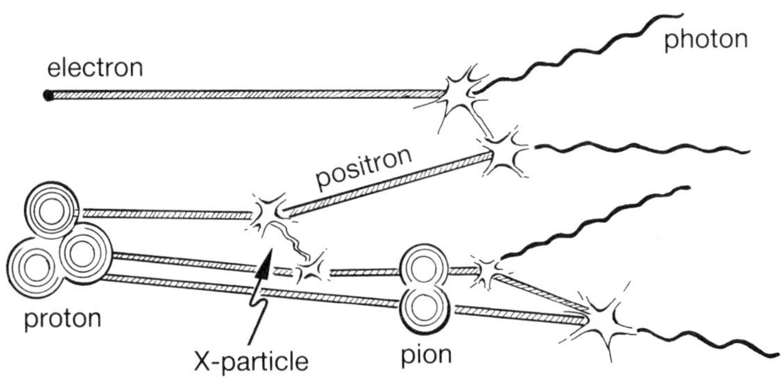

800

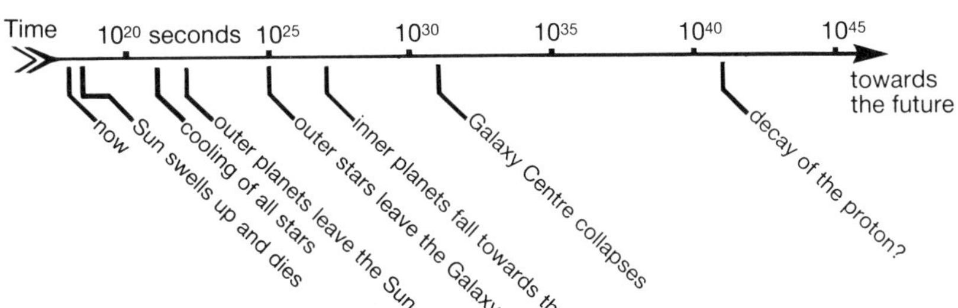

801

The Human Sense of Time

The Invention of History

Only a man harrowing clods
in a slow silent walk
with an old horse that stumbles and nods
half asleep as they stalk.

Only thin smoke without flame
From the heaps of couch-grass;
Yet this will go onward the same
Though dynasties pass.

Yonder a maid and her wight
Come whispering by:
War's annals will cloud into night
Ere their story die.
Thomas Hardy, *In Time of 'The Breaking of Nations'.*

The idea that in time everything changes, that everything therefore has a history, and that two things of the same kind will therefore be different at different periods, was one that did not develop before the Renaissance. Only very slowly were its implications understood.
Its gradual penetration of human understanding parallels the rise of secularism.

803 Lorenzo Valla and the Donation of Constantine, 1439
Lorenzo Valla, De falso credita et ementita Constantini donatione declamatio
In this work Valla, a leading humanist scholar and virulent controversialist and opponent of the church, employed linguistic tools to show the anachronism of the Constitutum Constantini stylistically, and that it could not therefore have been written in the 4th century.

In the instantaneity of God's view of creation, all time is eternally present. For the devout of the christian world, creation, the fall, and redemption were symbolically present in the cycles of church ritual that perpetuated the past.
In such a world there was no ambiguity in presenting events of the distant past in contemporary dress, for they were really present.

804 Jacob Cornelisz van Oostsanem, Calvary
Photograph
All the events of the passion are here presented in a single canvass sequentially yet simultaneously, linear time and God's instantaneous time co-habiting together, while the dress is that of the time that Cornelisz painted.

Photograph : Rijksmuseum, Amsterdam.

807

Because the modern world is imbued with the sense of history with the knowledge that events distant in time are different one from another, the producer of a period play today may present it -anachronistically - in modern dress, to insist that art transcends time and is eternally present.

805 A Modern Dress Production of Shakespeare
Photograph

Such an idea underlies a recent volume of criticism

806 Jan Kott, ***Shakespeare our Contemporary,*** 1964

The development of a sense of anachronism occurred at different periods for different subjects.

Although the philosophical historians of the enlightenment were well aware of the way in which institutions and social structures had changed across the centuries, no anachronism was felt in replacing the original movement of this Renaissance clock, for clocks were not yet matter for history.

807 The Monstrance Clock of Cardinal de Granvelle, 1564 and later
Augsburg.
A remarkable calendar and astronomical clock probably originally fitted with an astrolabe dial. Virtually nothing of the original movement survives. In 1769 the clock was entrusted for repair to Antide Janvier who despite the admiration he professed for the mechanism in one of his publications, nonetheless restored it. At a still later date, probably c. 1850 virtually the whole mechanism was replaced with a fine disregard for age and historic interest.
Musée du Temps et de l'Histoire, Besançon.

But for some scholars clocks were worthy of history.

808 An Early Study in the History of Horology, 1778
Daines Barrington, 'Observations on the earliest Introduction of Clocks : In a Letter to the Honourable Mr. Justice Blackstone'. *Archaeologia* V, 1779, 416-28.
That the lawyer, antiquary and naturalist Daines Barrington (1727 - 1800) should write a study of the origins of clocks is indicative of a new sense of the past that developed during the enlightenment. The rise of connoisseurship and of collecting in the applied arts together with the rapid development of technology (especially in horology) led to a recognition of the difference between past and present even in the realm of common everyday objects. That old time-measurers should come to be seen as historical monuments of time reflects an understanding of the past as something different, and of the past as a new element of the present to be explored.
Private Collection.

809 A Summary of the Time of Man, 1849
Calendar-clock signed 'Cette horloge est composée par Delvart, Curé de Zouafques et exécutée par Potez, horloger áu Audricq, dans le départment du Pas-de-Calais, l'an 1849'. (This clock was designed by Delvart, curé of Zouafques, and built by Potez clockmaker at Audricq in the department of the Pas-de-Calais in the year 1849.)
Brass and steel in painted wood cabinet with glazed top.
2m 400 x 1m 280 x 420 mm
On 18 dials the clock provides all the time indications needed for man's daily life and for ecclesiastical purposes. Hours and minutes, the hours with a double XII hour scale and indication of evening and morning, seconds.
Seconds.
Day of the week.
Date.
Zodiacal calendar.
Time of sunrise.
Time of sunset.
Equation of time.
Epochs of the world.
Comparison of the lunar and solar year with corrections for ordinary and leap years.
Lunar cycle.
Epacts.
Golden number, dominical letter and solar cycle.
Indiction.
Solar and lunar eclipses for 18 years.
Movable feasts of the Christian church.
Planetary system.
Six train movement in an iron frame, anchor escapement, three feet long pendulum seconds beating. Second signature by Potez on one of the wheels 'Executée et perfectionée par Potez horloger à Audicq Pas-de-Calais' (Made and perfected by Potez clockmaker at Audricq Pas-de-Calais).
Private Collection.

810 The Flow of Time, c. 1980
Water-clock by Bernard Guitton
Overall height 2 m.

Bibliography

Abridgements
Patents for inventions. Abridgements of Specifications Class 139 Watches, Clocks, and other Time-keepers, 1855 - 1930, 2 vols., Seattle, 1979.

Allexandre
Jacques Allexandre, *Traité général des horloges,* Paris, 1734.

Ambronn
L. Ambronn, *Handbuch der Astronomischen Instrumentenkunde,* 2 vols., Berlin, 1899.

Anon
Anon, 'Maezel's Metronome', *The Quarterly Musical Magazine and Review,* III, 1821, p. 303.

Arberry
Arthur J. Arberry, *The Koran Interpreted,* London, 1955, 1964.

Arendt, Broadway, English, Kemp Poulton & Symons,
J. Arendt, J. Broadway, J.E. English, M. Kemp, A.L. Poulton & A.M. Symons 'Pineal and Photoperiodicity : Practical Aspects', in W.Th.J.M. Hekkens, G.A. Kerkhof & W.J. Rietveld (eds), *Trends in Chronobiology,* Oxford, 1988, pp. 137 - 148.

Armstrong & Mckcamp
Joe E. Armstrong & John Mckcamp, 'Notes on a Water-clock in the Athenian Agora', *Hesperia,* 46, 1977, pp. 147 - 161.

Asahina
Tei-Ichi Asahina, 'Japan's Oldest', *Bulletin of the National Association of Watch and Clock Collectors,* 60, October 1955.

Backhouse, Turner & Webster
Janet Backhouse, D.H. Turner & Leslie Webster (eds.), *The Golden Age of Anglo-Saxon Art, 966 - 1066,* Ipswich, 1984.

Baillie
G.H. Baillie, *Watchmakers and Clockmakers of the World,* 2nd ed., London, 1947.

Bain
A. Bain, 'A Short History of the Electric Clock', in W.D. Hackmann, *Alexander Bain's Short History of the Electric Clock* (1852), London, 1973.

Barrington
Daines Barrington, 'Observations on the Earliest Introduction of Clocks : In a Letter to the Honourable Mr. Justice Blackstone', *Archaeologia*, V. 1779, pp. 416 -428.

Bartky & Dick (1)
I.R. Bartky & S.J. Dick, 'The First Time Balls', *Journal for the History of Astronomy,* XII, 1981, pp. 155 - 164.

Bartky & Dick (2)
I.R. Bartky & S.J. Dick, 'The First North American Time Ball', *Journal of the History of Astronomy,* XIII, 1982, pp. 50 - 54.

BCHE
Science Museum (Special Exhibition Gallery), London, *British Clockmakers' Heritage Exhibition, May 23rd to September 14th,* London, 1952.
Beek & Kurpershoek
Marijke Beek & Ernest Kurpershoek, *Amsterdam's Nieuwe Kerk,* Amsterdam, 1983.

Bedini
Silvio A. Bedini, 'The Scent of Time. A Study of the Use of Fire and Incense for Time Measurement in Oriental Countries', *Transactions of the American Philosophical Society,* NS, 53 pt 5, 1963.

Bedini & Maddison
Silvio A. Bedini, 'Mechanical Universe the Astrarium of Giovanni de' Dondi', *Transactions of the American Philosophical Society,* 56, 5, October 1966.

Beeson, Robinson & Ward
C.F.C. Beeson, T.O. Robinson & F.A.B. Ward, 'The Invention of the Anchor Escapement', *Antiquarian Horology,* VII, 3, 1971, pp. 225 - 228.

Bénézit
E. Bénézit, *Dictionaire critique des peintres, sculpteurs, dessinateurs et graveurs,* new ed., 8 vols., 1948 - 1955.

Benner & Forbes
A.R. Benner & F.H. Forbes (tr.), *The Letters of Alciphron, Aelian and Philostratus,* London-Cambridge, 1949.

Berryer & Lébioles
Anne-Marie Berryer & Liliane Dresse de Lébioles, *La mesure du temps à travers les ges aux Musées Royaux d'art et d'Histoire',* 2nd ed., Bruxelles, 1974.

Beschrijving
Beschrijving van de in Nederland gebruikelijke telegraaftoestellen, Afbeeldingen, Den Haag, 1863.

Biarne
J. Biarne, 'Le temps du moine d'après les premières règles monastiques d'Occident (IVe-VIe siècles), in *Le temps Chrétien de la Fin de l'Antiquité au Moyen-Age, IIIe - XIIIe siècles* (Colloques Internationaux du Centre National de la Recherche Scientifique, no 604', Paris, 1984, pp. 99 - 128.

Bigourdan
M.G. Bigourdan, 'Le jour et ses divisions', *Annuaire pour l'an 1914 publié par le bureau des longitudes,* Paris, 1914.

Billmeir
A Catalogue of Scientific Instruments from the 13th to the 19th Centuries from the Collection of J.A. Billmeir C.B.E., 2nd ed., Oxford, 1955.

Bion
Bion, *Traité de la construction et des principaux usages des instruments de mathématiques,* Paris, 1709.

Block
Wolfram Block, 'The Radellof Cross Beat Clock of 1660', *Antiquarian Horology,* 1972, pp. 700 - 703.

Bobinger
Maximilian Bobinger, *Alt Augsburger Kompassmacher,* Augsburg, 1966.

Boeke
K. Boeke, *Cosmic View : the Universe in 40 Jumps,* New York, 1957.

Boeren
P.C. Boeren, with A.S. Kortweg & G. Piket, *Catalogus von de liturgische handschriften van de Koninklijke Bibliotheek,* 's-Gravenhage, 1988.

Borchardt
Ludvig Borchardt, *Die alte Ägyptische Zeitmessung* (Die Geschichte der Zeitmessung, ed. E.V. Bassermann-Jordan, Bd I. 1g. B), Berlin & Leipzig, 1920.

Bos
H.J.M. Bos, *Descriptive Catalogue, Mechanical Instruments in the Utrecht University Museum,* Utrecht, 1968.

Bowra
C.M. Bowra, *Memories 1898 - 1939,* London, 1966.

Brachner
Alto Brachner (ed.), *G.F. Brander (1713 - 1783), Wissenschaftliche Instrumente aus seiner Werkstatt,* München, 1983.

Breugelmans & Dekker
R. Breugelmans & E. Dekker, 'Adriaan & Anthonisz and the Gregorian Calendar', in Ton Croiset, van Uchelen, Kurt van der Horst & Gunther Schilder (eds.), *Theatrum orbis librorum, Liber amicorum presented to Nico Israel on the occasion of his seventieth Birthday,* Utrecht, 1989, pp. 137 - 159.

Brieux (1)
Alain Brieux, *Instruments scientifiques, livres anciens, collection Léonard Linton, Point Lookout, N.Y.,* vente Drouot, 9 - 10 octobre 1980.

Brieux (2)
Alain Brieux, *Instruments scientiques du troisième siècle avant l'ère chrétienne à 1946, Paris, 1983.*

Brieux & Maddison
Alain Brieux & Francis Maddison, avec la collaboration de Ludvig Kallus, Ysuf Rghib & Jean-Pierre Verdet, *Répertoire des facteurs d'Astrolabes et de leurs oeuvres, Première partie : Islam* (in préparation).

Brown
G. Baldin Brown, *The Ruthwell and Bewcastle Crosses,* 2nd ed., 1921.

Brusa
Giuseppe Brusa, 'La Navicelle Orarie di Venezia', *Annali dell' Instituto e Museo di Storia dela Scienza di Firenze,* V, 1, 1980.

Bryden
David J. Bryden, *The Whipple Museum of the History of Science: Catalogue 6 Sundials and Related Instruments,* Cambridge, 1988.

Buchner
Edmond Buchner, *Die Sonnenuhr des Augustus,* Mainz-am-Rhein, 1982.

Bünning
E. Bünning, *Die Physiologie des Wachstums under der Bewegungen,* Berlin, 1939.

Capart
J. Capart, 'Horloges éqyptiennes', *Bulletin des Muses Royaux d'Art et d'Histoire,* 1938, no 3.

Cardinal (1)
Catherine Cardinal, *Ferdinand Berthoud, 1727 -1807, horloger mecanicien du Roi et de la marine,* La Chaux-de-Fonds, 1984.

Cardinal (2)
Catherine Cardinal, *La montre des origines au XIXe siècle,* Fribourg & Paris, 1985.

Chapiro, Meslin-Perrier & Turner
Adolphe Chapiro, Chantal Meslin-Perrier & Anthony Turner, *Catalogue de l'horlogerie et des instruments de precision du bout du XVIe au milieu du XVIIe siècle,* Paris, 1989.

Chayette-Sabrier
Hervé Chayette & Jean-Claude Sabrier, *Montres et pendules de collection,* vente Drouot, 4 juin 1981, Paris, 1981.

Christie's (1)
Christie's South Kensington, *Scientific and medical instruments including Barometers,* 10 July 1986.

Christie's (2)
Christie's South Kensington, *Scientific and Medical Instruments,* April 14, 1988.

Christie's(3)
Christie's South Kensington, *Fine Scientific and Philosophical Instruments,* 29 March 1990.

Cittert-Eymers
& Hagen
J.G. Cittert-Eymers & M.J. Hagen, *Zonnewijzers aan en bijgebouwen in Nederlanden, enige Astronomische (Toren) Uurwerken,* Zutphen, 1984.

Clanchy
M.T. Clanchy, *From Memory to Written Record, England, 1066 - 1307,* Oxford, 1979.

Clarke, Morrison-Low & Simpson
T.N. Clarke, A.D. Morrison-Low & A.D.C. Simpson, *Brass & Glass, Scientific Instrument Making Workshops in Scotland as Illustrated by Instruments from the Arthur Frank Collection at the Royal Museum of Scotland,* Edinburgh, 1989.

Clapham
A.W. Clapham, *English Romanesque Architecture Before the Conquest,* Oxford, 1930 (1964).

Clutton & Daniels
Cecil Clutton & Georges Daniels, *Clocks and Watches in the Collection of the Worshipful Company of Clockmakers,* London, 1975.

CNAM
Conservatoire National des Arts et Métiers, Catalogue du Musée, section JB, Horlogerie, Paris, 1949.

Combridge
J.H. Combridge, 'The 13th Century 'King Hezekiah' Water-clock', *Antiquarian Horology,* XII, 1980, p. 300.

Coomaraswamy
A.K. Coomaraswamy, *The Treatise of al-Jazarí on Automata,* Boston, 1924.

Daressy
G. Daressy, 'Le Mastaba de Mera', *Mémoires présentés à l'Institut Egyptien,* III, 531 - 574.

Dekker (1)
E. Dekker, 'Frederik Kaiser en zijn pogingen tot hervorming van 'het sterrekundig deel van onze zeevaart', *Tijdschrift voor de geschiedenis der geneeskunde, natuurwetenschappen, wiskunde en techniek,* 13, 1990, pp. 23 - 41.

Dekker (2)
Elly Dekker, 'From a World Ruling Time to a Time-Ruled World', this volume ch. 3.

Delphin
G. Delphin, 'l'Astronomie au Maroc', *Journal Asiatique,* XVII, no 2, 1891, pp. 117 -201.

Desroches-Noblecourt
Christian Desroches-Noblecourt, *Le Noblecourt, Le grand phaaron Ramsès II et son temps,* Montral, 1985.

Draaisma & Kaarls
Y. Draaisma & R. Kaarls, 'Tijd en Tijdregeling', *Zenith,* 4, 1977, pp. 122 -127.

Drachmann
G. Drachmann, 'Ktesibios's Waterclock and Heron's adjustable Siphon', *Centaurus,* 20, 1977, pp. 1 - 10.

Drover (1)
C.B. Drover, 'A Medieval Monastic Water Clock', *Antiquarian Horology,* i, 1953-56, 54-8.

Drover (2)
C.B. Drover, 'The 13th Century 'King Hezekiah' Water Clock, *Antiquarian Horology,* xii, 1980, 160-4.

Duncan
P.B. Duncan, *A Catalogue of the Ashmolean Museum,* Oxford, Oxford, 1836.

Dunsheath
P. Dunsheath, *A History of Electrical Engineering,* London, 1962.

Eden & Lloyd
H.K.F. Eden & Eleanor Loyd, *The Book of the Sun-dials,* London, 4th ed., 1900.

Electrische Tijd-
aanwijzing
Electrische tijdaanwijzing, Tentoonstelling in het Nederland Goud-, Zilver- en Klokkenmuseum, Schoonhoven, 1985.

Eliade (1)
Mircea Eliade, 'Time and Eternity in Indian Thought', *Papers from the Eranos Yearbooks,* 3, 1958, pp. 173 - 200.

Eliade (2)
Mircea Eliade, *Le mythe de l'éternel retour,* Paris, 1960.

Erbrich
K. Erbrich, *Präzisions Pendeluhren,* München, 1978.

d'Errico & Thevenin
Francesco d'Errico & André Thevenin, 'La notion du temps au paléolithique et le problème des calendriers lunaires', in Jean-Pierre Mohen (ed3Antiquarian Horology, XVIII, 1989, pp. 371 - 380.

Evans
L.T. Evans, *The Physiology of Flowering,* South Melbourne, 1969.

Farrelly
The Book of Bamboo, San Franscisco, 1984

Farré-Olivé
Edouard Farré-Olivé, 'A Medieval Catalan Clepsyda and Carillon', *Antiquarian Horology,* XVIII, 1989, pp. 371 - 380.

Faverger
A. Faverger, *L'électricité et ses applications à la chronométrie,* Neuchàtel, 1924.

Field & Wright
J.V. Field & M.T. Wright, *Early Gearing: Geared Mechanism in the Ancient and Medieval World,* London, 1985.

Foucault
L. Foucault, 'Démonstration physique du mouvement de rotation de la terre au moyen du pendule', *Comptes-rendus Hebdomataires des scéances de l'Académie des Sciences,* 32, 1851, pp. 135 - 138.

Frankfort
H. Frankfort, *The Cenotaph of Seti I at Abydos* (Egyptian Exploration Society Memoir 39), 2 vols., London, 1933.

Franklin
Alfred Franklin, *La vie privée d'autrefois, arts et métiers, modes, moeurs, usages des Parisiens du XIIe au XVIIIe siècle d'après des documents orginaux ou inédits. La mesure du temps,* Paris, 1888.

Gibbs
Sharon L. Gibbs, *Greek and Roman Sun-dials,* New Haven à London, 1976.

Gibbs, Henderson & Price
Sharon L. Gibbs, Janice A. Henderson & Derek de Solla Price, *A Computerized Checklist of Astrolabes*, New Haven, 1973.

Gingerich
Owen Gingerich, 'Zoomorphic Astrolabes and the Introduction of Arabic Star Names into Europe', in *From Deferent to Equant: a Volume of Studies in the History of Science in the Ancient and Medieval near East* (Annals of the N.Y. Academy of Science), 198, 89 - 104.

Gouk
Penelope Gouk, *The Ivory Sundials of Nuremberg, 1500 - 1700,* Cambridge, 1988.

Gould
Rupert T. Gould, *The Marine Chronometer, its History and Development,* London, 1923.

Green
Arthur Robert Green, 'Anglo-Saxon Sundials', *The Antiquaries Journal,* VIII, 1928, pp. 489 - 512.

Grunewald
Philippe Grunenwald, 'Un chronographe encreur signé Rieussec', *Chronométrophilia',* 14, 1983.

Gunther
Robert T. Gunther, *The Astrolabes of the World,* London, 3th ed., 1976.

Guyou (1)
E. Guyou, 'Application des chronomètres décimaux à la pratique de la navigation', in *Congrès International de Chronométrie, comptes-rendus des travaux, procès-verbaux, rapports et mémoires, Paris, 1902, pp. 116 - 121.*

Guyou (2)
E. Guyou, 'De l'extension du système décimal à la mesure de la circonférence, éphémérides et tables numériques en parties décimales du quart de cercle, préparés pour une application à la navigation', *Annales du Bureau des Longitudes,* VI, 1903, C3 - C89.

d'Haenens
Albert d'Haenens, 'La quotidienneté monastique au Moyen-Age', in *Klosterliche Sachkultur des Spätmittelalters,* Vienna, 1940.

Harding
Rosamond E.M. Harding, *The Metronome and its Precursors,* 3th ed., Frome & London, 1983.

Hassan
A.Y. Hassan (ed.), *Al-Jami' Bain al-'Um 'Amad al-Nâfi' fi Sina'at al-Hiyal,* Aleppo, 1979.

Haussherr
R. Haussherr, *Bible moralisée,* Paris & Graz, 1973.

Heilbron
J.L. Heilbron, 'Lectures on the history of atomic physics, 1900 - 1920', in *History of Twentieth Century Physics, Proceedings of the International School of Physics 'Enrico Fermi',* LVII, New York, 1977.

Hill (1)
Donald R. Hill, *The Book of Knowledge of Ingenious Mechanical Devices ... by Ibn al-Razzâz al-Jazarî,* Dordrecht & Boston, 1974.

Hill (2)
Donald R. Hill (ed. & tr.), *On the Construction of Water-Clocks, Kitâb Arshimîdas fî amal al-binkamât,* London, 1976.

Hill (3)
Donald R. Hill, 'Notice of an important al-Jazarî Manuscript', *Journal for the History of Arabic Sciences,* II, 1978, pp. 291-298.

Hill (4)
Donald R. Hill, *Arabic Water-Clocks* (Sources and Studies in the History of Arabic-Islamic Sciences (4), Aleppo, 1981.

Hirsch
A. Hirsch, *La pendule électrique de précision de M. Hipp,* Neuchâtel, 1988.

Hoefer
F.A. Hoefer, *Geschiedenis der openbare tijdaanwijzing,* Leiden, 1887.

Hoogeveen
W.P. Hoogeveen, *Elektrische Uurwerken,* De Bilt, 1938.

Hope-Jones
F. Hope-Jones, *Electrical Timekeeping,* 2nd ed., London, 1949.

Howard
Chester, W. Howard, 'A de Elavo Returned', *Bulletin of the National Association of Watch and Clock Collectors,* Inc., VII, no 5, pp. 296-300.

Howse
Derek Howse, *Greenwich time and the discovery of the longitude,* Oxford, 1980.

Hunter Blair
Peter Hunter Blair, *An Introduction to Anglo-Saxon England,* Cambridge, 1962.

Icke
Vincent Icke, 'Time scales', this volume, ch. 1.

Jacquet & Chapuis
Eugène Jacquet & Alfred Chapuis, *Technique and History of the Swiss Watch,* 2nd ed., London, 1970.

Jeon
Sang-woon Jeon, *Science and Technology in Korea Traditional Instruments and Techniques,* Cambridge (Mass.) & London, 1974.

Kaiser (1)
F. Kaiser, 'Eene nieuwe toepassing van het beginsel der noniën op den tijd, voor het waarnemen van plotselinge verschijnselen', *Tijdschrift voor de Wis- en Natuurkundige Wetenschappen, uitgegeven door de Eerste Klasse van het Kon. Ned. Instituut,* deel V, 1851, p. 8 en verder.

Kaiser (2)
F. Kaiser, 'De tijdseinen der Nederlandsche marine', *Verhandelingen en berigten betrekkelijk het zeewezen en de zeevaartkunde,* 1, afd. II, 1860, 139 - 225.

Kaiser (3)
P.J. Kaiser, *Atlas voor de sterrenkunde,* Leiden, 1873.

Kartaschoff
P. Kartaschoff, *Frequency and time,* London & New York, 1978.

King
David King, 'A Survey of Medieval Islamic Shadow Schemes for Simple Time-reckoning', *Oriens,* 32, 1990, pp. 191-249.

King & Millburn
Henry C. King & John R. Millburn, *Geared to the Stars. The Evolution of Planetariums, Orreries and Astronomical Clocks,* Bristol, 1978.

Krupps
Krupps, 'Astronomy, Pyramids and Priests', in E.C. Krupps (ed.), *In Search of Ancient Astronomies,* Harmondsworth, 1984.

Kurz
Otto Kurz, *European Clocks and Watches in the Near East,* London & Leiden, 1975.

de Laborde
A. de Laborde, *La Bible moralisée,* Paris, 1911 - 1927.

de Lalande
Jérôme de Lalande, 'Histoire abrégée de l' astronomie depuis 1781 jusqu'à 1802', in *Bibliographie Astronomique,* Paris, 1802.

Landes
David S. Landes, *Revolution in Time, Clocks and the Making of the Modern World,* London, 1983.

Lambrick
Georges Lambrick, 'Further Excavations in the Second Site of the Dominican Priory, Oxford', *Oxoniensia,* 1985, pp. 167 - 168.

Lane
E.W. Lane, *Account of the Manners & Customs of the Modern Egyptians (1836),* London, 1936.

Lloyd
H. Alan Lloyd, *The Collector's Dictionary of Clocks,* London, 1964.

von Lunardi
Heinrich von Lunardi, *900 Jahre Nürnberg. 600 Jahre Nürnberger Uhren,* Stuttgart & Wien, 1974.

McCluskey
Stephen C. McClusky, 'Gregory of Tours, Monastic Time Keeping and Early Christian Attitudes to Astronomy', *Isis,* 81, 1990, pp. 9 - 22.

Maddison (1)
Francis R. Maddison, *A Supplement to a Catalogue of Scientific Instruments in the Collection of J.A. Billmeir Esq.,* Oxford & London, 1957.

Maddison (2)
Francis R. Maddison, 'Medieval Scientific Instruments and the Development of Navigational Instruments in the XVth and XVIth Centuries' (Agrupamento de estudos de Cartografia antiga XXX Seccao de Coimbra), *Rivista de Universidade de Coimbra,* XXIV, 1968.

Maddison, Scott & Kent
Francis R. Maddison, Bryan Scott & Alan Kent, 'An early medieval Water-Clock', *Antiquarian Horology,* III, 1962, pp. 348 - 353.

Maddison & Turner
Francis R. Maddison & A.J. Turner, *University of Oxford, Museum of the History of Science, Catalogue 2, Watches,* Oxford, 1973.

Marshak
Alexander Marshak, *The Roots of Civilisation ...,* London, 1972.

Mauchline
J. Mauchline, 'The Gezer Calendar', in D. Winton Thomas (ed.), *Documents from Old Testament Times,* London, 1958.

Mayer
L.A. Mayer, *Islamic Astrolabists and their Works,* Genève, 1956.

Meis
Reinhard Meis, *Zakhorloges, van halshorloge tot tourbillon,* Haarlem & Antwerpen, 1981.

Menninger
Karl Menninger, *Number Words and Number Symbols. A Cultural History of Numbers,* London & Cambridge (Mass.), 1969.

Mercer
Vaudrey Mercer, *The Life and Letters of Edward John Dent, Chronometer Maker and Some Account of his Successors,* London, 1977.

Miclet
B. Miclet, 'Paul Pouvillon M.O.F. et son horloge à planétaire', *Bulletin d'Ancaha,* 43, 1985, pp. 23 - 28.

Milham
Willis I. Milham, *Time & Timekeepers, Including the History, Construction, Care and Accuracy of Clocks and Watches,* 4th ed., New York, 1942.

Mills
A.A. Mills, 'The Mercury Clock of the *Libros del Saber', Annals of Science,* 45, pp. 329 - 344.

Millas-Vallicrosa
J.M. Millas-Vallicrosa, 'Translations of Oriental Scientific Works (to the end of the Thirteenth Century), in Guy S. Metraux et François Crouzet', *The Evolution of Science ...,* New York, 1963, pp. 128 - 167.

Momigliano
Arnaldo Momigliano, 'Time in Ancient Historiography', *History & Theory VI,* 1966, pp. 1 23.

Van der Molen
E.J. van der Molen, *Klokken : Zaans Uurwerk Museum,* Zaandam, 1976.

Moll
G. Moll, 'An Account of Experiments on the Velocity of Sound made in Holland', *Philosophical Transactions of the Royal Society,* 114, 1824, pp. 425 - 456.

Montagu
Jeremy Montagu, *The World of Medieval and Renaissance Musical Instruments,* London & Vancouver, 1976.

Montet
Pierre Montet, *Eternal Egypt,* London, 1964.

Morpurgo
Enrico Morpurgo, *Nederlandse Klokken en horlogemakers vanaf 1300,* Amsterdam 1970.

Morrison & Eames
P. Morrison & R. Eames, *Powers of Ten,* San Francisco, 1982.

Mouliérac
Jeanne Mouliérac, 'La Collection Marcel Destombes', in Turner (4), pp. 77 - 126.

Naffah
Christiane Naffah, 'Un cadran cylindrique ottoman du XVIIIe siècle, in Turner (4), pp. 37 - 51.

Needham
Joseph Needham with Wang Ling, *Science & Civilisation in China 3. Mathematics and the Sciences of the Heavens and the Earth,* Cambridge, 1959.

Needham, Ling & Price
Joseph Needham, Wang Ling & Derek de Solla Price, *Heavenly Clockwork. The Great Astronomical Clocks of Medieval China,* 2nd ed. with supplement by John H. Combridge, Cambridge, 1986.

Needham, Gwei-Djen, Combridge & Major
Joseph Needham, Lu Gwei-Djen, John Combridge & John S. Major, *The Hall of Heavenly Records : Korean Astronomical Instruments and Clocks, 1380 - 1780,* Cambridge, 1986.

Negoro
R.M.A.A. Tjondro Negoro, 'De Koperen Zonnewijzer van Gresik', *Tijdschrift van Bataviaasch Genootschap van Kunsten en Wetenschappen,* 27, 1882, pp. 47 - 68.

Neugebauer
Otto Neugebauer, *The Exact Sciences in Antiquity,* 2nd ed., New York, 1969.

Noble & Price
J.V. Noble & J.D. Price, 'The Water-Clock in the Tower of the Winds', *American Journal of Archaeology, 72, 1968,* pp. 345 - 355.

North
J.D. North (ed. & tr.), *Richard of Wallingford. An Edition of his Writings with Introductions, English Translation and Commentary,* 3 vols., Oxford, 1976.

Okasha
Elizabeth Okasha, *Handlist of Anglo-Saxon Non-Runic Inscriptions,* Cambridge, 1971.

Page
R.I. Page, 'The Bewcastle Cross', *Nottingham Medieval studies,* IV, 1960, pp. 36 - 37.

Parker
R.A. Parker, *The Calendars of Ancient Egypt,* Chicago, 1950.

Perdrizet
Paul Perdrizet, *Le calendrier de la nation d'Allemagne de l'ancienne université de Paris,* Paris, 1937.

Poulle (1)
Emmanuel Poulle, *Les instruments de la théorie des planètes selon Ptolémée : Equatoires et horlogerie planétaire du XIIe au XVIe siècle,* Hautes Etudes Médiévales et Modernes, 42), 2 vols. Genve & Paris, 1980.

Poulle (2)
Emmanuel Poulle, *Les instruments astronomiques du Moyen-Age,* Paris, 1983.

Price
D.J. de Solla Price, 'Portable Sundials in Antiquity including a new example from Aphrodisias', *Centaurus,* 14, 1989, pp. 242 - 266.

Prospérité
Un grand pas en avant dans l'art du chronométrage : le chronographe', *Prospérité,* avril-mai, 1931.

Rasquin
Victor Rasquin, 'Les instruments non mécaniques', in *La mesure du temps dans les collections belges,* Bruxelles, 1984.

Roeder, Strömgen & Strömgren
Hans Roeder, Elis Strömgren & Bengt Strömgren, *Tycho Brahe's Description of his Instruments and Scientific Work as given in the Astronomiae Instaurata Mechanica ... (1598),* Copenhagen, 1946.

Riccardi
Pietro Riccardi, *Biblioteca mathematica Italiana dalla orgine della stampa ai primi anni del secolo XIX,* new ed., Milano, 1952.

Riché
Pierre Riché, *Ecoles et enseignement dans l'Occident chrétien de la fin du Ve siècle au milieu du XIe siècle, Paris,* 1979.

Roberts
Derek Roberts, *An Exhibition of Amazing Clocks,* Tunbridge Wells, 1987.

Robertson
J. Drummond Robertson, *The Evolution of Clockwork,* London, 1931.

Rohr
René R.J. Rohr, *Les cadrans solaires anciens d'Alsace,* Colmar, 1971.

Rooseboom
Maria Rooseboom, *Bijdrage tot de Geschiedenis der Instrumentmakerskunst in de Noordelijke Nederlanden tot omstreeks 1840,* Leiden, 1850.

Schatten
Schatten van de Koninklijke Bibliotheek : Acht eeuwen verluchte handschriften, 's-Gravenhage, 1980.

Schell
C.W.L. Schell, 'Honderd jaar tijdsein', *Electro-Techniek,* 1952, pp. 463 - 467.

Schneiders
B.G.G. Schneiders, 'De Kristalafdeling van het Dr. Neherlaboratorium', in *Studieblad PTT,* 1962, no 2, pp. 45 - 47.

Schweingruber
F.H. Schweingruber, *Der Jahrring : Standort, Methodik, Zeit und Klima in der Dendrologie,* Bern, 1983.

Secular Spirit
The Secular Spirit : Life and Art at the End of the Middle Ages New York, 1975.

Serrurier
L. Serrurier, *Bibliothèque japonaise, Catalogue raisonné,* Leiden.

Sleeswyk (1)
André Sleeswyk, 'The Celestial River : A Reconstruction', *Technology & Culture,* 19, 1978, pp. 423 - 449.

Sleeswyk (2)
André Wegenar Sleeswyk, 'The 13th Century King Hezekial' water-clock', *Antiquarian Horology,* XI, 1979, pp. 488 - 494.

Sloley
R.W. Sloley, 'Ancient Clepsydrae', *Ancient Eqypt,* June 1924.

Smart
W.M. Smart, *Textbook of Spherical Astronomy,* 6th ed., Cambridge, 1977.

Stebbings
Frederick A. Stebbins, 'A Roman Sundial', *The Journal of the Royal Astronomial Society of Canada,* 52, 1958, pp. 250 - 254.

Stchoukine
Ivan Stchoukine, 'Un manuscrit du traité d'al-Jazarî sur les automates', *Gazette des Beaux-Arts, 1*934 - 1935, pp. 134 - 140.

Stimson
Alan Stimson, *The Mariner's Astrolabe, a Survey of Known, Surviving Sea Astrolabes,* Utrecht, 1988.

Sully
Henry Sully, *Règle artificielle du temps ...,* nouvelle éd. ... par Julien Le Roy, Paris, 1737.

Tardy
Tardy, *French Clocks, the World Over, Part Two, from Louis XVI Style to Louis XVIII -Charles X period,* 5th ed., Paris, 1981.

Taylor & Taylor
Joan Taylor & Harold M. Taylor, 'Architectural Sculpture in pre-Norman England', *The Journal of the British Archaeological Association,* 3rd Ser., XXIX, 1966, pp. 3 - 51.

Toulouse & Piéron
Edouard Toulouse & Henri Piéron, *Technique de psychologie expérimentale de Toulouse, Vaschide et Piéron,* 2 vols., Paris, 1911.

Turner (1)
A.J. Turner, 'Christopher Wren and the Wadham Clock', *Antiquarian Horology,* VII, 3, 1971, pp. 229 - 230.

Turner (2)
A.J. Turner, *Water-clocks, Sand-glasses, Fire-clocks* (The Time Museum, Catalogue of the Collection I. 3), Rockford, 1985.

Turner (3)
A.J. Turner, *Early Scientific Instruments : Europe 1400 - 1800,* London, 1987.

Turner (4)
A.J. Turner (ed.), *Etudes 1987 - 1989* (Astrolabica No 5), Paris, 1989.

Turner (5)
A.J. Turner, *Mathematical Instruments in Antiquity and the Middle Ages : an Introduction* (in preparation).

Turner (6)
A.J. Turner, *Sun-dials, Meridians and Nocturnals : a History* (in preparation).

Turner (G)
G 1'E Turner, *Nineteenth Century Scientific Instruments,* London, 1983.

Ungerer
Alfred Ungerer, *Les horloges astronomiques et monumentales les plus remarquables de l'Antiquité jusqu'à nos jours,* Strasbourg, 1931.

Van Cittert
P.H. van Citters, *Astrolabes, A Critical Description of the Astrolabes, Noctilabes [sic] and Quadrants in the care of the Utrecht University Museum,* Leiden, 1954.

Van Herk
G. van Herk, 'De sterrenhemel zelf (IX)', in *'Hemel en Dampkring,* 40, 1942, pp. 172 - 180.

Voorhoer
P. Voorhoer, *Handlist of Arabic Manuscripts,* Leiden, 1957.

Ward
F.A.B. Ward, *Ministry of Education, Science Museum, Handbook of the Collections Illustrating Time Measurement, Pt II, Descriptive Catalogue,* 2nd ed., London, 1950.

Weber
G.A. Weber, *The Naval Observatory, its History, Activities and Organisation,* Service monographs of the United States Government, no 39, Baltimore, 1926.

Weitzmann
Kurt Weitzmann, *Late Antique and Early Christian Book Illumination,* New York, 1977.

Welcomme
Jacques Welcomme, 'Les instruments mécaniques', in *La mesure du temps dans les collections Belges,* Bruxelles, 1984.

Winter
H.J.J. Winter, 'A Shepherd's Time-Stick Nagari Inscribed', *Physis, Rivista Internazionale di Storia della Scienza,* VI, 4, 1964.

Whitrow (1)
G.J. Whitrow, *The Natural Philosophy of Time,* London & Edinburgh, 1961.

Whitrow (2)
G.J. Whitrow, *The Nature of Time,* Harmondsworth, 1975.

Whitrow (3)
G.J. Whitrow, *Time in History. Views of Time From prehistory to the present day,* Oxford, 1988.

Yamaguchi
Ryuji Yamaguchi, *The Clocks of Japan. A Study of the Antique Japanese Clocks During the Period of Tokugawa Shogunate,* 2nd ed., Tokyo, 1950.

Zába
Zbynek Zába, *'L'orientation astronomique dans l'ancienne Egypte et la précession de l'axe du monde'* (Ceskoslovenská Akademie Ved : Archiv Orient. Supplementa II), Prague, 1953.

Zinner
Ernst Zinner, *Deutsche und Niederländische Astronomische Instrumente des 11. - 18. Jahrhunderts,* 2nd ed., München, 1967.

Index

Index of makers of clocks, watches and instruments exhibited

List of color-illustrations